Magnetic Venture

Magnetic Venture

The Story of Oxford Instruments

Audrey Wood

OXFORD
UNIVERSITY PRESS

OXFORD

UNIVERSITY PRESS

Great Clarendon Street, Oxford OX2 6DP

Oxford University Press is a department of the University of Oxford.
It furthers the University's objective of excellence in research, scholarship,
and education by publishing worldwide in

Oxford New York

Athens Auckland Bangkok Bogotá Buenos Aires Calcutta
Cape Town Chennai Dar es Salaam Delhi Florence Hong Kong Istanbul
Karachi Kuala Lumpur Madrid Melbourne Mexico City Mumbai
Nairobi Paris São Paulo Shanghai Singapore Taipei Tokyo Toronto Warsaw
and associated companies in Berlin Ibadan

Oxford is a registered trade mark of Oxford University Press
in the UK and in certain other countries

Published in the United States
by Oxford University Press Inc., New York

British Library Cataloguing in Publication Data

Data available

Library of Congress Cataloging in Publication Data
Wood, Audrey, Lady.
Magnetic venture : the story of Oxford Instruments / Audrey Wood.
    p.   cm.
Includes bibliographical references and index.
1. Oxford Instruments—History.   2. Scientific apparatus and instruments industry—Great
Britain—History.   3. Medical instruments and apparatus industry—Great Britain—History.
4. Nuclear magnetic resonance—Industrial applications—History.   I. Title.
HD9706.6.G74 O94   2000   338.7'681'0941—dc21   00-040059
ISBN 0-19-924108-2

10 9 8 7 6 5 4 3 2 1

Typeset by Hope Services (Abingdon) Ltd.
Printed in Great Britain
on acid-free paper by
T. J. International Ltd., Padstow, Cornwall

This book is dedicated to scientist-entrepreneurs just starting down the road we have travelled, in the hope that they will discover as much excitement and satisfaction as we have found.

Acknowledgements

I would like to express my gratitude to the people who have helped with this story. First, I must thank Martin, my husband, for making time in his busy life for reading all the chapters at various stages, and for giving me help and honest criticism. I am grateful to: Sir Peter Williams, the late John Woodgate, and Professor Mike Brady, who all read the full manuscript and came up with valuable comments and suggestions; to Nigel Keen for a few corrections, and his views on the Company in Chapter 23; to Sir Rex Richards, who gave me a lot of help with the chapter on Nuclear Magnetic Resonance; to Richard Kennett for allowing me to quote from a letter; to Michael Leask, who helped to check the glossary for accuracy as well as understandability; and to other people in the Clarendon Laboratory, including the late Nicholas Kurti, for their information on the earlier years.

My thanks go also to the many Oxford Instruments people who have answered my questions, both historical and technical, have corrected passages for me, and have looked out photographs and other material. These include Dave Andrews, Simon Bennett, Nigel Boulding, Brad Boyer, Paul Brankin, Barry Bright, John Burgoyne, Emma Comer, Tony Ford, Allan Goodbrand, Alex Goodie, Della Gorham, Tony Groves, Mia Hale, David Hawksworth, Robin Higgons, Ron Jones John Kearns, Nick Kerley, Neil Killoran, Jiro Kitaura, Marcel Kruip, Martin Lamaison, John Lewis-Crosby, Andrew Mackintosh, Steve McQuillan, Jane Methven, Robert Milne, Katja Pollanz, Victor Regoczy, Alistair Smith, Richard Thomson, Kevin Timms, and Jim Worth.

I must not forget to thank Oxford Instruments itself, the subject of this work, and its subsidiary companies, for allowing me to quote passages from the Annual Report and Accounts and other company publications, and to use private minutes and memoranda, diagrams, and photographs. I must include Oxford Magnet Technology, whose managers have provided me with photographs and two cartoons and given me permission to reprint the piece in Chapter 17 that first appeared in an advertising supplement for Siemens AG. I thank The Oxford Trust for permission to use both published and private material and photographs, and Paul Bradstock for looking at the chapter on the Trust for me.

I am grateful to the following bodies: International Business Machines Corporation for permission to reprint passages of text and a Venn diagram from the *IBM Journal of Research and Development* issue on X-ray lithography (May

1993); The Wellcome Trust for permission to reproduce passages and borrow the title from *Making the Human Body Transparent: The Impact of Nuclear Magnetic Resonance and Magnetic Resonance Imaging* from Wellcome Witnesses to Twentieth Century Medicine (September 1998); Newsquest (Oxfordshire) Ltd, a Gannet Company, for permission to publish a number of photographs that have appeared in the *Oxford Mail* or the *Oxford Times* over the years; 3i plc for generously arranging for me to use the Mancoff cartoon in Chapter 13; Fig. 1.3 is © (British) Crown copyright, 1999 Defence Evaluation and Research Agency, reproduced with the permission of the Controller, Her Majesty's Stationery Office.

I want to thank: Jane King, Carole Wheeler, and Julia Mackenzie, who, in succession, have written letters and dug up information for me, and have saved me from unexpected happenings on my computer; Madeleine Murphy for working on the text and preparing the references; Cyril Band and Simon Emmett for helping with photographs and drawings; and Anthony Keats for drawing many of the figures and preparing the plates.

I would like to thank Richard Coopey for using his deep knowledge of post-war industrial history in the postscript, which places Oxford Instruments in a wider context; David Stanford for giving me advice on publishing; and David Fishlock for providing me with many media contacts. Finally I would like to express my gratitude to David Musson, Sarah Dobson, and Hilary Walford for advising and guiding me through the processes towards publication.

Contents

List of Plates

List of Figures

List of Boxes

Abbreviations

ACCOSP	Advisory Council on Central Oxfordshire Science Parks
ACOST	Advisory Council on Science and Technology
AERE	Atomic Energy Research Establishment
AGM	Annual General Meeting
AIM	Alternative Investment Market
ALF	Advanced Lithography Facility
APL	Air Products Ltd.
ATG	Accelerator Technology Group
BAe	British Aerospace
BESSY	Berlin Electron Storage Ring for Synchrotron Radiation
BHF	British Heart Foundation
BOC	British Oxygen Company
BSCCO	bismuth strontium calcium copper oxide
CBI	Confederation of British Industry
CEBAF	Continuous Electron Beam Accelerator Facility
CEO	Chief Executive Office
CERN	European Centre for Nuclear Research
CFC	continuous flow cryostat
CLAS	CEBAF Large Acceptance Spectrometer
COO	Chief Operating Officer
COPUS	Council on the Public Understanding of Science
CSC	Coventry Steel Caravans
DHSS	Department of Health and Social Security
DIAC	distributed intelligent data acquisition and control system
DM	Deutsch Mark
DRAM	dynamic random access memory
DTI	Department of Trade and Industry
EC	European Community
ECG	electrocardiogram
ECR	electron cyclotron resonance
EDX	energy-dispersive X-ray
EEC	European Economic Community
EEG	electroencephalogram
EGM	Extraordinary General Meeting
EMG	electromyography
ERM	Exchange Rate Mechanism
FOT	Furukawa Oxford Technology

FT	*Financial Times*
FT	Fourier Transform
GDPM	goal-directed project management
GE	General Electric (USA)
GEC	General Electric Company (UK)
HTS	high-temperature superconductor
IAG	Industrial Analysis Group
IBM	International Business Machines
IC	integrated circuit
ICFC	Industrial and Commercial Finance Corporation (now 3i)
IGC	Intermagnetics General Corporation
IMI	Imperial Metal Industries
IPR	intellectual property rights
IT	information technology
JEOL	Japan Electro Optics
KK	Kabushiki Kaisha
LHC	Large Hadron Collider
LTS	low-temperature superconductivity
MAG	Microanalysis Group
MAT	moving annual total
MATR	miniature analogue tape recorder
MD	Managing Director
MIT	Massachusetts Institute of Technology
MMIC	microminiature integrated circuit
MRC	Medical Research Council
MRI	magnetic resonance imaging
MSS	magnetic surgery system
NbSn	niobium tin
NbTi	niobium titanium
NbZr	niobium zirconium
NCS	National Committee on Superconductivity
NMG	Nuclear Measurements Group
NML	National Magnet Laboratory
NMR	nuclear magnetic resonance
NMRC	National Magnetic Resonance Collaboratorium
NRDC	National Research and Development Corporation
NRPB	National Radiological Protection Board
NTT	Nippon Telegraph and Telephone
OA	Oxford Automation
OAI	Oxford Analytical Instruments
OAP	Oxford Airco Partnership
OD	Oxford Dynamics
OEE	overall equipment effectiveness
OEM	original equipment manufacturer
OI	Oxford Innovation
OIA	Oxford Instruments Analytical
OIKK	Oxford Instruments KK
OIM	Oxford Instruments Medical

OION	Oxford Investment Opportunity Network
OIS	Oxford Instruments Superconductivity
OM	Oxford Medical
OMD	Oxford Medical Devices
OMS	Oxford Medical Systems
OMT	Oxford Magnet Technology
OPEC	Organization of Petroleum Exporting Countires
ORS	Oxford Research Systems
OST	Oxford Superconducting Technology
P/E ratio	price/earnings ratio
PET	Positron Emission Tomography
PT	Plasma Technology
PwT	PriceWaterhouseCoopers
R & D	research and development
RAL	Rutherford and Appleton Laboratory
RF	radio frequency
RI	Research Instruments
RIE	reactive ion etching
RM	Research Machines
RPI	Retail Price Index
RRE	Royal Radar Establishment
RSNA	Radiological Society of North America
SBIR	Small Business Innovation Research
SBU	strategic business unit
SERC	Science and Engineering Research Council
SMART	Small Firms Merit Award (DTI)
SMEs	small and medium-sized enterprises
SOR	Synchrotron Orbital Radiation
SPG	Special Projects Group
SRAM	static random access memory
SRC	Science Research Council
SQUID	superconducting quantum interference device
SSC	Superconducting Supercollider
STEP	Science and Technology Enterprise Project
TAC	Technology Advisory Committee
TBAC	Thames Business Advice Centre
TDC	Technical Development Capital
TEC	Training and Enterprise Council
TFMH	thin film magnetic read–write heads
TMR	topical magnetic resonance
TOT	The Oxford Trust
TQM	total quality management
UEI	United Engineering Industries
UKAEA	UK Atomic Energy Authority
USM	Unlisted Securities Market
VG	Vacuum Generators
VLSI	very large-scale integrated (circuit)
WDX	wavelength-dispersive X-ray

XRF	X-ray fluorescence
XRL	X-ray lithography
XTG	X-Ray Technology Group
YBCO	yttrium barium copper oxide

It is not imaginable to such as have not tried, what labour an historian (that would be exact) is condemned to. He must read all, good and bad, and remove a world of rubbish before he can lay the foundation.

(John Evelyn, in a letter to Samuel Pepys, 28 April 1682)

The historian 'superinduces upon events the charm of order'.

(Sir Arthur Quiller-Couch, *The Art of Writing* (1916))

Prologue

At the turn of the millennium Oxford Instruments was forty years old. In those four decades it had grown from its small beginnings, as almost the first Oxford University 'spin-out' company, making specialized copper-wound electro-magnets for a minute market, to the status of an internationally known public company quoted on the London Stock Exchange. Recognized for many years as world leader in applied superconductivity, the Group went on to achieve that coveted rank in other technologies. Innovation is its lifeblood, and a substantial percentage of its turnover is yearly ploughed back into research and development.

At March 1999 Oxford Instruments employed some 1,800 people in ten man-ufacturing units based in the UK and the USA, and in overseas sales companies and offices. At that date they had, for some years, been grouped under three loose divisions covering superconductivity, analytical instruments, and medical systems. In addition, a large and successful joint venture made magnets for med-ical imaging, while agencies in some other countries contributed to the Group's strong distribution network.

As for much of the manufacturing industry of the UK, the last two years of the 1990s were a difficult time. Several regions of the world were suffering from political and economic uncertainties, leading to shrinking markets in which an overvalued pound made it hard to compete. In the autumn of 1999 Oxford Instruments embarked on a major reorganization, aimed at simplifying its oper-ational structure and improving its efficiency.

The beginning of the twenty-first century is a time of great changes in tech-nologies. This restructuring will prepare the Company for taking emerging opportunities in this changing environment for a fresh surge of growth. Its unique products and creative ideas, driven forward by a beneficial amalgam of old-timers and new faces, should power this growth, and soon enable it to tran-scend the difficulties it encountered at the end of the 1990s.

This book is the story of the up and down road Oxford Instruments has trav-elled over those forty years. The Marquise du Deffand's remark 'La distance n'y fait rien; il n'y a que le premier pas qui coûte' is commonly quoted, but she did not get it right; perseverance and the refusal to give up under adversity are at least as significant for a long journey. But for that first step in any project there must be an idea in the mind of an 'adventurer', together with the will to act on it.

Oxford Instruments was the brainchild of Martin Wood, an unlikely entrepreneur who came from a staid professional middle-class family. But a look at his ancestry shows that there were indeed adventurers and entrepreneurs in the past. Robert Wood, a traveller and a politician, led an expedition that rediscovered Palmyra in 1751. He was also appointed Undersecretary of State under Pitt in 1756. On the other side of the family, a young son of a Yorkshire woollen industry patriarch was ordered by his father to get out of England following an affair with a milkmaid. He answered an advertisement put out by Peter the Great, who wanted technicians, and he established a company making woollen uniforms for the Russian army. His Thornton descendants built up the largest woollen mill in Russia on the banks of the River Neva in St Petersburg; at its height it employed over 3,000 people, with its own school and hospital. But the family lost it all after the 1917 Revolution.

In starting and driving the Company forward in its early years, Martin was aided and encouraged by me, his wife, the author of this account. I, too, was descended from 'adventurers'. My grandparents had been missionaries in the wilder parts of southern Africa in the 1870s, and my parents covered a forty-year span as missionaries in China.

At the end of the old millennium Martin Wood was still Deputy Chairman of the Company we founded. He did not often think of retirement, either from Oxford Instruments or from the many other bodies that have recruited him to their deliberations for his wide experience. These have included scientific committees and councils, Government advisory boards, local economic associations, and small companies in the Oxford area. In 1985 we set up The Oxford Trust, a charitable foundation whose aim is to encourage the study and application of science and technology. Among its many activities it operates an 'incubator', where some thirty fledgling science-based companies are currently finding their feet with a 'big brother' organization to give them advice and to ease their early years. It is a very different world for a small company today than when Oxford Instruments took the first steps of its journey.

1 First Steps

How Martin Wood came to start a company making magnets

'Mighty things from small beginnings grow.'
(John Dryden, *Annus Mirabellis* (1667))

Martin Wood's first encounter with industry came when he was 15 years old. It was wartime, and in his long school holiday he was asked to take over the job his sister had just left at a small agricultural machinery factory near his home in rural Oxfordshire. His task was to take a curved flat wooden part from the man on his right, drill six holes in it through a metal template, and pass it on to the man on his left. He soon found that he could drill two of the wooden parts at once, then three. Four was too many. Before the mid-morning tea break the man on his right was sweating to keep him supplied, and a small mountain of parts was encroaching on the territory of the man on his left. Both were distinctly cross. At the end of the day he was called into the foreman's office:

I expected a pat on the back for raising my productivity by 300 per cent, but all he did was to tell me that, as my sister was over 18, and I was only 15, I would receive seven pence halfpenny an hour instead of her eleven pence. I slouched home fuming at the unfairness of life, kicking stones along the gutter to relieve my frustration.

But he had encountered some basic issues of industrial management, which he never forgot. One was that there are few jobs where a little thought cannot improve efficiency, but that this depends on the whole team; and another, that treating employees as cogs in a machine, and failing to give encouragement and praise where they are due, stifle enthusiasm and creativity.

Martin's next encounter with industry was on a different scale. Just after the war ended he spent three years as a 'Bevin boy', enjoying his national service down an old-fashioned south Wales coal mine instead of going into the 'boring' army. He found the underground environment an exciting challenge. This was before nationalization of the pits, and he worked with an older collier who, from the piecework earnings of the two, paid his young assistant what he thought fit over and above the basic weekly boys' rate. Martin usually earned something under £3 a week, which was not bad in those days. Some boys were fairly treated, some not. He became a union representative for his coalface, and led a mild 'work-to-rule' protest to highlight the injustices. He soon discovered where power lay when he was relegated to the lowest wettest coalface in the whole mine. But his mind began to reflect on how the industry might be better run.

There followed six years of engineering education, in Cambridge, and Imperial College, London. This period included six months spent working his way round the USA in a variety of jobs from foundry work in a sewing-machine factory in Illinois, to scaring eagles away from a turkey farm in California. Still fascinated by mining, he then worked for a further spell as a management trainee in the vast inefficient battleground that was the British coal industry in the mid-1950s.

Disillusioned once more, this time with the seeming impossibility of improving anything in the coal industry, Martin sought advice from an old family friend, who had an idea that the Professor of Engineering in Oxford University was looking for a young engineer. He took time off from his underground work and went to see Professor Alexander Thom, who offered him the job. Not long before he was due in Oxford he received an urgent phone call from the mortified professor saying the University Chest, the treasury, which he had not previously bothered to consult, had refused to provide the new salary. If Martin still wanted to come south to Oxford, he could spend enough time teaching engineering drawing and tutoring undergraduates to keep him from starving while he looked around for another job. He had already burned his boats by giving in his notice to the Coal Board, and had no desire to rebuild them, so he moved south. In the meantime Professor Thom had discovered that an engineer was needed in the Clarendon Laboratory over the road, part of the Physics Faculty. Martin got on well with those he met 'over the road', not realizing until the end of the visit that he was actually being interviewed for an advertised position. On such chance happenings his future life's work was to depend.

So, in 1955, recently married, Martin started working as a Senior Research Officer in the Clarendon Laboratory under Dr (later Professor) Nicholas Kurti, who sadly died in 1998 at the age of 90. This laboratory was, and still is, what is now called a 'world-class' centre for research in solid state physics. Among other methods of investigation, these studies involved experiments at very low temperatures and in very high magnetic fields (see Box 1.1).

Martin's assignment was to develop better magnets as well as managing and improving the existing high-field installation with its water-cooling system and its huge old 2 MW DC generator. Nicholas Kurti had found this generator before the war, and it now had a new lease of life providing electricity for magnets after its days powering the trams of Manchester. Magnets, their design, construction, and operation, became Martin's daily preoccupation, one shared by very few other people anywhere. The Clarendon was one of only ten or twelve laboratories worldwide capable of generating magnetic fields above 3 T,

Box 1.1. What is a Magnetic Field?

The *Oxford Concise Science Dictionary* defines a magnetic field as 'a field of force that exists around a magnetic body or a current-carrying conductor'. I once heard a lecture in which Sir Lawrence Bragg described magnetism as 'electricity looked at sideways'. An electric current produces an associated magnetic field, and a moving magnetic field in its turn induces an electric current in a conductor. This was Michael Faraday's great discovery (see Fig. 1.1).

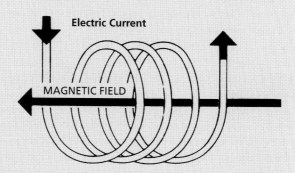

Fig. 1.1 The link between electric current and magnetic field

The magnetic field round a single straight wire carrying a current is very weak, but when an electric current flows through a coil of wire it generates a stronger magnetic field at its centre. The strength of the field depends on the strength of the current, the number of turns and the size and shape of the coil. Magnetic fields are normally measured in gauss or tesla. 1 tesla (T) = 10,000 gauss, or 10 kilogauss. The earth's magnetic field, which swings a compass needle to line up in a north–south direction, varies between 0.2 and 0.6 gauss, or 200 to 600 microtesla.

which they did by forcing huge currents of electricity through water-cooled coils made of copper conductor. D.Phil. students and postdoctoral research physicists from many countries, whose chosen investigations needed these high fields, migrated to Oxford University. The laboratory scintillated with exciting research; in the winter months it certainly shone out at night, the only time the large generator was permitted to run, as it took nearly 10 per cent of the output of the 25 MW generator that then served Oxford City. The power-station manager had to be contacted before the machine could be switched on.

Martin found his interesting new work in the Clarendon Laboratory demanded a range of abilities. There was the straight engineering needed for running the large electrical and water-cooling plant; then he had to learn all about magnet design and construction and to improve on the existing magnets; and he needed to know something of the experiments the physicists planned to put in the equipment. Later, he organized the design and building of a new laboratory with eight magnet workstations.

New universities were at last being built in the UK following the relaxation of post-war austerity. A few physicists leaving the Clarendon to take up new posts wanted to continue their lines of research, and sought advice on setting up similar, if more modest, facilities in their emergent departments. Martin helped with information on the commercially available items such as power supplies, water purifiers, and control instrumentation, and he sketched designs for hydraulic hoists and other parts. Magnets themselves were more difficult, but, to help these new laboratories along, the department agreed to its technicians using the workshops out of hours to make bits and pieces of hardware, including magnet coils. This system operated without strain while it did not interfere with the work on research equipment for the resident Clarendon physicists, but demand crept up, and it started to become an embarrassment to the laboratory. The 'customer' physicists could assemble most of their facilities, but there was nowhere else to go for high-field magnets—an unsatisfied need in a minute market.

Although he had been active and happy for the past few years in the stimulating, congenial, and open world of the University, Martin began to revisit his dormant ideas on industry. How could the industrial experience be improved beyond the poor encounters in his past? He had occasionally had dreams of running a much better company, one that would be more efficient in technical ways, and, at the same time, harness rather than confront the human spirit. He developed a proposal for a more-or-less commercial private initiative that would satisfy the needs of the tiny high-magnetic-field community, and would also relieve the Clarendon of this embarrassment. He suggested this to Nicholas Kurti, whose instant response was 'What can I do to help you?' A Hungarian Jewish physicist, who had also worked in Germany, and came to Oxford to escape the Nazis, Nicholas retained a continental perspective, recognizing the value of close interactions between university and industry. This attitude was quite uncommon in British universities at the time and is still not universal. Nicholas was to give us a lot of help and support over the years, with introductions to great men of science and warm mentions of the Company in his lectures—all immensely important as we were developing our early research market.

Professor Brebis Bleaney, head of the department, had himself collaborated successfully with industry during the war, and, with his support, and the enthusiastic backing of Nicholas Kurti, Martin put in a request to the University 'Visitatorial Board'—Oxford often has unique names for its committees. The Board sanctioned this extramural activity, granting permission for him to start a company and spend a part of his time on its work. Nicholas Kurti's one stipulation was that he should continue in his tenured position in the University; so we would still be able to eat, however small the market! As Martin's partner in the business, I would have to relieve him of as much work as possible. I had a scientific education, but limited physics and no engineering, so this would have to be in unskilled labour and administration; luckily I had already taught myself to type while tied to the house with a young baby. As we worked together, we found we had complementary abilities, the foundation for a lasting partnership.

Looking back more than forty years, we can see that the strength of the Company over its formative period lay in this partnership and the mutual support it provided through hard times. For its first ten years Martin was to continue his university work, but this often overlapped his Company activities. To begin with all the scientific input and engineering design came from Martin, and he had many other strengths. He had a vision of what he wanted to achieve in the Company, a flair for leadership and strategic thinking, and the ability to persuade people to his viewpoint; but he would never describe himself as a day-to-day manager. In the early phases that was more my role, and I developed an aptitude for finance, legal matters, and administration. After a while I started to adopt a few current management techniques. Martin is a born optimist, and I used to say that one of my tasks was keeping his feet on the ground while I thought up ways and means of reaching as far forward as possible. We made a good negotiating team; Martin expounded his ideas for the Company and I produced the business projections, giving potential senior employees or financial supporters the comfort of feeling that we were not just 'boffins'.

A new organization has to have a name. 'Wood Enterprises' would hardly be suitable for a company based on electrical conduction. We looked towards a prestigious role model. The Cambridge Instrument Company, then a major force in the scientific-instruments field, owed its rise to prominence to Horace Darwin, the youngest of Charles Darwin's five sons. He was an engineer who designed instruments for several scientists. In 1881 he became one of the partners in a small jobbing workshop with a few mechanics who had been making equipment for newly developing experimental laboratories in Cambridge. A description of this firm in *Horace Darwin's Shop*, the history of the Cambridge company, foreshadows how our new company would function: 'Darwin would, if necessary, design a complete piece of apparatus to meet a customer's needs. If a piece of experimental equipment proved successful it might become standard apparatus in the laboratory' (Cattermole and Wolfe 1987: 26). 'The Oxford Instrument Company' would be a name to give us the flexibility to expand or to turn in other directions in the future. Both our families were strong in the medical field, and one brother had often talked of medical equipment needs in hospitals in Kenya and Tanzania where he worked.

So in 1959 the new Oxford Instrument Company was formed. I say 'new' because there was an earlier Oxford Instrument Company, started in the early 1930s by Professor Lindemann (who became Lord Cherwell) to make various electrical gauges and devices. Later on it made photomultipliers for the original Baird television broadcasting system. The Company was then run by Dr Bolton King, a colourful character who, when in Berlin, had learned how to make photomultiplier tubes, and took on a young glass-blower and a few technicians. Alfred Sommer, who had connections with the Company, told me of Diana, Bolton King's large yellow dog, which more than once got among the vacuum equipment in the basement at No. 6 Keeble Road, where the firm was located, and smashed several vacuum pumps by wagging her tail. The Company was not very successful and had closed down by 1940. But, because of its previous existence, we were not immediately allowed to register the new firm as a limited liability company. By November 1961, having traded under the name for two years, we were finally permitted to incorporate the Company. Perhaps some fixed number of years had elapsed since the demise of the older firm. Only twice did we encounter any trace of it; in the early 1970s a large brass pressure gauge dating from the 1930s was sent to us with a complaint that it no longer worked. Our technicians followed the simplest course of mending it and sending it back with apologies. A little later a solicitor, executing a will, asked for all recent reports and accounts as he had to value a single share issued in 1932.

For a few months the work of the *new* Oxford Instrument Company was almost all in design or consultancy. An extra table in the spare bedroom sufficed for spreading out 'blueprints' in the evening, and we already owned two typewriters. But soon we found ourselves in discussions with old contacts at the Royal Radar Establishment (RRE) in Malvern, and at the Harwell Laboratory of the UK Atomic Energy Authority (UKAEA), who wanted complete magnets. These were for 'pancake-type' magnets on the same basic design as the standard ones in the Clarendon. The one for Harwell was quite fancy, as it was intended for use in a small atomic reactor and had demanding specifications with a transverse slot through the middle for a beam of neutrons. During 1960 orders arrived for both these magnets. (See Box 1.2 and Fig. 1.2.)

With the beginnings of serious manufacture, more space had to be found—study, spare bedroom, and ex-coal cellar were no longer enough. At the bottom of our large garden in North Oxford we installed a 'garden shed' (see Plate 4) which had once been half of a post-war 'prefab' home. With adequate electricity connected this became a good small workshop, some four by eight metres. The Clarendon was generous with loans of its special winding machine for the pancake-type coils, and Joe Milligan, a versatile retired Clarendon technician, who had actually developed this machine, came to help whenever needed. Martin, of course, knew all the suppliers from his university work, and the cost of the materials was manageable. The magnet cases and spacers were made of resin-bonded fibreglass—one of his Clarendon innovations. There was plenty of

Box 1.2. Magnets of Different Types

What does the word 'magnet' conjure up for the general reader? In spite of the widespread advance of magnetism into medicine over the past fifteen years, he or she probably still visualizes the small red horseshoe magnets of the schoolroom. These toys, which one can use for hanging pins or paper clips in a string, or arranging iron filings into the pretty patterns determined by the 'lines of force', are made of magnetized iron; they achieve magnetic fields several hundred times more powerful than the earth's field (see Fig. 1.2).

When a much larger horseshoe-shaped iron yoke is wound round with copper wire carrying a current, as in laboratory electromagnets, the iron becomes magnetized. Michael Faraday made early magnets of this type; there is one in the Royal Institution (see Fig. 1.2, and King 1973: fig. 16). In the small gap between the two 'poles' the scientist may achieve a magnetic field of up to some 50,000 times the earth's field—say 2 T. The limitation is the magnetic saturation field of the iron. Electromagnets are widely used in industry, and some are huge; in the scrapyard they move whole cars to their fate in a crusher—as can be seen in the James Bond film *Goldfinger*. There, a car made partly with secret gold components, and containing a dead villain, is lifted into the air by a giant electromagnet and dumped in the crusher, coming out the other end as a compact block to be lifted into another villain's truck. The magnet could not have lifted a car made entirely of gold, which is non-magnetic.

Solenoids are cylindrical coils of wire; a current in the wire produces a magnetic field at the center of the coil. Many small low-field solenoids are used in industry for activating switches, valves, and circuit-breakers—there are scores in a car and they are common in household appliances such as washing machines, telephones, and video recorders.

Higher fields can be obtained by using more power in large solenoids and removing the resulting heat with cooling water. Very high-field solenoids for research were fairly new in the 1950s when Martin joined the Clarendon Laboratory. At the time the standard solenoids there were made of four or six stacked pancake coils. Each was formed from inch-wide copper strip, wound round with nylon fishing line for electrical insulation and to create cooling channels, and coiled up like a slice of Swiss roll. These magnets achieved fields of about 6 T, which was limited by the characteristics of the generator. They took as much power as 2,000 single-bar electric fires, and the resultant heat was removed by purified water pumped at high pressure through the gaps between the adjacent turns of the copper conductor, held apart by the fishing line.

In the Massachusetts Institute of Technology (MIT) in Cambridge, the US National Magnet Laboratory (NML) made solenoids to a pattern devised by Professor Francis Bitter. Here round flat plates, punched with many small cooling-water holes, and with a larger hole in the centre, were connected to make a flat spiral separated by thin sheet insulation that contained corresponding holes. Although successful with the power supply at the NML, this design was not suitable for the old generator in the Clarendon, which produced power at 400 V—as required for trams. In the 1960s and 1970s Oxford Instruments was to make a number of Bitter coils for laboratories with suitable power supplies.

Martin pioneered a new type of high field solenoid in the Clarendon at the beginning of the 1960s. It came to be known as a polyhelix coil, and was made of several concentric copper helices like a nest of different diameter 'slinky toys', the spring-like coil that walks downstairs in a series of somersaults. This design could pack more current-carrying copper than the pancake coils, in the same volume, and achieved fields of over 10 T within the solenoid.

1928: the elctromagnet installed in the Belle Vue laboratory in Paris

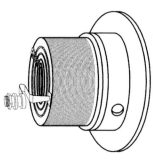

1960s onwards: polyhelix-type magnet in the Clarendon, 30–50 cm diameter

Horseshoe magnet

1845: the electromaget Faraday used for experiments in the Royal instituiton

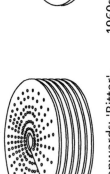

1950s onwards: 'Bitter'-type magnet, 30–60 cm diameter

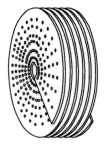

1950s onwards: Clarendon pankake-type magnet, 30–40cm diameter

Fig. 1.2. Different types of magnet

unpleasant unskilled labour for me involved in 'laying up' the fibreglass, and Joe came to machine the parts to shape on an old lathe Martin had picked up in an auction.

In the late summer of 1961 the first magnet was finished, assembled, and ready to go to Malvern (see Fig. 1.3). There was no way of testing it away from the RRE with its particular power supply—a hall full of submarine batteries—and I had to wrestle with bureaucratic delivery forms in septuplicate, and argue with administrators over the impossible clauses about established performance data and proven reliability. Fortunately it exceeded its specifications, providing a steady field of 8 T, and it remained in use for many years. We charged the modest sum of £1,250 for this advanced magnet, unobtainable elsewhere (equivalent to about £17,000 at the beginning of 2000). We made a little profit; apart from the materials, the outgoings were small and the wages bill only the occasional invoice from Joe Milligan. The Harwell magnet was finished and delivered not long afterwards. The bill for this more-difficult magnet was £1,850. After all the work, and arguments with the administration on cover for 'consequential damage', this magnet was never used. The Harwell safety officers passed the equipment as fit for service, but the insurance inspectors became alarmed at the

Fig. 1.3. The first magnet made by Oxford Instruments, delivered in 1961, in place at RRE, Malvern

amount of electrical current for the magnet, which would have to pass through the shell of a reactor.

During 1961 we received an order from the Nuclear Fusion Research Group of the UKAEA, then still located at Harwell. This was for thirty-nine helically wound copper coils, for modulating the magnetic field in the first substantial 'mirror machine' or 'magnetic bottle' to be constructed there. For this job we had to develop a new type of winding machine, and we needed an oven to anneal the copper coils after the stresses of winding, and we would also need an acid-dipping bath to clean off the oxidation deposits after the heating process. This job would take up most of the workshop, and there were other projects under discussion. We were already growing out of the garden shed.

Premises for small companies were very difficult to find in 1961; there were no science parks, few industrial estates, and scarcely any suitable workshops for small new companies to rent. Even if one could find a possible old building, say an old church hall or dairy, it was hard to obtain planning consent for a change of use for industrial purposes. The word 'industry' always seemed to conjure up visions of noise and smoke and factory-gate disputes. We combed North Oxford and, in a back street, we discovered a small disused stables and slaughterhouse, clearly untouched since the last beast had departed in 1939 (see Fig. 1.4). We were able to rent No. 3 Middle Way, Summertown, from John Lindsey,

Fig. 1.4. The old stables and slaughterhouse at Middle Way

the son of the original butcher. We succeeded in getting provisional, temporary, personal, planning permission—no precedent was to be established—and we set about burning the decades-old straw, clearing out the pre-war butcher-boy bicycles and rusty poleaxes, removing the stable partitions and mangers, attacking the extensive woodworm, and arranging for the roof to be mended.

By the autumn of 1961 the first two big magnets had gone. There was no hurry for the thirty-nine coils for Harwell and no other urgent business. We 'shut up shop' and left for Boston to attend the first international conference on high magnetic fields, at MIT, in Cambridge, Massachusetts. Martin was to give a paper there on his polyhelix magnet, newly developed in the Clarendon Laboratory. We planned to follow this with visits to several other laboratories using magnets of various sorts. Apart from offering our wares, we needed to do some primitive 'market research', and to learn at first hand what was going on in our small sphere across the Atlantic. The USA was later to become our largest market.

2 The Superconductor Breakthrough

How the Company rose to the challenge of a new technology

Today we are entering a new era which may come to be known as the decade of high-magnetic-field research. Perhaps the present conference will be remembered as a milestone in this era.

(Henry H. Kolm and Benjamin Lax. 'Preface' in Hulm *et al.*, *High Magnetic Fields: Proceedings of the International Conference on High Magnetic Fields* (1962))

Fortune helps those who dare.

(Virgil 70–19 bc, *Aenead* 10.284)

Saturday, 4 November 1961, was a momentous day in the annals of Oxford Instruments, a day that was to determine the future course of the Company. On the Tuesday we arrived in Boston for the conference at MIT in Cambridge, Massachusetts. Through the next three days of papers and discussions on current ways of making and using high magnetic fields, the concourse buzzed with talk of the recent breakthrough in superconductivity. The excitement was about the newly identified 'Type II' superconductors. Like the long-known 'Type I' materials, these could carry an electric current with absolutely no resistance, provided they were kept extremely cold—in practice in liquid helium within a few degrees of absolute zero. The significant new property of the Type II materials was their ability to carry much higher currents and to remain superconducting in much higher magnetic fields. Dr John Hulm, of the Westinghouse Research Laboratories, had put it graphically, in an interview for the Raleigh *News and Observer*, North Carolina (22 Sept. 1961): 'The breakthrough is that this superconducting wire does not lose its superconductivity because of the magnetism it creates . . . Until now this has been the bugaboo in trying to make useful magnets out of superconducting materials.' How far could these advances go? Would it become feasible to use these new materials instead of copper for winding high-field magnets?

Since the conference had first been planned, teams had started working on the new superconductors in several US laboratories. Some had been fabricating and testing many permutations and combinations of the metals involved. The best, niobium tin (NbSn), was hard to make in a useable form; but in the previous few months one material, niobium zirconium (NbZr), had got to the point where it could be used for winding small magnets. John Hulm's team had claimed the first magnet. The conference organizers had decided to tack on an extra session on the Saturday afternoon for submitted papers, and for a discussion on progress with these promising new materials. It was at this Saturday afternoon session that superconductivity really 'arrived' on the magnet-making scene. Up to the last minute scientists were asking for time to report on their experiments; to accommodate them in ten-minute slots, the session was put forward by an hour—drastically curtailing lunch. Some delegates had planned to go early, but the lecture theatre was still crowded, with people even sitting in the stairways, for a most unusual tail-end-of-conference session.

Scheduled talks by scientists from the Lincoln Laboratories, the Bell Telephone Laboratories, Atomics International, Oak Ridge National Laboratory, and, of course, Westinghouse, described their experiences in making small magnets from the new NbZr wire. From the research laboratories of the Bell Telephone company (now AT&T), Dr Kunzler, who had achieved the original breakthrough, talked of his latest fantastic results with a tiny single crystal of NbSn. Professor Kolm's team at MIT had actually succeeded in making the first ring magnet from this promising but intractable material. The participants scribbled to get down the experimental details and results. Cameras clicked at every slide. A race seemed to be on for the highest field in a magnet; back in the laboratories were teams of scientists working round the clock to beat the previous record before the conference ended. They kept ringing Boston with the latest results, prompting dramatic two-minute interpolations; Bell Telephone was even humorously accused of blocking the lines from competing laboratories. Westinghouse could not say what the upper limit of its latest magnet was because it had not got a power supply giving more than 100 A. Someone had a newer record with niobium tin; someone else had put more current into a NbZr magnet and produced a field of 6 T—comparable to the field in our far larger copper-wound magnets. It was all new and surprising and very important. We were seeing scientific history in the making (see Box 2.1).

A few quotations from the proceedings of the conference, which came out the following February, will give the flavour of this early stage of the technology:

These data are for short lengths of wire. However, assuming that such performance can be duplicated in very long lengths, it is obvious that given about one pound of suitable wire, one should be able to construct a solenoid . . . that would generate fields approaching 70,000 gauss [7 T] in a small volume . . . we observe a wide variation in the superconducting behavior according to the particular metallurgical history . . . At the lower fields Ic [critical current] was very much lower in the solenoid than in the free wire. This effect is not yet understood.   (Hulm *et al.* 1962: 334–6)

Fig. 2.1.
H. Kammerlingh
Onnes, seated, and
J. D. van der Waals
in the cryogenics
laboratory, Leiden
University

Box 2.1. The History of Superconductivity

A superconductor is a material that, under the right conditions, has absolutely no resistance to the flow of a steady electric current. The phenomenon has been known since 1911, when Professor Kammerlingh Onnes, of Leiden University in the Netherlands (see Fig. 2.1), found that mercury lost all resistance when it was cooled by liquid helium to a few degrees above absolute zero (0.0 Kelvin (K) or −273 °C). Like many other physicists, Kammerlingh Onnes wanted to measure the electrical resistance of metals at low temperatures. Scientists had known for a long time that the resistivity of a metallic conductor reduced with the temperature, but there was speculation on what would happen as they got down towards absolute zero. James Dewar suggested the electrical resistance of a pure metal would also go to zero, but Lord Kelvin—he of the Kelvin scale—thought it would start rising again at some point and become infinitely great at absolute zero. This question was pivotal to the research aimed at probing the atomic structure of matter, but the technology for testing these theories took a long time to come.

Before 1883 there was no liquid nitrogen (which boils at 77 K at atmospheric pressure) or liquid oxygen (at 90.2 K). In 1898 James Dewar liquefied hydrogen, which has a boiling point of 20.5 K. Only small amounts of these liquids could be produced then, and the equipment needed for experiments on the resistivity of metals was difficult to construct. Intermediate temperatures were obtained by reducing the pressure, which lowers the boiling point, much as, on a very high mountain, a kettle of water will boil at a temperature that provides only tepid tea. As the scientists succeeded in making measurements at lower and lower temperatures, the resistances went on decreasing with temperature (see Fig. 2.2).

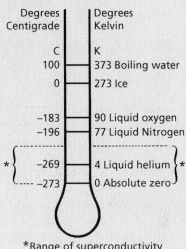

Fig. 2.2. Temperature scales
Note: not to scale.

*Range of superconductivity

In 1908 Kammerlingh Onnes succeeded in liquefying helium, which has a boiling point of 4.2 K at atmospheric pressure; this could be reduced to 1.8 K by lowering the pressure. There are drawings and photographs of the Leiden Laboratory from this period—the liquefaction equipment was an unbelievably complicated maze of glass tubing and valves (see Fig. 2.3). It needed a great deal of nurturing to keep all the valves tight and stop the tiny amounts of liquid from evaporating too fast. With his excellent chief technician, Mr Flim, Kammerlingh Onnes set out to test the different theories on the resistivity of metals. At first he sided with Lord Kelvin rather than James Dewar. Using platinum wire the resistance levelled off at 4.3 K, which was interpreted as meaning either that Lord Kelvin was right, or that impurities were disturbing the measurements—all the scientists were using the purest metals they could obtain. The same thing happened with gold.

Kammerlingh Onnes then turned to mercury, which could be purified to a high degree by distillation. In October 1911 the team worked on a thread of mercury in a capillary tube. When the resistance suddenly disappeared at 4.2 K,

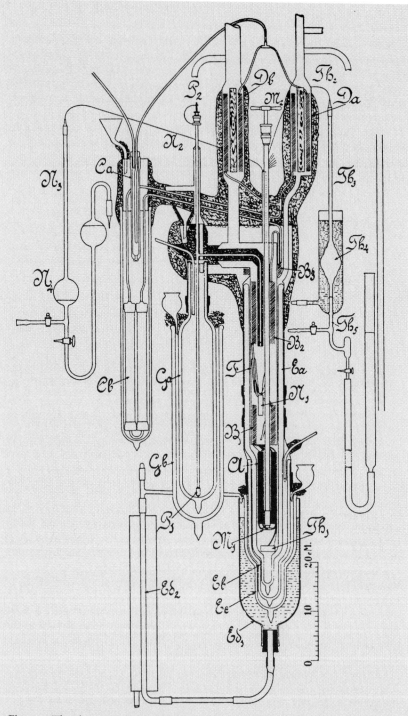

Fig. 2.3. The glass apparatus used by Kammerlingh Onnes for the first liquefaction of helium.

they suspected a short circuit. They checked the equipment over, and soon it was clear that the resistance was at least not going to rise as they approached absolute zero. Kammerlingh Onnes had already revised his ideas, and now supposed the resistance was really going down steadily to absolute zero. But in November, as they experimented further, it became clear that the resistance really did drop to zero, and suddenly, at as high a temperature as 4.15 K, and remained there as the temperature went down further. In his communication describing these results (Kammerlingh Onnes 1911: 13–15) he called the new state of the metal one of 'superconduction'. He received a Nobel Prize for this work in 1913.

When Kammerlingh Onnes found that other metals, including lead and tin, exhibited the same remarkable property, even when they contained impurities, the possibilities seemed limitless. He immediately proposed a high field superconducting magnet for his laboratory, wound from lead wire, and put in an application for 'modest' funds for its construction. With other scientists and commentators, he began predicting a resistance-free future for the electrical industry; superconducting motors, generators, and transmission lines would save huge amounts of energy lost to heat in conventional copper windings. But these dreams were soon to fade. In subsequent experiments he found that the superconducting capacity of these metals could be destroyed by carrying quite a small electric current, and by low magnetic fields, as well as by a rise in the temperature—John Hulm's 'bugaboo'.

Between 1911 and 1950 superconductivity remained largely an interesting curiosity with no practical applications. Work aimed at more useful superconductors continued in several laboratories, but it was limited by the difficulty of producing the necessary liquid helium. The next step might have come sooner had progress not been interrupted by two world wars and by the elimination of one promising line of research when the Russian scientist Shubnikov was executed in a Stalinist purge. The phenomenon defeated theoretical physicists including Einstein, who, in 1922, declared that the experimentalists would have to provide more data before a sound theory could be devised. With the increasing availability of liquid helium, more scientists started working on the problem in the 1950s, and their results led J. Bardeen, L. Cooper and J. Schrieffer to propose the 'BCS' theory in 1956. This was widely accepted, and these three scientists jointly received a Nobel Prize, but the phenomenon is still not fully understood.

One of the scientists active in the field in the 1950s was Berndt Matthias, who was working at the Bell Telephone Laboratories in New Jersey. He concocted thousands of possible superconducting materials, some of which he tested, but only at a low field, and some of which he passed to colleagues to work on. He was a great believer in experiment and intuition, and a friend once referred to him as 'the only living sixteenth century alchemist'. The crucial experiment came in December 1960 when John Kunzler, a colleague of Berndt Matthias, took a new crystal of niobium tin and found it would remain superconducting in a field of 8.8 T while carrying a high-current density. This was the first high-field Type II superconductor to be identified.

The onset of resistance in the coil at the peak field was sudden and resulted in immediate quenching of the current, an audible report, and a slight boil-off of liquid helium . . . Much larger scale superconducting magnets of similar character . . . should be feasible. For example, a 12-inch-i.d. magnet . . . would use about 90 pounds of Nb-25 at. % Zr wire per foot of solenoid length . . . Such a solenoid with a length of two feet or more should generate a field of $\approx$ 60 kgauss [6 T].   (Hake *et al*. 1962: 342–3)

The brittle nature of the niobium-tin core of this wire requires that it not be bent after the heat treatment which forms the active component, Nb$_3$Sn . . . The currents carried by the coils were down by factors of 2 or 3 from the best short-sample results which have been obtained; it thus appears that non-uniformity in the wire as drawn, or some part of its subsequent treatment reduced its performance.   (Salter *et al*. 1962: 344–6)

Leaving to catch our plane before the famous last session of the conference ended, we flew to New York. Right through the flight friends were discussing the new superconductors and whether they would challenge the old ways of making high magnetic fields. The liquid helium, essential for cooling these materials, was costly and hard to obtain in most places, and the new ill-understood superconductors were inconsistent and unreliable. But if these problems could be overcome, any laboratory would need only a coil made from this superconducting wire, a superior vacuum flask for the liquid helium, and an electric current supply, similar to that from a car battery, and its scientists would have a high-magnetic-field facility. No huge generator or complicated engineering installation would be needed, and no minor waterworks for cooling copper windings. The market would grow enormously, perhaps from ten to 10,000 users, and this was the market we had started out to serve two years earlier.

Later that evening, on the New York subway, we looked again at the implications. The future of our infant high-magnetic-field industry clearly lay with superconductors. Were we bold enough to face these challenges? Before we got to Brooklyn we, the company of two, had reached the momentous decision to embark on this exciting new technology and make superconducting magnets. Thirty-eight years later, at the end of the century they still formed the largest part of the Group's production.

After the conference we visited laboratories around the USA from New York State to Oak Ridge, Tennessee (see Fig. 2.4); to Los Alamos in New Mexico; to Berkeley, California; and to the Argonne Laboratories in Chicago. Meetings were almost all about conventional ways of getting to high fields. These laboratories needed special magnets of various shapes and specifications, and these were normally constructed in their own workshops. Equipment had to be reliable and it would be some years before superconducting magnets could be considered; but there was much talk of the new materials.

Surprisingly, one type of superconducting wire had come out of the laboratory and was already on the market. A metallurgist, Dr Jimmy Wong, had been working at the Wah Chang Corporation on a US Government development project. He had been adding small amounts of the metal zirconium to pure niobium to give it the mechanical strength needed for the radioactive

Fig. 2.4. A discussion on magnet design at the Oak Ridge National Laboratory, Tennessee, in November 1961: Martin and Audrey Wood with David Coffey and R. L. Brown

shielding of a small reactor intended for powering an aeroplane. The project—to build a plane designed to orbit the earth continuously—did not get far; but John Hulm and Stan Autler had found that zirconium vastly improved the long-known superconducting properties of pure niobium. They knew Jimmy Wong from student days, and kept phoning him to ask for samples of wire made of niobium with various amounts of zirconium added. They soon settled for 25 per cent zirconium, as it was more malleable than 33 per cent, and still produced an effective superconductor. This alloy was still relatively hard but, unlike NbSn, the first Type II superconductor identified, it could quite easily be drawn into fine wire. This material was being sold by the pound weight, at £200 a pound (equivalent to about £2,500 in 2000). It cost more than price-controlled gold at that time.

On our return home, the tiny Oxford Instrument Company ordered one pound of NbZr wire—our first investment in real research and development (R & D) not financed by a customer. Martin planned to make the largest magnet he could from the material. Apart from the various conference reports, there were no books or manuals on how to design a magnet using superconductors; how to make the current leads and the terminals; what tension to put on the wire; and so on. He made his calculations using an existing computer program for fixing the overall dimensions, and his engineering intuition for all the other factors in the design. At that time we were allowed to use the huge university computer, a mass of valves and switches and wiring, in air-conditioned rooms, occupying a large Victorian house in South Parks Road. Martin punched the parameters onto narrow paper tape, stuck the program tape to it, and handed it in for processing. The next day, if he was lucky, he collected the print-out. By the end of the century, as well as the mainframe computers for complicated tasks, virtually every desk in the Company bore a personal computer (PC), each of which had far more computing power than that old leviathan. But we were very glad of it at the time. Although our magnet calculations were relatively easy, trying all the variables by any other method could have taken months.

By March 1962 the design was complete and Martin had made the experimental former for the coil. This was rather like an outsize cotton reel, with a central brass tube, and round plates, made from an insulating material, stuck onto each end. We wanted to produce the first magnet in Europe, so, as soon as we heard the wire had arrived in London, Martin went up to fetch it. He borrowed a winding machine and very carefully wound the fine wire onto the former in many even layers. This was a tense process with so much invested in the material, and the first time he tried, one of the end plates started to slip; he had to unwind it carefully, use a stronger glue, and start the winding again.

After a night and a day the magnet was finished, ready for the next stage—the crucial tests. Nicholas Kurti and others in the Clarendon were eager to see the new superconductors in action. The laboratory provided the necessary liquid helium and liquid nitrogen, and the glass cryostats—like large nesting vacuum flasks—to stop the cold liquids evaporating too fast. Martin wired the magnet to his car battery through a simple rheostat for controlling the electric current, and meters for measuring resistance and current. He pre-cooled the magnet in the much cheaper liquid nitrogen—which boils at the 'relatively high' temperature of minus 197 °C—and then immersed it in the liquid helium in the inner cryostat at minus 269 °C, about four degrees above absolute zero, or 4 Kelvin (K) (see Fig. 2.5).

Through the glass the liquid helium could be seen bubbling hard as it cooled the magnet. As the coil got colder, the resistance reading on the instrument fell steadily, showing normal behaviour for a metal. After a while slower boiling could be seen, indicating it was nearly cold right through. Suddenly the resistance dropped to zero—an astounding thing to see for the first time. The wire had become superconducting. Martin started to wind up the current through the rheostat; as it rose slowly to 2 amperes (A), then 4 A and on up, the resistance remained at zero. At 17 A—above the design current—the field in the 1.8-centimetre bore of the magnet was 4 T—way above anything previously possible in that laboratory without the major engineering, power, and cooling installations. A little beyond 17 A the NbZr reached its critical current and returned suddenly to its non-superconducting and highly resistive state, boiling off a lot of liquid helium as the current now heated the metal. This almost explosive

Fig. 2.5. Martin Wood testing our first superconducting magnet

transition is known as a 'quench'. There was no liquid helium left for further trials that day, but the experimental prototype had performed perfectly.

On 12 April 1962 a small report and picture of this magnet (see Plate 5) appeared in the journal the *New Scientist*, and a little later the journal *Cryogenics* published Martin's paper entitled 'Some Aspects of the Design of Superconducting Solenoids' (Wood 1962). This was the first magnet to be made from the new superconductors outside the USA, and it attracted a lot of interest. Within weeks we were talking to prospective customers. Tests with the magnet went on and Martin started refining the design of terminals for transferring current from copper leads to superconducting wire, and developing superconducting switches (see Box 2.2).

Box 2.2. Superconducting Switches

A switch consists of a superconducting link between the terminals of a magnet that is 'opened' with a tiny heating element round it, which keeps it in the non-superconducting or 'normal' state while the current is being increased in the magnet. On reaching the desired current and field, the heating element is switched off and the current flows across the now-superconducting link. The power source can then be removed and the current goes on flowing and keeping the magnetic field in place indefinitely—provided the magnet remains cold enough. This allows energy to be stored in a magnet and is the nearest thing to date to perpetual motion.

During the summer of 1962 we showed the magnet at the first exhibition the company attended. This was at a Physical Society meeting in Harrogate, where the tiny Oxford Instruments' stand was placed next to the large and professional display from Cambridge Instruments. For this occasion we printed our first company leaflet. A little later Martin and Nicholas Kurti were invited to a Royal Society Conversazione to give a demonstration of the storage of energy in a superconducting magnet (see Fig. 2.6). This was an elegant evening-dress affair in Burlington House, harking back to the days when the fashionable world was interested in scientific experiments. Martin and Nicholas developed quite an act, removing the battery with a flourish and demonstrating the persistent field in the magnet with the help of nails and spanners stuck tightly to the outside of the glass cryostats. When they finally 'opened' the superconducting switch, the current was released into a second external circuit, where it lit up small light bulbs, rang a bell, and powered a tiny fan. With the loss of the current the magnetic field decayed, and the nails and spanners dropped away one by one to cheers from the floor.

By this time the old stables at Middle Way were taking shape as pleasant and sunny workshops. Martin and Joe Milligan examined lathes and milling machines discarded by government departments and bought two or three of these machine tools at auctions, while the Company invested in a new coil-winding machine suitable for superconducting wire. On the strength of our one-magnet experience, Oxford Instruments accepted a couple of real contracts for these magnets, and we ordered more niobium zirconium wire with the same specification as the first reel. But these early materials were variable, unstable, and easily damaged. We had been very lucky with that first pound of superconducting wire. I hate to contemplate whether the Company would be

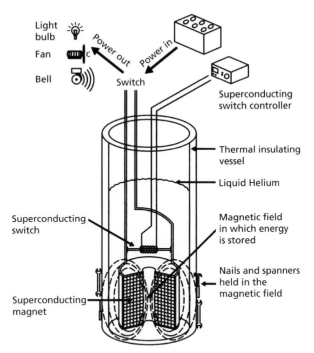

Light bulb

Fan

Bell

Power out

Power in

Switch

Superconducting switch controller

Thermal insulating vessel

Liquid Helium

Magnetic field in which energy is stored

Superconducting switch

Nails and spanners held in the magnetic field

Superconducting magnet

Fig. 2.6. The principle of the storage of energy in a superconducting magnet

in the same business today had it been like the second batch; this did not function nearly as well, and we encountered many problems getting these orders into specification.

Recurrent technical problems with almost all the various forms of superconducting materials succeeding each other on the market have been a feature of our industry from the beginning. There was one period of real famine when no metallurgical company seemed able to make useable superconductor. This is not just a matter of making an alloy with the correct proportions of all the right elements. As in cooking, where the coolness of the milk, the freshness of the eggs, the time taken to beat the batter, and the temperature of the frying pan determine the quality of the pancakes; so the quality of the finished superconductor is determined by the actual processes of preparing the billet of component materials, swaging it, extruding it, drawing it down through successive dies into finer and finer wires, and periodically taking it up to just the right temperature to anneal it for the next process. Not all these factors are well understood even now.

Oxford Instruments had enjoyed an easy start in its tiny market for high-field copper solenoids. Because of this, possibly unique, technical and commercial experience with high magnetic fields, and our temerity in diving into the new technology, the fledgling company led Europe in applied superconductivity in the early 1960s. Since the 1980s it has led the world. It was certainly a difficult industry for a novice company to be in, but a small group can learn fast, and can usually be more flexible than a large hierarchical company. All the competitors in our field have had the same problems with the technology over the years. The large electrical companies were very excited in the early days. John Hulm of Westinghouse had told his interviewer in 1961, 'The supermagnet promises to revolutionize almost every aspect of man's use of electricity.' He foresaw applications 'in generating and distributing electricity, in building new atom-smashing machines, magnetic containers trying to draw peaceful power from the H-bomb reaction, and light-weight power plants for interplanetary space ships' (*News and Observer*, 22 Sept. 1961). For these larger American companies, which had led the emergent technology, the work later became a small marginal activity to keep them up to date; most of them lost interest as problems proliferated and these visionary goals retreated into the future. The small companies that came into the field to make laboratory-scale magnets were mostly underfunded and, over the years, several were taken over or succumbed to the costly technical headaches. In Oxford Instruments, through our early and growing experience, we always managed to get over our difficulties in the end. And as we emerged into the light at the end of these dark tunnels the landscape contained fewer competitors.

3 The Juvenile Company

Success with the new magnets brought growth and a supply problem, solved by starting a sister company

> The defining characteristic of enterprise is the drive to make things happen, to recognise opportunities and find the means to realise them, in short to devise and carry through an agenda for action.
>
> (Elizabeth Garnsey, 'Science-Based Enterprise: Threat or Opportunity?' (1997))

> Necessity is the mother of invention.
>
> (An old proverb)

Like heat or pressure or X-rays, magnetic fields are a basic tool in scientific research. Physicists need ever-higher fields to extend their experiments, and the superconductor breakthrough was opening up new possibilities. The brand new technology for making these magnets was little understood and difficult, but Oxford Instruments was among the very few organizations or people with *any* previous experience of providing high magnetic fields. We were in a new business at the right time, and accepted the risks inherent in this rare opportunity. The Company soon started to grow.

Martin talked to potential customers, designed superconducting magnets for their particular experimental needs, and made the time to wind them in the garden shed. Joe Milligan made parts and magnet formers, but was still very busy on the conventional copper-wound-magnet business, which had not come to a full stop. In fact an early duplicated letter to potential customers shows that this business was growing. In announcing a 'range of standard superconducting solenoids producing fields of from 20 to 50 kilogauss [2 to 5 T]', it talks of these magnets as 'supplementing the high power solenoids which we have been making for some years . . . At the moment we have solenoids under design or construction with power requirements ranging from 10 kW to 10 MW.'

For commercial superconducting magnet systems the peripheral equipment needed to be more sophisticated than for the early tests. We would have to provide metal cryostats to contain the liquid helium instead of fragile glass dewars (vacuum vessels); and electronically controlled low voltage power supplies were required for delivering the smoothed electric current we found necessary for reliable operation. With little experience in these matters we again turned to the Clarendon Laboratory, where Eric Tilbury, head of the Low Temperature Workshop, agreed to make a few metal cryostats for us in his spare time. Several other technicians, working in their garages, were happy to assemble the power supplies designed for us by David Smith, who was in charge of the electronics workshop there. But it was becoming clear that we needed our own full-time staff.

By the end of 1962 we had received many enquiries and ten actual orders for the new magnets. Early in 1963 we delivered our first commercial superconducting magnet system—a small but quite complicated 'split coil' for Dr Butt of Birkbeck College, London (see Fig. 3.1). Its photograph featured in the *Times Science Review*, in a perceptive article in which Professor Eric Mendoza foresaw a few exciting possibilities for large superconducting magnets, some of which have since become a reality:

No solenoid has yet been constructed which has quite reached 100 kilogauss [10 T]. But it is already possible to purchase solenoids, manufactured in this country, giving 40 kilogauss over a few cubic centimetres . . . Certainly nobody has yet constructed or made a solenoid much bigger than a few inches long, let alone one occupying a few cubic feet. But the possibility exists and is exciting.   (Mendoza 1962: 14)

We succeeded in finding an excellent recruit as our first full-time employee, and, in February 1963, John Rackstraw started work as a technician/design engineer. He had been a prize-winning apprentice at the Harwell research laboratories—subsequently to be the source of many of our best engineers. John, who was tragically killed in a car crash after giving seven valuable years to the Company, described the move as 'leaving the cosy security of a government laboratory for the excitement of a brand new industry'. John was soon followed by another technician and a 'boy', and, in the early summer, by Vernon Brook, who had worked in the Clarendon and became our Workshop Supervisor. The team grew further, and in the autumn of 1963 we were joined by Frank Thornton, a physicist who had rare previous experience with superconductors. He came as Design and Production Manager, and was later to become the Company's first Managing Director (MD).

At the time Frank joined the Company Martin wrote to an adviser: 'It marks the point at which we become a proper firm, with a substantial wage bill of £3,400 a year [equivalent to about £41,000 at the beginning of 2000] instead of a group of semi-amateurs.' But he was wrong. The virtues of enthusiastic amateurism carried us forward for some years, and it took major business problems and subsequent management changes to introduce real professionalism. By the 1990s there was much more business training and advice available to small companies than in the early 1960s, and our staff, preoccupied with achieving techni-

Fig. 3.1. Our first commercial superconducting magnet— a small split-pair made for Birkbeck College, London

cal success in this exciting and demanding field, disliked paperwork. But administration there had to be. There was no provision for an office in the converted stables in Middle Way, so early in 1964 a caravan was hoisted by crane into the stable yard (see Fig. 3.2). One end formed an office-cum-meeting-room for Frank Thornton and the other end was fitted with desks for a part-time secretary and for me.

The small management and administrative team started to learn about running a real business. We had to study personnel selection and salary levels; payroll tax and National Insurance; purchasing and storing materials at appropriate levels, and keeping the necessary records; communicating with the

Fig. 3.2. The caravan/offices arriving at Middle Way

factory inspector and complying with fire regulations; and importing superconductor and other necessities and exporting magnet systems. On the production
side we began to discover the woes of being 'pig in the middle', as poor superconductors, or delays in the delivery of other vital materials, resulted in agitated
customers waiting to start their experiments. Our management systems were
still in their formative stages; this was long before the days of the office computer, and there were frequent pitfalls. We soon found how easily confusion can
arise over dimensions, usually between customer and designer, but sometimes
even between designer and machinist. Magnets were planned for particular
experiments and there were few duplicates. There was often a dilemma for us;
should we get a working system to a clamouring customer fast, or should we
take the time to make sure of the best possible performance and reliability? In
all aspects of the business we felt the pressure of deadlines:, deadlines for tests,
deadlines for deliveries, deadlines for exhibition materials, and deadlines for

publication of information sheets. Deadlines are actually a fact of life and are always with us.

The wages bill had gone up, but, although the Company was growing fast, we did not yet feel a serious need for outside finance. The overheads were still low, and the operation sufficiently small for the costs not to get out of hand unawares. Beyond the limited finance we could provide ourselves, and a few loans from the family, our bank allowed the Company an overdraft, secured on our house and other possessions, and they let it increase as the assets of the business grew. Entrepreneurs today often refuse to risk everything for their companies. At that stage we were full of confidence in the future, and we needed the working capital for growth. If you have taken a personal risk for your company, it certainly concentrates the mind and gives you an extra incentive to 'make it work'.

We started out as ignorant of the financial side of a company as of its general management. In retrospect, we were rather arrogant, imagining that it would not be difficult for people with a scientific training to grow a company successfully. The day-to-day book-keeping entries were not onerous; the annual stocktake for the audit seemed straightforward; as equipment was delivered the invoices went out and, eventually, the proceeds came back in. The 'job cards', on which all time and materials expenditure for each project was supposed to be recorded, formed the basis for future quotations. We knew by then how inconsistent the expensive superconductor could be, and how easily it could be damaged by a quench, so, for the more problematical magnets, the designers normally doubled the cost of this material before sending out a quotation. Even so, we probably charged much too little for our almost-unique magnets, but we still made a small profit from our base at Middle Way.

Jimmy Wong was still involved in making NbZr wire, now in a new company, Supercon. It was set up with ambitious investors who were excited by the general hype about the future for superconducting motors, generators, and power transmission lines. After a year or two the metallurgists found that superconductors became more stable when bonded to a significant amount of high-conductivity metal like copper or silver. NbZr did not bond well, and using a lot of copper also reduced the 'packing factor', and so the magnetic field inside the coil. But for a while our NbZr magnets, made from quite thinly copper-plated Supercon wire, performed well. By 1964 an early form of NbSn conductor was being offered by another American company, RCA. This was more difficult to use than NbZr, but we started experimenting with it as NbSn has higher magnetic-field characteristics.

There was one major obstacle to progress—liquid helium. This was needed for testing every magnet and cryostat. The British Oxygen Company (BOC) offered a limited and expensive supply in seventeen-litre, vacuum-insulated containers, but it had to be collected from their site in south-west London. More distant users had to make arrangements to get the suspicious-looking vapour-producing containers put on a train, where heat, vibration, and delays could easily reduce the volume of liquid by half by the time it was collected at the other end. This unsatisfactory liquid-helium situation was deterring many potential magnet customers, and was generally applying a brake to the progress of British research in low temperatures or high magnetic fields.

The Clarendon Laboratory, among the few in the UK with *any* kind of helium liquefier, could produce two litres of liquid an hour from its laboratory-designed machine. Oxford Instruments was permitted to buy small quantities of this, but only when all academic needs had been satisfied. Magnets were sometimes held up for weeks as unexpected experimental problems in the Clarendon put back our scheduled test date. When we did get a small supply for these vital tests, we had to return the expensive helium gas to the Laboratory. Having no suitable compressor for filling cylinders, we returned the gas in large white meteorological balloons, which were stuffed into a van. The huge balloons were anchored inside, but, if overfilled, would push out of the van's back doors trying to rise up in the air as the van was slowly driven the mile back to the Laboratory past staring pedestrians.

We had to find a solution to the liquid-helium problem, which had been a worry from the moment we had decided to make superconducting magnets. In 1962 we considered building our own liquefier with the help of the Clarendon, aiming to supply both ourselves and our magnet customers. We even sought funding from a Government development corporation, which welcomed the idea of a British manufacturer of liquefiers. But the months dragged on through discussions and letters and unexpected problems, and, by the end of 1963, we were obviously getting nowhere. Meanwhile magnets were waiting to be tested, deliveries were delayed and customers were becoming more and more unhappy. The solution of building our own machine was clearly, by then, too long a process to satisfy these needs, but something had to be done. In 1962 we had got as far as forming a sister company called Oxford Cryogenics. This was planned as a service company to supply liquid helium and other special cryogenic materials to our equipment customers as well as to Oxford Instruments. It was set up as a separate company, because we expected to need outside equity finance for such a costly item as a helium liquefier, even if we could build one ourselves, and Oxford Instruments seemed, at the time, to be adequately financed with family loans and the overdraft.

The only source of commercial liquefiers was the USA, where Arthur D. Little, a Boston company, built machines to a design by Professor Sam Collins of MIT. It had recently introduced a semi-automatic, eight-litres-an-hour model; it cost £35,000 (equivalent to over £400,000 at the beginning of 2000) a lot more than the total investment in Oxford Instruments to date. We decided to buy one, the first of the new machines to be sold outside the USA. To our surprise, the bank agreed to fund it on overdraft, on the security of the plant itself with limited new family guarantees, so we did not need equity finance. The next problem was where to *house* Oxford Cryogenics and the new liquefier, as the old stables at Middle Way were bursting at the seams. After failing to get plan-ning permission to use a redundant chapel on the nearby airfield, and rejecting a barn on a brother-in-law's pig farm, we settled for a disused laundry (see Fig. 3.3). This we were able to rent on a temporary basis from Oxford City Council pending development decisions for the area.

Fig. 3.3. The old laundry in its heyday, with pre-war washing machines

In May 1964 Albert Hutchins, 'Hutch' to everyone, joined Oxford Cryogenics to run the new plant. He set about preparing the draughty leaking old building in Abbey Place—the site of a medieval abbey—near Paradise Square, for the reception of this shining space-age equipment. The liquefier was flown over from Boston and arrived in June. Milton Streeter of Arthur D. Little, known as 'Milt' to everyone in the cryogenic world at that time, came over to install it with the help of our own technicians. Liquid helium was flowing twelve days after delivery, which was probably a record (see Fig. 3.4). Many years later Milt reminded us of his first visit to Oxford Instruments, when he had privately con-sidered the possibility of selling us a liquefier the least likely to succeed of any prospect he had ever had.

In July, with the staff augmented by another technician and two drivers, Oxford Cryogenics advertised its 'milk round'. We offered liquid helium, delivered to the door, throughout Great Britain, at a lower price than BOC charged ex-works. A van made a regular circuit round customers' laboratories

Fig. 3.4. A change in technology: the helium liquefier in place of the washing machines

delivering the seventeen-litre containers of liquid and collecting the recompressed gas in cylinders. In the first four months the Company sold some 3,000 litres of liquid helium—probably a lot more than the customer laboratories had used in the whole of the previous year. The new service gave a boost to low-temperature research in Great Britain, which, for some years, remained ahead of most of Europe in this field. BOC was taken by surprise at the launch of the milk round, and its service soon became more customer friendly. With the newly competitive situation the price of liquid helium soon came down from £5 a litre (£58 in 2000 terms) to around £3, depending on quantity. In 2000 the price for relatively small contracts is between £2.50 and £3 a litre—in depreciated pounds. At the end of the century the Oxford Instruments Group, including the joint venture, Oxford Magnet Technology, was using well over two million litres a year, and was probably the largest user in Europe. The two principal liquefiers on sites in the Group could, between them, produce 300 litres an hour.

Unlike the state of affairs in the Middle Way workshop, there was no shortage of space—of a kind—in the ex-laundry. When the old concrete washing machine plinths and canvas belts dangling from their pre-war drive shaft had been cleared away, more space became usable. Soon the testing of magnets and cryostats moved down to Abbey Place, to be followed by the emerging development activity, which included work on cryogenic improvements and on the new

NbSn conductor. The small electronics team later got squeezed out of Middle Way and found another possible home for its work where the old ironing machines had been.

On a cold November day in 1964 we held a party in the old laundry for the 'formal' opening of the liquefier and the milk round. Everyone worked through the weekend to improve the tatty walls, curtain off the worst unused spaces, and set up a little exhibition. On display were magnets, cryostats, and electronic equipment made by Oxford Instruments as well as the shining liquefier and items from the 'Cryospares' supply service—special tubing, valves, sensors, and other bits and pieces needed by cryogenic experimentalists. The press, customers, and eminent scientists were invited to this celebration along with our professional supporters such as the bank manager, accountant, and solicitor. Staff and wives provided excellent food, and good mulled wine took off the chill from the draughty building, which was impossible to heat adequately. Nicholas Kurti spoke about the beginnings of helium liquefaction in the Clarendon in the 1930s—using an apparatus known as a Simon Bomb. The press loved the contrast between futuristic equipment and ramshackle building, and next morning we read one or two enthusiastic accounts with headlines such as 'Superconduction in an Old Laundry' (*Guardian*, 11 Nov. 1964).

Helium consumption rose fast over the next two years and in 1966, partly for volume but even more for security, Oxford Cryogenics installed another similar liquefier—bought second-hand this time. While Hutch continued to run the expanded plant, Ken Gamlen moved in to manage the enlarged operation. There were always problems. Helium molecules are very small and light and can escape through the tiniest pore in a tube wall to float upwards through the earth's atmosphere to outer space. Leaks are a never-ending headache for the cryogenic engineer, and helium is expensive. Our accountant called it a 'floating-away asset'. Then there was the problem of contaminated gas. Many gases freeze solid well above the boiling point of liquid helium, and, in spite of a purifier for the returned helium gas, the liquefiers regularly became blocked. Then there were vacuum problems in the storage containers and transfer lines, and valve problems, as well as all the normal problems of a growing business.

Oxford Cryogenics served its purpose in supplying Oxford Instruments' needs and making liquid helium readily and competitively available for our equipment customers in Great Britain. In spite of growing price and service competition with BOC, the Company managed to achieve a small profit in each of its four years under our control. But we did not really want to remain in the liquid-gas business, however esoteric our particular niche in it. Beside this, the scene was changing; helium gas was beginning to be used, mixed with oxygen, as a breathing gas for divers in the new North Sea oilfields. Soon companies in the USA, where natural gas wells are comparatively rich in helium, started experimental shipments of the gas in liquid form in huge vacuum-insulated containers, instead of shipping it in large numbers of heavy gas cylinders. In 1968 we sold our liquid-helium enterprise to Air Products Ltd. (APL), the second major gas company in the UK, thereby ensuring a competitive supply for Oxford Instruments, and for others, in the future.

Meanwhile Oxford Instruments was growing fast. The availability of liquid helium on the doorstep brought a surge of new orders for superconducting magnets and cryogenic equipment. There were still many technical problems—most of the larger orders were really development projects—but at least the vital testing of magnets and cryostats was no longer held up for lack of liquid helium. Workshop and office space at Middle Way and a shortage of good technicians were our main constraints. As early as April 1964 we were looking seriously for new premises. The one or two possibilities we investigated proved to be 'contrary to the County development plan', and an article in the *Oxford Mail* (10 Apr. 1964) headed 'Go Ahead Firm Looks for a New Home' brought in no suggestions. In those days the UK produced hardly any embryo companies 'spun off' from university research. Apart from the Oxford University Press, which claims to be more than 500 years old, Oxford Instruments was probably already the largest the University had spawned in its eight centuries in existence.

Throughout the country there were very few suitable small factory units for scientific companies. In the USA—the 'land of opportunity'—planning rules were less rigid and speculative factory building and local-authority encouragement made life easier for companies such as ours. There, the blurring of the boundaries between university and industry was leading to little clusters of science-based companies near campuses, where the results of research could be exploited, and academics could continue their investigations while also serving as part-time directors or consultants in these new enterprises. At that time the most famous of these clusters was in the hinterland of Boston, Massachusetts. There the oft-cited route 128 ring road was lined with the modern factories of growing firms like High Voltage Engineering, spawned from MIT or Harvard University. The Boston authorities, and the banks, had consciously encouraged and supported these science-based 'spin-off' companies as the old heavy industries declined.

During the 1990s, three decades after these developments, a few areas in this country were just about getting to the same stage. Why is it taking us so long? In the 1960s there were groups of scientists working on committees set up to look at the reasons for the country's failure to exploit its excellent research in pure science. They recognized the need to get university, government, and industrial scientists to talk to each other. There were committees or working parties under Professor (now Lord) Dainton of Nottingham University, Sir Gordon Sutherland of Cambridge, Sir Neville Mott, head of the Cavendish Laboratory (the Cambridge Physics Department), and Professor Michael Swan from Edinburgh. Since then there have been several reports on these problems, some of them from these committees; and there have been many more articles in papers and journals commenting on the situation.

David Fishlock, then Science Editor of the *Financial Times*, wrote an article entitled 'From Scientist to Manager' (30 June 1967). He looked for reasons why the first-rate British research did not spin off more small companies as it did in the USA. He retold a story from Dr Jeremy Bray, then Joint Parliamentary Secretary to the Ministry of Technology: 'A young research chemist . . . wrote to him at the height of the Torrey Canyon affair with a proposal—an excellent one, incidentally—for disposing of oil. But his letter concluded by saying that, of course, he didn't want to make any money out of it. Why the hell not? asks Bray.' Mr Fishlock saw part of the problem as a long-standing cultural disdain for manufacture among the UK élite, and scorn for 'making money from trade', as Jane Austen's characters would have put it.

R. Hobart Ellis, then editor of the US journal *Physics Today*, came to Great Britain to talk to scientists about why so few British physicists moved from academia to industry, or started their own companies. In January 1968 he published his findings (Hobart Ellis 1968). He identified a number of reasons for the differences between the two countries: the swing in the UK against science in schools, combined with early specialization, which left potential scientist entrepreneurs ill equipped for management; a failure of society generally to appreciate scientists and to value and use them properly in traditional industry; the dearth of scientists and engineers in top management positions; prejudice against engineering among those involved in pure science; the big gulf in status between university degree courses and the technical college route to qualifications, often achieved through part-time study while holding down an engineering job; the much smaller home market, than in the USA, for any new device; and, finally, the difficulties in obtaining adequate finance for growth, in this climate of suspicion of science and invention.

As David Fishlock wrote in his article, 'Finance is one of the biggest obstacles—not least, in the view of several scientists who have ventured, because those with money to lend tend to take a pride in refusing to understand technical matters' (*Financial Times*, 30 June 1967). Since success builds on success, I would also suggest that not many even *thought* about venturing because there were so few role-model entrepreneurs here, and there was no 'critical mass' of similar companies to join.

Some of these barriers to the formation and growth of spin-off companies have remained, but in the 1980s the Conservative Government under Margaret (now Baroness) Thatcher made great strides in changing the culture and encouraging enterprise. During the 1990s there developed a wider understanding of the link between a healthy sector of growing innovative companies and national economic wellbeing. All political parties have, since then, encouraged the growth of this type of company. But there are still too few people in the government, the financial sector, or the media with enough scientific knowledge to make informed judgements on some important issues, or to encourage the public's understanding of science and technology, and of statistics. The latter is important for developing in the community a better discrimination on important topics. Newspapers sell on sensationalism. The worst of them inflame anti-science campaigns and the irrational fears roused by matters like food scares. I will give an example of this disregard for statistics: while smoking and road accidents continue to kill many thousands of people every year, and are rarely newsworthy, the suspected human form of BSE had, by the end of the century, accounted for only around fifty. Yet every CJD-variant death these papers hear of hits the headlines and increases the public's terror of BSE, while millions, unperturbed, go on smoking and driving cars dangerously.

Back in 1964 our main problem was finding possible premises not too far from the University, a problem then often overlooked in the various committee reports. But early in 1965 we found advertised in the *Oxford Times* a 3,000-square-foot boat-building shed, standing on two-thirds of an acre of land, by the River Thames at Osney Lock (see Fig. 3.5). There was no access except by river, but a newly designated industrial estate for the relocation of old-established Oxford companies, such as printers and stonemasons, was being pegged out on the neighbouring meadows. There would soon be road access. We could not qualify as an old-established firm, but were able to buy the boathouse for £13,650 (equivalent to £157,000 at the beginning of 2000). This was double the size of the old Middle Way stables, and later we were able to increase the site to an acre for a further £6,000. The funds for the new premises came again from

Fig. 3.5. The boathouse as we found it

our generous bank manager in the form of an overdraft. The facility was large enough to cover the extensive work needed to make the very basic structure habitable, and to provide a small prefabricated office unit to be added shortly. The building work progressed painfully slowly, and, as conditions became more and more congested at Middle Way, small groups migrated to corners of the cold and damp but spacious building at Abbey Place.

Several engineers and technicians joined the Company in advance of the move to Osney Mead. Tony Groves remained with the Company for over thirty years and finally became Production Director of the operation making whole body magnets for medical imaging, before taking early retirement. Michael Cooper is still with the Group in 2000. John Pilcher worked for many years in the original company, which became the Research Instruments Division, of which he was MD until his retirement in 1998. These exceptionally long-serving members of staff have a unique place in our history; they have supported the Company through thick and thin down the long path from its juvenile days to its present position as an international public company.

Early in 1965 our outside accountant and auditor, Will Penfold, completed the year's accounts to the end of September 1964. On sales of £41,000 (equivalent to £480,000 at the beginning of 2000), we had made a net profit before tax of £2,460 (£29,000 in early 2000), about 6 per cent. *The Economist* published an article entitled 'Getting Down to Nothing' (22 May 1965). It began, 'One of the smallest and choicest companies in the country, working in an entirely new field, is now making a profit; which could serve as a definition for the starting point of an industry. If so, a lot could spring from the tiny shoulders of Oxford Instruments.' Prophetic words; apart from its own growth, Oxford Instruments has indeed been the core from which a cluster of cryogenic and magnet companies has arisen, mostly in the Oxford–Abingdon–Witney triangle. Entrepreneurial employees have seen a niche in the market, and left to set up on their own, sometimes with the blessing of the Company. But we have also experienced more painful defections.

In 1965 our experiences of this kind of breakaway were still to come. That September another financial year closed before the Company was able to move. Sales had grown to £94,000, equivalent to well over a million pounds at the end of the century, and the pre-tax profit at over £7,000 (£78,500 at the beginning of 2000) had risen to nearly 7.5 per cent of turnover.

4 Triumphs and Trials

In spite of technical successes, the Company's survival was threatened by growing pains in the business

Nothing in the world can take the place of persistence . . . Talent will not: . . . Education will not: . . . Persistence and determination are omnipotent.

(Calvin Coolidge, from a speech in the 1920s)

Come what come may
Time and the hour run through the roughest day.

(Shakespeare, *Macbeth* (1606)

One weekend in October 1965 the Company finally moved to the not-quite-ready ex-boathouse on Osney Mead. The small office block had not even been started, so the caravan offices, now two in number, were heaved by crane out of the slaughterhouse yard and moved to the rough site in front of the boathouse. At the date of the move Oxford Instruments employed twenty-five people including the directors, Frank Thornton, by then the MD, Martin, and myself.

We expected the new premises, with over twice the floor area of the old stables, to give us enough space for several years. With a freehold factory and a growing reputation we were able to recruit more scientists, engineers, and technicians to cope with the swelling order book. More support staff were then needed; secretaries to deal with more letters and other paperwork; an extra book-keeper to keep track of the increasing number of purchases and sales and the rising wage bill; a storekeeper for the expanding numbers of parts and materials; a lady to provide sandwiches; and so on. It seemed no time at all before the staff had expanded to fill the new building, and we were talking about extensions and more temporary huts (see Fig. 4.1).

Soon after the move John Woodgate joined the Company. He had been working on a linear accelerator at the Rutherford and Appleton Laboratory (RAL), south of Oxford, one of the research laboratories owned by the Science Research Council (SRC). He was to fill several important technical and management positions in the Group over the next thirty-three years. He played a

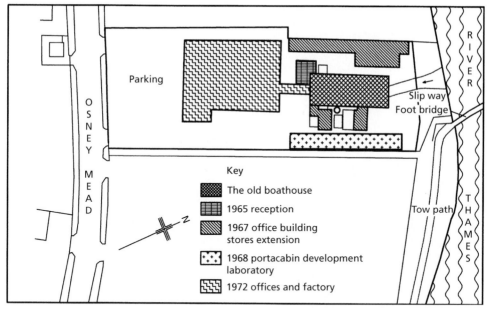

Fig. 4.1. Development of the Osney Mead site

pivotal role when, as its MD, he transformed the large division, which makes the magnets for body scanners, into one of the twenty-five most advanced manufacturing units in the UK. John was a Group director for many years; he finally retired through ill health in 1998. Sadly, he died in 1999.

Magnets, cryostats, and complete cryomagnetic systems were growing in size and complexity. The electronics team developed new instruments for monitoring the liquid-helium levels and for measuring and controlling the temperature inside cryostats—physicists often want to carry out the same measurement over a range of temperatures. There were some attempts to standardize magnet systems as well as cryostats and electronics, but the business remained obstinately project- rather than product-based. Systems, designed in close liaison with a customer-scientist, were often innovative and involved a large element of creative thinking and development, although, in response to the research-funding pattern, it was usually necessary to quote on a fixed-price basis. This was a risky matter, but it was on this work, as well as on our own few developments, that the Company's unique experience was founded.

Our own in-house developments were at the 'cutting edge' of the possible; if successful, these prototypes could always be sold. In 1966 the Board allocated £5,000 (equivalent to over £54,000 at the beginning of 2000), for the risky development of a 10 T magnet, using the high-performance but difficult superconductor NbSn (see Fig. 4.2 and Plate 5). Success brought an immediate sale, and orders for more of them—but always a little different in dimensions, field,

Fig. 4.2. Winding a 10T magnet from early niobium-tin tape

temperature requirements, or in the access needed for a beam of neutrons, or X-rays, or whatever.

The most notable development contract the Company was working on at that time came from the Atomic Energy Research Establishment (AERE) at Harwell. This was for the development of a 'helium three/helium four dilution refrigerator' based on a new cooling cycle proposed by Dr Heinz London, which was expected to produce extremely low temperatures. Harwell administrators had been to see us as early as 1962, when we were still using the garden shed. They said they would be choosing between Philips of Eindhoven, Arthur D. Little in the USA, and Oxford Instruments—a pretty wide spectrum. By 1964 our operation was a bit more substantial, and they decided to entrust the development to a tiny company close by rather than one overseas.

At that time, Professor Eric Mendoza of Manchester University was in the third year of a contract with AERE Harwell. The aim was to prove the principle of the cooling cycle and to get to maintainable temperatures below 0.1 K. This would be three times lower than the 0.3 K that had been achievable previously by evaporating liquid ³He, the lighter isotope of helium. He was making some progress, as also was a group in Leiden University, in the Netherlands, but he was soon to move to the University in Bangor. Harwell asked Oxford

Instruments to carry the project forward, working with Dr London, under a new two-year £10,000 contract. For this scientific development, to start in March 1965, we recruited Dr Dafydd Philips, a physicist who had been working with Professor Mendoza. He soon started to design a large apparatus to investigate the problems encountered in Manchester.

In the spring of 1965 Professor Henry Hall, of Manchester University, while discussing the original project, realized how he might make the cooling cycle work more effectively. In the summer he got his new apparatus to achieve temperatures well below 0.1 K (100 millidegrees above absolute zero). The cycle was well and truly proven, and shortly afterwards Professor Hall published in detail exactly how to build a similar refrigerator.

Harwell held patents on the cooling cycle and had offered Oxford Instruments a three-year sole licence for its commercial exploitation. The Company decided to go ahead straight away, using our own resources, to develop a commercial version based on Professor Hall's experience. This was to proceed in parallel with the scientific work with Heinz London (see Box 4.1 and Fig. 4.3). Early in 1966 our first prototype achieved a temperature of 0.067 K,

Fig. 4.3. John Woodgate and Peter Blowfield at work on an early dilution refrigerator

Box 4.1. The ³He/⁴He Dilution Refrigerator

Before the new cooling cycle was confirmed, the lowest possible steady temperature was 0.3 K, which was achieved by evaporative cooling of liquid ³He, the light isotope of helium, by lowering the pressure above it. The commoner isotope, ⁴He, boils at 4.2 K under normal pressure. This can be reduced to less than 1.0 K by evaporative cooling in the same way. When ⁴He is cooled to 2.18 K, another phenomenon occurs: it loses all viscosity and flows without friction; it has become 'superfluid'. It will now flow up the walls of its cryostat or through the finest crack or pore.

Dr Heinz London's first idea, proposed in 1951 at the second Low Temperature Physics Conference, held in the Clarendon Laboratory, was that a cooling effect would result from diluting ³He with superfluid ⁴He. The ³He atoms would 'not notice' the inflowing superfluid ⁴He atoms and the former would behave like an expanding gas, which always produces a cooling effect. Then in 1956 American scientists noticed that a mixture of the two isotopes separated out into two distinct phases below 0.87 K, the upper phase consisting largely of the lighter ³He (the concentrated phase) and the lower mostly of the heavier ⁴He (the dilute phase). In 1962 Heinz London and others suggested that an additional, and stronger, cooling effect, analogous to evaporative cooling, would result from ³He molecules crossing the boundary between the phases, because they would expand like a gas into the superfluid ⁴He.

In a practical closed cycle refrigerator, ³He is circulated. Coming from room temperature it is first condensed to a liquid at 1.3K and then flows down through heat exchangers to the 'mixing chamber', where the two phases are separated. As it crosses the phase boundary it cools, creating the lowest temperature in the cycle, and then flows back through the heat exchangers, is evaporated to a gas, and is circulated again by powerful vacuum pumps. (See Fig. 4.4.).

and in April we showed the second at the Physics Exhibition in Alexandra Palace. This stimulated a lot of interest, with articles by science correspondents intrigued by this 'coldest place on earth'. The Company soon received orders for this unique product, and by October 1966 we were offering Mark III versions at prices between £6,000 and £7,000 (equivalent to £67,000 to £75,000 at the beginning of 2000). The only competition for several years came from do-it-yourself physicists building their own equipment from Professor Hall's description. But it was not as easy as it might have appeared—some who tried it came back to buy from us later. These dilution refrigerators have been in demand ever since, and, between 1966 and 2000, the Company made over 500. To the layman, the prospect of getting from 0.3 K to below 0.1 K may not sound a big step, but the temperature is down by a factor of three, and a great many important events take place in the temperature range below 0.1 K.

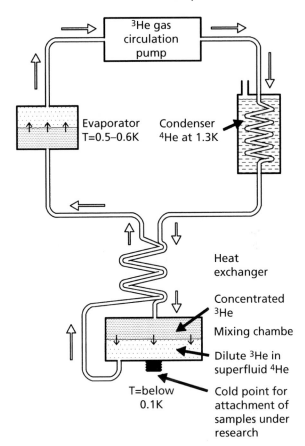

Fig. 4.4. The cooling cycle of the ³He⁴He dilution refrigerator

John Woodgate became the manager in charge of the production refrigerators, and, by 1967, had achieved a temperature of 0.03 K. By 1976 it had got down to 0.012 K, and our minimum has now been at 0.0035 K since the early 1990s. For most customers, ease of use and cooling power have become more important than reaching the ultimate lowest temperature. By the 1990s there were several standard versions in the Company's 'Kelvinox' range, and many special refrigerators have been made for particular experimental purposes (see Plate 9).

In 1967 the Company won its first Queen's Award, for technical innovation in developing the 10 T magnet and the dilution refrigerator. The Queen's Awards (see Fig. 4.5) were only in their second year, and were still very newsworthy. From 1,300 applicants Oxford Instruments was the smallest of eighty-five to win. The Queen's representative in Oxfordshire, Sir John Thompson, the Lord Lieutenant, came to present the trophy at a great party, where we showed round distinguished scientists, the press, local dignitaries, important suppliers, our bank manager, and other supporters, as well as the families of our hard-working staff.

The Queen's Award generated some publicity. Articles with headlines such as 'Award for Firm that Began in a Slaughterhouse' (*Oxford Mail*, 21 Apr. 1967), and 'The Saint and a Deep Freeze Both Winners' (*The Times*, 21 Apr. 1967), roused the curiosity of other journalists (see Plate 2). At that time there were hundreds, if not thousands, of small electronics companies in the UK, but in magnets, superconductivity, and cryogenics Oxford Instruments was unique (see Fig. 4.6). Over the next year several scientific and business journalists came to see the small company that was tackling such esoteric technologies, and wrote about our progress. The refrigerator featured on BBC television's scientific programme *Tomorrow's World*, and appearances on a couple of business programmes enhanced the Company's reputation. One of these was to be entitled 'Today's World Success Stories', but when it was transmitted we were disconcerted to find it had turned into 'How to Make Your First Million'. Technical success was not achieved without perpetual headaches and challenges to our ingenuity, and, as we 'outgrew our strength' in the late 1960s, survival was more of a preoccupation than very questionable future wealth. Most of the other companies that featured in this series have disappeared, either through failure or through takeover.

In the late 1960s the physics of superconductivity was not fully worked out, and even today there are gaps in the fundamental understanding of the phenomenon. The metallurgists used experience and pragmatism as well as knowledge of the materials employed. Small changes in the wire manufacturing process could

Fig. 4.5. The Queen's Award logo

Fig. 4.6. Ron Boyce fitting a large split-coil magnet into its cryostat

have unexpected repercussions. We had learned to live with frequent problems, but one episode, which lasted for many months in 1966 and 1967, put the Company under extreme pressure. The outside investors in the US Company Supercon had expected an overnight success, and after two or three years without seeing any spectacular new applications for their materials, they lost interest. Supercon started to fall apart; Jimmy Wong left, and it was taken over by a subsidiary of the Norton Group, NRC, which was already involved in developing superconductors.

After a while, unbeknownst to us, the engineers made 'improvements' to the production process for the superconducting wire to enhance one parameter of performance. It had other insidiously disastrous effects. The short-sample specifications on the reels of wire were good. Magnets were wound, successfully tested, and shipped to customers. Then, unaccountably, a number started to come back to us as below specification. The mystery deepened, and many large projects in production were held up while our scientists tried to get to the bottom of the problem. NRC-Supercon was the only current supplier of this material, although one or two other companies were working on new types of superconductor. Why were some magnets all right, while others, wound from wire with the same specifications, were developing this baffling fault? It took six months, the help of several physicists and metallurgists, and many thousands of pounds to solve the problem.

The cause of the trouble turned out to be a fundamental one. The small changes in the metallurgical process had resulted in an alloy whose crystal structure changed slowly at room temperature. Magnets that had been kept permanently in liquid helium did not suffer the same fate, but, for the many magnets that were not in use continuously, the problem was very serious. After the diagnosis, Jimmy Wong again joined NRC-Supercon to steer the Company to new superconductors. But it was not until March 1967 that we were able to obtain production quantities of good material. By then a better superconductor, niobium titanium (NbTi), had finally emerged from development, and was put on the market by the British company Imperial Metal Industries (IMI) of Birmingham.

In spite of the Company's prowess in finding solutions to these technical trials it was still quite naïve in business matters. At Osney Mead the turnover was growing fast; the monthly wages bill and 'bought-ledger' payments were larger, and the overheads were inevitably higher. The need for more working capital was masked, at first, by the increased overdraft facility for the purchase and conversion of the premises. This was meant to cover the completion of the much-delayed office building as well as the boathouse renovations. At Middle Way simple management methods and financial controls had been enough to keep track of the business. With a larger operation, and a management preoccupied with technical matters, they had now become inadequate. The financial situation soon deteriorated, and we started to bump up against our overdraft limit.

From the time of the move in 1965 we had known the Company would soon need permanent capital. In technical terms we were 'over-trading' like many other small companies; we had grown for too long on bank overdrafts, with periodic rescues from family loans and guarantees, which were now reaching their limit. Between 1965 and 1967 we looked at some eight or ten possible sources of finance: large companies interested in our technology; merchant banks; insurance companies; and even rich private individuals who would today be called 'business angels'. The term 'venture capital' was then almost unknown in the UK. The months went by and we failed to find the right partner, or to be offered acceptable terms. The superconductor 'famine', which started in 1966, compounded our problems, cutting into our cash flow alarmingly. Work had to be switched from new magnet production to investigations of the magnets coming back under guarantee. The monthly output dropped. Our bank manager became very anxious—he had never really understood the technical risks in the business. Our September 1966 accounts had to be withdrawn before publication owing to the impossibility of putting a value on our superconductor stocks, or on our liabilities under guarantee.

This was not the best scenario for negotiating for long-term finance with experienced business men, already rather nervous of the recondite technology. Somehow we kept going. We were still producing the occasional large copper-wound solenoid for the highest magnetic fields, and the cryogenic work, independent of magnets, was becoming a relatively substantial business. Beside this, the Harwell development contract on the dilution refrigerator provided for regular monthly payments for labour and materials—with a reasonable over-head—and we were able to switch more effort into this area.

Many a US technology company in similar circumstances has been reprieved by government development contracts to grow into a substantial enterprise. Since 1982 the US National Science Foundation's programme Small Business Innovation Research (SBIR) has directed all departments with R & D budgets of over $500 million to spend 2–3 per cent of these budgets with small companies. This well-structured incremental programme was laid down by an Act of Congress. Through funding agreements, many small firms have developed their technologies until able to compete with larger rivals and to grow and thrive. This has certainly helped the USA to build a substantial sector of growing science-based companies.

In the winter of 1966–7 the country was in general gloom. Interest rates were rising, and the UK was trying, unsuccessfully, to join the European Economic Community (EEC). Pressure from our bank was escalating. Early in 1967 we started negotiations with IMI, the company working on the new superconductor. The managers involved in this operation saw a minority investment in Oxford Instruments as 'vertical integration'—having a stake in the end product as well as in the superconductor. A further possible stage could be a joint venture for making the very large magnets being discussed for particle physics. IMI would bring money and superconductor to the party and Oxford Instruments our extensive know-how. But, after enthusiastic meetings with the managers of the interested division, the IMI Board drew back, feeling their exposure to this 'risky and capital-intensive new industry' was already too great.

We turned rapidly to the next possibility on the list. APL, the British subsidiary of an American company, was supplying Oxford Cryogenics with helium gas and had, for some time, shown an interest in closer ties. But it soon became clear that any relationship that brought us adequate funding would mean an eventual takeover, and we wanted to remain independent. Independence enables a small company to reach decisions fast and to act fast; it is in control of its own destiny. Such freedom has its own dangers, but you know where you are. A large potential owner may be subject to changes; divisions can be reorganized; contacts may move to another section; the parent company itself may be taken over. Once control is lost to a larger company, a helpful and supportive big brother may give place to a preoccupied and uninterested parent, banning risk and stifling action. And the company that is taken over is also liable to change for the worse.

A takeover, or 'trade sale', is a common destination for a growing and successful small company, especially one that is running out of cash and support.

Sadly, many of the companies acquired fail to achieve the results expected by the new parent. Somehow, at the end of the negotiations the hard-pressed management relaxes; access to the much-needed new resources, and the novel dependent relationship, seem to dampen the urgency and drive for achievement. For both parties, with apologies to Shakespeare, 'Our remedies oft in ourselves do lie, which we ascribe to the other party.'

Nowadays, with secondary markets, the capital-hungry small company may be able to procure a listing on an exchange even before it achieves a profit, especially in the high-risk/high-reward areas like biotechnology, where years of research must precede income. A flotation may well be urged on a company by early individual or venture-capital investors wanting an 'exit route'; they know that the investing public is tempted to gamble on the chance of the high reward. In 1967 Oxford Instruments, with a turnover equivalent to more than £2 million at the end of the century and profits before tax of about 8 per cent, did not dream of a public listing. There was no secondary market, and the rule of thumb for a full flotation was a profit of £1 million before tax (equivalent to nearly £10 million at the beginning of 2000).

APL was an excellent company, so, in spite of our misgivings, and under strong pressure from the bank, we went on negotiating—for several months. Nothing could be decided by APL without reference to the parent company across the Atlantic, Air Products and Chemicals Inc., but by June 1967 we were close to a settlement. We had even initialled Heads of Agreement, when the discovery of a transatlantic misunderstanding, over how debt was to be treated, led to a hiccup. The effect was to reduce by such a large sum the value of the offer, as we had understood it, that negotiations were suspended while each side thought again. We were so hard pressed that the break might have been temporary, had a different way out of the maze not opened up for us.

The four months of negotiations between February and June 1967 had seen great changes in the Company's underlying prospects. Oxford Instruments' staff had been working flat out to improve both technical and financial situations. Good superconductor had finally started to arrive from IMI in March, followed by the rapid completion of the queue of delayed projects. This led to good profits in May and June. With a settlement over the old faulty superconductor stock we were clear to issue new accounts, which now extended to March 1967. They showed a profit for the eighteen months of £28,000 (equivalent to nearly £300,000 at the beginning of 2000), and the order book was substantial and growing. There was the first Queen's Award, announced that April, and Oxford Instruments was being cited in the press as the kind of company the country needed.

Although these improvements swelled the debtors' list and foreshadowed successful trading ahead, it would be some time before they made any impression on the cash flow and the overdraft. Our sorely tried bank manager, sitting in an ancient institution in London, who had been so supportive over our earlier growth, was fast reaching the end of his patience. He could not see round the corner we were turning. He had been counting on a rapid end to our problems through the APL negotiations, and, as they dragged on and on, he appointed an outside accountant to keep him informed and to assist with the Company's financial affairs pending their successful conclusion. This man was quite helpful in convincing the bank that a solution was attainable, but one day he mentioned casually that he normally acted as a receiver. To forestall any really uncomfortable action our bank manager might unexpectedly take, we decided to seek new banking facilities locally. We were frank with the Manager of the Midland Bank in Oxford, who recognized that we were pulling out of our worst technical problems. He agreed to take over the debenture and fund the overdraft, giving us time to complete our negotiations for long-term finance without a gun at our heads.

Oxford Instruments' reprieve was to come from virtually the only 'venture-capital' organization then working in the UK. During the previous year, the Industrial and Commercial Finance Corporation (ICFC, now 3i) had taken over full ownership of a small investment operation, Technical Development Capital (TDC). At the press conference marking the 'merger', ICFC's Chairman, Lord Sherfield, said: 'It is our aim to ensure that no worthwhile technical development fails to be exploited in this country merely through lack of financial backing at the commercial stage.' Our earlier contacts with ICFC had been at a fairly low level, and had not been satisfactory, but the new subsidiary, TDC, was *seeking* technology investments. In fact, Sir Frank Turnbull, one of its directors, had been to see the Company in February, when we had been deep in other negotiations. He had wanted Martin's ideas on how to get exploitable results of research from the laboratory into commercial production. Sir Frank had invited us to approach TDC should our other prospects for long-term capital come to nothing.

As soon as the APL negotiations faltered, we got in touch with Sir Frank. By then we had a much better picture to present; there was the recent Queen's Award, the resolution of our worst technical problems, the good current trading profits, and the reasonable eighteen months' accounts. On the wider stage, Harold Wilson was still seeking the revitalization of British industry through the 'white heat of the scientific revolution' (MacArthur 1993: 336), and there was growing disquiet in the country at the queue of high-technology companies being taken over by US firms. At our first meeting with ICFC and TDC in London, Lord Sherfield and his team were clearly predisposed to be helpful. Then and there they agreed to make an investment offer, and 'Heads of Terms' were signed within a fortnight. They bought 20 per cent of the Group, including Oxford Cryogenics, in preferred ordinary shares, for £45,000 (equivalent to nearly £420,000 at the beginning of 2000), and they put a similar sum into the Company as loan capital under a debenture. Oxford Instruments would remain independent, with enough capital to carry us forward. How glad we were that we had resisted the bank's pressure on us, and dug in our heels when the APL negotiations turned out to be valuing Oxford Instruments as so much less than we considered reasonable. The relationship with ICFC (and later 3i) has remained almost consistently good over the thirty plus years since then. At the time of our public flotation in 1983 one of their managers told us that the 3i investment in Oxford Instruments was possibly the most profitable in its history.

In the negotiations in 1967, ICFC asked for more equity investment from the very few other shareholders, to confirm our commitment. We took this opportunity to transfer low-price 'rights-issue' ordinary shares to some score of senior employees and a few others who had given us generous support through difficult times. Oxford Instruments was one of the first companies to realize the value of employees sharing in ownership. Now, after various new schemes over the years, almost all those who work in the Company are shareholders, and many who have been with the firm a long time own substantial stakes in the Group.

With adequate finances in place, Oxford Instruments surged forward. The staff of both operations grew from thirty when we left the old stables at Middle Way to seventy-six by March 1967 and to 105 by March 1969. Although Oxford Cryogenics was sold in 1968, taking a handful off the payroll, employee numbers still went smoothly up (see Fig. 4.7). With the loss of the old laundry

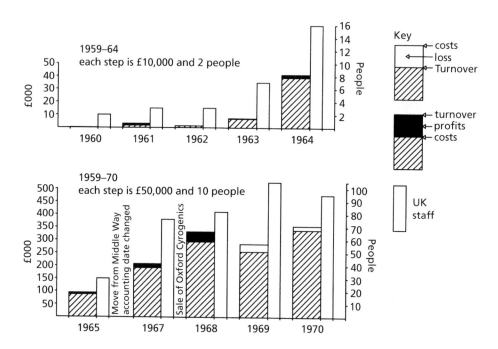

Fig. 4.7. Oxford Instruments' progress in the 1960s

through that sale, more room had to be found. The prefabricated office building had, by then, been finished, and extensions and temporary buildings started to appear like mushrooms (see Plate 4). There were still very few modern factories for rent, but soon the Company managed to negotiate a lease on a small factory half a mile away. This building had once housed the horseradish-sauce production for Coopers, who are better known for their Oxford marmalade.

Exports were growing. In 1967 the Company appointed a newly established firm to represent us in Switzerland, Germany, and France. Cryophysics was founded by an American physicist, Ken Geiger. To begin with, Oxford Instruments accounted for most of his sales. Ken and his small staff worked effectively at marketing and sales and produced a good newsletter, describing achievements in research as well as the equipment he was selling. This gave the Company a firm base in Europe from which to grow. In 1968 we set up a small subsidiary on the west coast of America, run by Dr Jay Templeton. The US market is a very hard nut for a small European company to crack. Although Jay generated a reasonable number of sales, like many another US office of a European company, the subsidiary lost money, and underwent several changes over the next decade. We had quite a few enquiries from Japan, and in 1969 Martin and I spent three weeks in Tokyo, looking for an agent and trying to understand the culture. After visits to several agencies and scientific companies where we found it difficult to judge whether there was the necessary enthusiasm, we appointed Columbia Import and Export as our Japanese agents. Mr Tobita, its founder, had been brought up in the USA, and with him we could meet in mutual understanding. This relationship lasted for many years.

David Roberts was then in charge of the sales office in Oxford, but the sales team really included all our scientists and engineers, who visited and corresponded with customers around the world. The project engineers discussed the proposed experiments and designed the equipment to do the job; when an order came, the engineer concerned would follow it through production and testing to delivery.

Frank Thornton had been MD since 1965, and in the late 1960s John Rackstraw, our first employee, served as Production Manager. Colin Hillier was then in charge of Design and Neil Munro headed the Electronics section. Dr Barry McKinnon was recruited late in 1968 to run a more formalized Development Group and Dr John Williams joined the Company from MIT to work on large magnets. He left his mark designing and directing the building of the world's first commercially made superconducting quadrupole magnet, the largest magnet the Company had ever made (see Fig. 4.8). This was an experimental magnet for the European Centre for Nuclear Research (CERN) in Geneva. It was designed for focusing beams of charged particles from an accelerator, and was funded by the British Government.

Confident in the ICFC support, the Company expanded fast; but preoccupied with technical developments and manufacture, the management still did not give enough attention to the mechanisms of good business practice. Peter

Fig. 4.8. John Williams supervising the assembly of the world's first commercial superconducting quadrupole beam-focusing magnet

Austin, ICFC's manager in its Reading office, came to discuss progress from time to time, but it was then the company's policy to back people, and let them get on with the job. With the shortage of outside experience in both board and management, we would really have benefited from more 'hands-on' guidance. The three directors learned a lot in a short space of time from a one-day management training session, with case histories, which ICFC organized for its investee companies. At this seminar we were also given a mauve-coloured booklet that gave clear, simple, useful advice. It remained my management bible for two or three years, but, alas, it seems to be extinct, and I cannot now locate a copy.

There were, by then, quite a few one- or two-day management courses put on by the British Institute of Management and other bodies. Our managers were too busy to attend longer courses, but a few spent the odd day on this sort of training. As a company grows it needs to use more formalized management techniques. For hard-working managers, one day away from the normal routine at such a course sets the mind thinking along new paths. Participants can compare their experiences with those of others, and listening to clearly analysed case histories does wonders for both technique and confidence. Today the need for such training is generally understood, and it is readily available in courses of all kinds, not to mention the advice to be found on the crowded business shelves of bookshops—the problem is what to choose.

In 1968 and 1969 Oxford Instruments suffered from more bouts of technical difficulty arising from changes in type and composition of superconductors. Management time was focused on trying to solve these problems, and some of the controls already in place started to slip. It was later to become clear that the cost records for some finished projects were not complete. Without a sound basis for making new quotations for similar projects, the Company was in danger of under-quoting. There was, at that time, one British competitor for laboratory scale magnet contracts, the British Oxygen Company (BOC). It had set up a division to work on the technology, no doubt in anticipation of a future market in much larger devices. In this competitive climate, with BOC not particularly worried about making a profit in this development division, prices were driven down. This competition, combined with insecure costing for difficult equipment, and problems with new forms of superconductor, were leading towards actual losses on quite a few long-term contracts. Failures in recording some of the expenditure on these projects hid this trend for a year or two. The Company's cash flow started to suffer, but, as sales were expanding at 80 or 90 per cent a year, we believed the need for more working capital could account for this. The losses were not fully identified until 1971.

The bank was patient over the new technical problems that were not of our own making, but became concerned as the overdraft continued to climb. ICFC remained supportive, but pointed out some administrative deficiencies. Towards the end of 1969 Martin left the University to spend more time in the Company, both on its organization and marketing, and on its technical and financial problems. As we had recruited a Chief Accountant a year earlier, my main work at that time was in marketing and publicity. In my Board role I tried to analyse monthly progress and to encourage improvements in management reporting and in training. By Christmas 1969 the position was becoming serious. Early in 1970 Martin, as Chairman, instituted more formal weekly meetings of the three directors to try to resolve the difficulties, and to get better management accounting and control systems in place. Monthly output budgets and cash-flow forecasts rarely seemed to coincide with the actual outcome, and swift corrective action was now necessary if the patience of the bank was to last. Martin and I became apprehensive for the Company we had started, and for its many loyal employees. It was too late now for management training; we needed rapid advice and help from outside. We decided to bring in management consultants from ICFC.

The ICFC consultancy, scheduled to start in June 1970, was to be a valuable exercise that taught the Company a great deal. But before it could begin, Frank Thornton, the MD, who had contributed so much to the early days of the

Company, decided to leave. He said he had ideas for setting up his own non-competing business. He preferred a more intimate management style, with an enthusiastic small group of friends, to a regime with additional paperwork and other changes, which he felt we were forcing on him. Frank's way of running things appealed to several members of staff with whom he had been working closely through the previous year. These included the accountant and three or four other senior people. They left, one after the other, to join him in the late summer in a new company, Thor Cryogenics. Despite the unfortunate aspects of this breakaway, we remain grateful for the early pioneering work of the people who left us at this time.

In the event, Thor took up a building ten miles from Oxford, and immediately started to compete with many Oxford Instruments' products. Knowing all our customers, it soon began to take a few contracts, often at very low prices. For some years Thor remained a source of irritation or worse, but it was never to achieve lasting success. In 1970, as it became clear that they were offering some standard products on what were basically Oxford Instruments' designs, Martin and I were taken by the Company Solicitor to consult legal council. He advised us that the courts were always reluctant to stop someone from earning their living by what they had learned in a previous job. Non-competition clauses in contracts of employment would probably not be enough, and, even if one could prove that company drawings had been used, the outcome would not be certain. Litigation costs a lot of money, and, perhaps even more disturbing, it costs time and emotional stress. We decided to get on with the business of running the Company and getting it through this difficult period.

This sort of breakaway can happen for a variety of reasons, some good, some bad. One example is Vacuum Generators (VG). Bernard Eastwell had been working in Mullards and, with another engineer, he founded VG in 1962 to manufacture high vacuum components. To begin with these were not dissimilar to those made by Mullards. Others joined them from Mullards and from AEI, and VG grew to become a flourishing group of scientific companies that was ahead of the world in some technologies. I do not know what the situation was in Mullards before this breakaway, but sometimes teams of entrepreneurial managers or scientists, frustrated in their ideas for new developments by the sluggish companies they work for, decide to 'go it alone'. One has sympathy with this route and any resulting competition may be beneficial to both sides as well as the national economy. There are also less laudable cases where a team keeps secret its new ideas or gadgets, developed in an employer's time, and leaves to set up in competition. In the USA 'spin-outs' of all kinds are commoner than in Europe, and are more generally accepted.

The history of virtually every organization has its bad times, its boardroom battles, its schisms, its threats of litigation, its financial danger points, its partings from old colleagues in strained circumstances. For Oxford Instruments this was the unhappiest time, and I am not going to dwell on it for long. It was a time for spending every waking moment doing what had to be done to keep afloat.

The finances were stretched almost to breaking point; the suppliers were wondering whether they would ever be paid; rumours of our imminent demise were rife; and one or two customers were hesitating before placing orders. But the Company had a lot of momentum and plenty of contracts in hand. No irreplaceable people had gone. The latest superconductor problem was yielding to our engineers. Magnets and cryostats continued to be delivered and, eventually, paid for, and we discovered a fund of new qualities and firm loyalties in those who remained. In some ways the small exodus was a blessing in disguise: it cleared the decks of dissatisfied people in preparation for the next stage in the Company's development. But at the time we could not see that far ahead.

The business had to be run. In the vacuum that followed Frank's departure Martin took over the reins of technical and people management and I gave up most of my other responsibilities to manage the financial and administrative functions. We still had confidence in the future. Oxford Instruments was a sizeable organization with a good reputation technically, a turnover of £350,000 (equivalent to about £3.4 million at the beginning of 2000) and a substantial pool of talented and experienced engineers, physicists, and technicians. Confidence was all important. It was absolutely vital to keep up the confidence of the staff, of the essential suppliers, of our financial supporters, and of our customers. Martin talked through the necessary reshuffle of executive jobs and reassured staff and customers. I took charge of the administrative office and the chaotic financial situation I found, negotiated with our creditors, wrote monthly cash-flow budgets, and kept the bank up to date with progress.

As the Company started to stabilize we looked to the future. Like horses for courses, most growing companies need a new style of management for each major phase of development. Oxford Instruments had got to the stage where it needed more-professional management techniques if it was to prosper again. Ian Cuff, a senior executive in ICFC, carried out a management audit and was impressed with the calibre of the executive managers and potential managers who remained. But there was no one immediately capable of taking the top executive job, so, with his help, we set in motion the recruitment of a new MD. After the preliminary interviews held on our behalf by ICFC, the short-listed candidates came to Oxford to see the Company. Within half an hour of our first introduction we recognized that one candidate stood head and shoulders above the rest. He had a depth of penetration and clarity of exposition of business matters quite new to Oxford Instruments. He took the job.

The Company was extremely fortunate to find Barrie Marson, who was almost ideally equipped to be our MD (see Fig. 4.9). A physicist, he had experience in aerodynamics, in electronics, and in infra-red technology. As a director of Kent Instruments, he had built up a new Digital Systems Division from nothing to an operation comparable in size to Oxford Instruments. He knew how to grow a science-based company, but he also knew the disciplines necessary in larger organizations. His main hesitation was his lack of experience in cryogenic and superconductor technology, but we were already looking towards diversifi-

Fig. 4.9. Barrie Marson

cation for growth, as the business faced a new recession, and the obvious area for this was in electronics. Barrie joined at a low point in our fortunes and suffered from a frustrating shortage of working capital for several difficult years. His prudent realism and business insight formed a good counterbalance to the enthusiastic optimism and eagerness for technical risk-taking of some of the senior staff. A gentle and reasonable man, Barrie set very high standards for hard work and commitment—for himself as well as for others. As Chief Executive he achieved recovery and diversification until, as Chairman, he took the Company public with a stock-market flotation in 1983. In his time Oxford Instruments grew from a staff of around 100 to 1,300; from a turnover of half a million pounds to nearly a hundred million; and from a loss of £20,000 to a profit of nearly twenty million. By any measure this was a Herculean accomplishment. We all owe much to him.

5 The Slow Climb from the Morass

Recovery, technical progress, and diversification against a difficult economic background

Now here you see it takes all the running you can do to stay in the same place. If you want to get somewhere else you must run at least twice as fast as that.

(Lewis Carroll, *Through the Looking Glass* (1872))

The 1970s were a difficult decade, confronting British industry with many challenges. Changes of government in 1970, 1974, and 1979 produced a switchback of political and economic philosophies. Industry faced severe fluctuations in interest rates, in exchange rates, and in taxes. Labour unrest led to major strikes through much of the decade, the powerful and militant unions constantly pushing up wages—in one year by an average of 33 per cent. The unemployment rate climbed and there were many business failures. In 1973 some of the principal oil-producing countries united to drive oil prices up, eventually by a factor of four. This first 'oil price shock' resulted in an international recession in the industrialized world, which, unusually, was accompanied by severe inflation. Inflation was indeed the worst enemy in the decade; between 1970 and 1980 prices increased by 350 per cent—you needed £3.50 in 1980 to buy an item that had cost £1 in 1970.

On the more positive side, in 1973, after General de Gaulle finally gave up the French Presidency, the UK, under the premiership of Edward Heath, was at last able to join the EEC. This was confirmed by a 72 per cent 'yes' vote in a referendum under Harold Wilson's next Labour Government, although the Labour Party had previously been rather negative about the idea. Before the end of the decade North Sea oil started to flow and began to improve our chronically bad position in the balance of international payments; but the pound then started to rise, uncomfortably for exporters.

In January 1971, against an already difficult background, Barrie Marson took up his appointment as MD of Oxford Instruments. He soon started to introduce better management systems and controls. The financial position of the Company was considerably worse than anyone had realized. The lack of fully reliable records on work in progress and stock usage, and confusion in some other areas, had led to difficulties in preparing the 1970 accounts. These were late, and had to be qualified by the auditors, as they could not certify that proper procedures had been used in every case. But, as far as could be assessed, the Company had still made a small profit that year. The cash flow was suffering, but that was known to be due, at least in part, to production delays caused by our recurrent technical problems, and, of course, to our growing need for working capital as the Company expanded.

By the time the 1971 accounts were to be prepared the position had become much clearer, and it presented a pretty dismal picture. Barrie wanted to get all the bad news behind us so that the Company could go forward into a new phase unfettered by the problems of the past couple of years. The 1971 accounts, when finally published, showed a loss for the *year* of over £20,000, but further losses of nearly £50,000 were declared for the previous year—for writing off inflated work in progress and other dubious assets, and for losses in the small US subsidiary. The total declared loss of nearly £70,000 (equivalent to £576,000 at the beginning of 2000), left a weakened balance sheet and a shocked bank manager. He took several years to recover his confidence in the Company and kept it so short of credit for working capital that future progress was put in jeopardy.

The ICFC office in Reading was, by this time, headed by Dr David Ellis, a physicist, who retained his confidence in the management under Barrie, and in the Company's technology and prospects. He was always a great support to the Board, and was to become a director of the Company twelve years later. On two occasions, over these difficult early years, David organized additional loans from ICFC for specific purposes.

The first of these loans was to enable the Company to take over a product, a line of chart recorders, from Telsec, a local electronics company. After the losses of the past few years, Barrie was understandably anxious about the projects involving difficult one-off superconducting magnets. This leading-edge technology was unfamiliar to him, and, although an esoteric business such as this can be profitable, the Company needed 'bread-and-butter' turnover to balance the risks, contribute to the fixed overheads, and bring in a steady income. The flatbed recorder we acquired used one or two pens, that inscribed onto moving chart paper the traces caused by incoming electrical signals. This instrument could be used for recording many sorts of signal, such as variations in the voltage from a generator or the small electrical signals from the heart. It came to Oxford Instruments with a few employees, its stocks, orders, and customer lists, and provided a steady turnover of more than £100,000 in the first year. Although it needed upgrading and the margins were not very good, it was virtually risk-free. This chart recorder formed the basis of a new Electronics Division. It produced a modest profit for several years until all resources were needed for 'Medilog', a much more innovative and profitable new electronic product in the medical field. Then the ageing chart recorder was gently phased out.

Although the small, prefabricated office building had been finished for a year or two before Barrie started work, the odd extensions and temporary buildings and the rented factory half a mile away made for inefficient operation. Within a few months of his arrival he negotiated 'sale and lease-back' finance from ICFC for a 12,000-square-foot two-storey building. The new offices and workshops were to stand in front of the old boathouse, and the work soon began. The main benefit would be the future improvements in efficiency, but in May 1971 the artist's impression of the smart new premises, on the front of a newsletter (Fig. 5.1), served to reassure customers that Oxford Instruments was very definitely here to stay.

Fig. 5.1. Artist's impression of the factory and offices built in 1971

Our first newsletter had been issued a month earlier, in April, to restore confidence and to fill the gap left after the Cryophysics newsletter no longer carried news of Oxford Instruments' products. Cryophysics had been our European agent for several years, but its founder and owner, Ken Geiger, decided to serve Frank Thornton's new company instead. This left a worrying hole in our selling effort, and Martin and Barrie urgently opened negotiations with Klaus Schaeffer, who had his own German sales company selling scientific instruments. The two parties formed a joint venture, Oxford Geräte GMBH—meaning Oxford Equipment. The direct German translation of Oxford Instruments sounded too much like a well advertised aphrodisiac; care is often needed to avoid this sort of pitfall in another language. The new company got going very fast with a German sales manager and the infrastructure of Klaus Schaeffer's organization. This worked well for a while, but after three years of growth it became clear that the operation needed its own premises and Oxford Instruments was able to acquire the rest of Oxford Geräte, which then moved to Darmstadt.

The year 1971 saw other moves to improve marketing and sales. The mailing list was reorganized onto punched cards—it was before the days of the now-ubiquitous PC with database software. The Company recruited an Export Office Manager, and, in 1972, a new Marketing Manager, Bill Vince, a physicist and an experienced sales engineer. The sales staff, project engineers, and directors continued to travel widely to conferences and exhibitions, as well as visiting customers in their laboratories to discuss magnet systems or dilution refrigerators for particular experimental needs. Our very specialized business could not have survived on UK sales alone; from the beginning our market had been worldwide, and exports played a vital role. In 1972 the Company won its second Queen's Award, this time for a 'rapid and sustained' rise in exports, which had reached 70 per cent of turnover. This award boosted morale at a time when the challenge and excitement in the business had been somewhat dulled by financial stringency and by the dire state of the national economy. The presentation party was combined with the formal opening of the new building (see Fig. 5.2 and Plate 4). It gave our scientists and engineers an opportunity to demonstrate to families, friends, and professional supporters some fascinating scientific sideshows as well as the Company's progress in new products and its increasing professionalism.

Although Barrie was hampered in some of his intended improvements by the shortage of funds, the management was steadily becoming more professional. Our stopgap part-time accountant, who had struggled gallantly since late in

Fig. 5.2. The Lord Lieutenant, Sir John Thompson, with Lady Thompson and Martin and Audrey Wood after the presentation of the Queen's Award in 1972

1970 to bring order to the financial records, left in 1972. He was followed by John Lawrence, an experienced company accountant, who was to hold the position of Group Finance Director for many years. Because funding was so tight, overheads had to be kept to a minimum; priority had to be given to orders for products that could be produced rapidly to help the cash flow, and to larger contracts that provided for stage payments. Production was becoming more streamlined, but suffered some disruption from strikes in other industries. A long miners' strike led to the rationing of electricity in February 1972, when, for a difficult period, companies were allowed to operate machinery on only three days a week. Another miners' strike late in 1973 was soon followed by an electricity workers' strike and by the first great leap in the price of oil. This led to a state of emergency, and to a general election that spelt the downfall of the Conservative Government.

Oxford Instruments stayed afloat through these turbulent years, and, although old, unreliably costed, long-term projects were still causing losses, later quotations were sounder and made more realistic allowances for inflation. In the year to March 1972 the Company still made a loss, but only of £17,000. Its net worth had by then sunk to £35,000 (equivalent to £277,000 at the beginning of 2000). Oxford Instruments did not make any further annual losses through the rest of the century. In 1973 it emerged into profits of £38,000 on a turnover of £615,000. By 1974 the turnover had climbed to £750,000 and the profit to £55,000.

In spite of the squeeze on spending, the Company had not been standing still technically. With the help of Dr Gordon Davey of the Oxford University Engineering Department, the Company developed a new family of cryostats for all sorts of experiments over a range of temperatures—the continuous flow cryostats or CFCs (see Box 5.1 and Fig. 5.4).

The Company designed different versions of the dilution refrigerator to cater for those who wanted the lowest possible temperature and for those who wanted maximum cooling power at a slightly higher temperature—maybe a tenth of a degree above absolute zero rather than within the bottom hundredth. On the magnet front, attention was given to studying past projects in order to formulate more standardized systems to cater for a few established experimental techniques. Among these products were 'Spectromag' systems for various forms of spectroscopy; systems for measuring the 'magnetic susceptibility' of materials; systems for experiments with 'polarized neutrons'; and nuclear magnetic resonance (NMR) systems. The latter were to become an important business, and will feature in later chapters.

Magnets were benefiting from the development of a new form of NbTi conductor in which many extremely fine filaments of the superconductor are embedded in a matrix of copper. This material was conceived and first developed at RAL near Oxford; it is much more stable than the single core material

Box 5.1. Continuous Flow Cryostats

The older standard cryostats, also called dewars, contained a reservoir of liquid helium for cooling the magnet or other equipment. The new continuous flow cryostats (CFCs) were connected by flexible umbilical transfer tubes to a liquid helium storage container, and the liquid was drawn continuously through the system by pumps (see Fig. 5.3). This provided a rapid cool-down facility and hence the ability to change experimental samples fast. As they contained no reservoir of liquid, these cryostats could be used in any orientation. Several standard CFC models were launched for different areas of research, such as for spectroscopic work where the tail of the cryostat contained windows, or for experiments between the poles of an electromagnet.

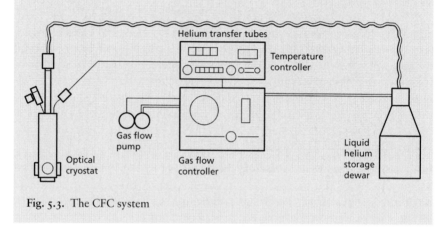

Fig. 5.3. The CFC system

and capable of producing higher magnetic fields. Multifilamentary NbTi has been the workhorse superconductor ever since then. NbSn had always been difficult to form into conductors, but eventually the metallurgists also succeeded in producing multifilamentary forms of this material. In the case of NbSn, the superconducting layer on the filaments has to be developed by reaction at a high temperature actually *after* the magnet has been wound. Fully superconducting joints have been difficult to achieve with both these multifilamenatry materials, but are vital for long-term stability in the types of magnet that have to stay permanently at field. The Company participated in development projects with RAL and with IMI in order to solve the problems of making these joints. This took some years, but the scientists were eventually successful, even with NbSn, where the method of production presented particularly difficult challenges.

The largest magnet system we made in those years gave a great deal of anxiety, and its progress became quite a saga. It was a hybrid magnet commissioned for upgrading the magnetic field capability of the Clarendon Laboratory. The

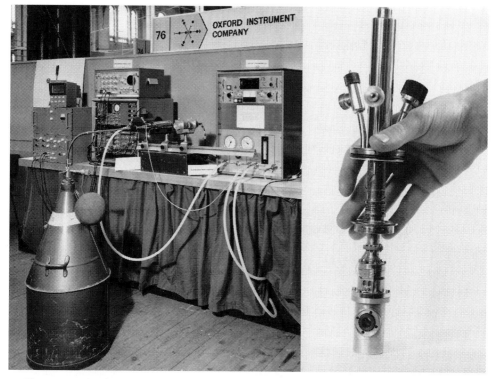

Fig. 5.4. A CFC being demonstrated at the 1971 Physics Exhibition, and (right) an optical CFC, held in a hand, with its outer cover removed

design was for a large outer superconducting magnet containing within its bore a compact resistive magnet, which was to take the total output from the 2 MW generator (see Fig. 5.5). The Company had made several resistive magnets through the 1960s, but all of these were either stacked pancakes or flat-plate 'Bitter'-design magnets. The only suitable high-field design to match the output from the Clarendon generator was the polyhelix configuration (see Box 1.2 and Fig. 1.2). This had been Martin's innovation in the Clarendon, and had since been taken further by a group of scientists and engineers in Canberra, Australia. This group had adapted the design to receive an enormous pulse of power from the famous flywheel-driven 'homopolar' generator at the Australian National University. The inner magnet for the Clarendon would be much smaller, but would, like the large early stacked pancake magnets, take the equivalent of the power feeding 2,000 single-bar electric fires; all this heat would have to be removed by rapidly pumped cooling water.

The team of scientists and engineers in Canberra, led by Dr Peter Carden, had formed a small company to make various parts for high-performance scientific

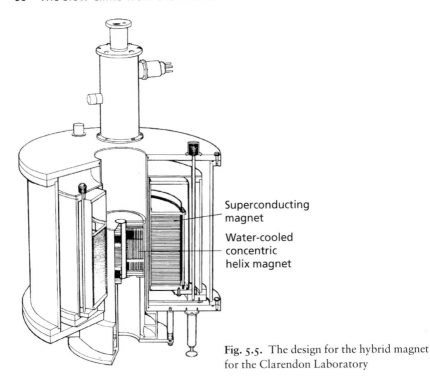

Superconducting
magnet

Water-cooled
concentric
helix magnet

Fig. 5.5. The design for the hybrid magnet
for the Clarendon Laboratory

equipment. They had developed several new techniques, such as a new way of
bonding insulation, and had used other new materials in constructing the large
Canberra magnet. Because of this experience, Oxford Instruments commissioned them to work on the details and to manufacture this critical central resistive magnet. Eventually the design was completed and manufacture began. The
Canberra team made slow progress, as some of the materials, including the very
special copper alloys, were on long delivery. Communications with Australia
were then by post, telephone, or telex—there were no fax machines or e-mail to
facilitate discussions on design details. There are always additional points to be
settled in this type of project; misunderstandings arose, some over quite obvious
matters like dimensions. The inner magnet eventually arrived but revealed a
number of problems on its preliminary tests. Peter Hanley, then our chief magnet engineer, decided it would have to go back to be altered.

The worst moment in this long saga came when a tiny inner helix, put on a
plane in Australia, failed to arrive back in England. After many enquiries we concluded that it had been stolen in Singapore Airport. The light-fingered porter
must have been puzzled by his singularly useless prize; but it had cost the
Company thousands of pounds. This caused a further delay of many months and
yet more costs, as the small Australian Company had been lax on insurance mat-

ters. We had to settle for a compromise in our demands in order to get a repeat helix made at all. The magnet was finally finished, tested in the Clarendon, and accepted early in 1975—before they lost patience altogether. After the long chapter of accidents it was a great success, and performed well for many years. The magnet provided a steady field of 16 T, which was occasionally pushed to almost 20 T for short periods, by overrunning the generator on a cold day, when the cooling water was more efficient at removing heat from the coils. This was over twice the laboratory's previous maximum, using only the generator. The outer 30 cm bore magnet could also be used on its own for experiments needing a large volume at a field of 6 T. The generator was finally condemned by the safety officer in the early 1990s. By that time the Company was able to make 20 T superconducting magnets for small-volume experiments, so high-field copper solenoids fed from the generator were no longer necessary.

At the beginning of the 1970s the world market for superconducting magnets was not large. The UK market, even when research laboratories were not suffering from funding cuts, was only a few hundred thousand pounds (equivalent to between one and two million pounds at the start of the new century). From 1970 Thor Cryogenics competed for orders round the world; it was able to get into the business rapidly because of its ex-Oxford Instruments' employees, who knew the technology and the market. BOC was also still competing, and still anxious to obtain contracts for complicated systems, even at the risk of losing money. Large one-off systems needed a lot of discussion with the research scientists and a considerable design effort before a quotation could be made. When an order was lost, these costs were wasted. This expensive duplication was cutting into the profits of all three groups.

BOC, with less experience than the other two companies, was certainly losing money on its magnet operation, and the market for the large superconducting machines, anticipated for the electrical industry, seemed as far away as ever. It decided to pull out of the field and offered to sell its magnet division to Oxford Instruments. BOC had some good technology, which complemented ours, particularly in the impregnation of magnet coils with resins. It wanted to hand over the assets of the Division just as they were, including stocks and equipment and problematical half-made systems—quite difficult to finish to someone else's design. Negotiations took many months as we weeded out equipment we did not want, obsolete stock, and other 'assets' likely to prove more of a liability. Had it not been for the existence of Thor Cryogenics we might have turned down the overtures, but we were aware that competition from a joint BOC/Thor operation might become quite formidable. Compromises were reached in March 1973. The Company bought the Division in exchange for 20 per cent of the shares in the expanded Group. The deal was, even at the time, rather generous to BOC, and it certainly proved to be so after Oxford Instruments became a publicly quoted company. However, deals like this may have hidden benefits. Some years later the Group was to profit from BOC's ownership of Airco Superconductors Inc. in the USA, which had a superconducting wire manufacturing operation. This Oxford Instruments eventually came to own in the 1980s to the great advantage of our magnet manufacturing divisions—and as a steady generator of profits.

The purchase of the BOC magnet unit did not include any premises and none of the production staff wished to move to Oxford. The only addition to our payroll was one physicist, Dr Paul Brankin, who was to perform several very important roles in the Group over the years. In the early 1990s he played a seminal part in the conception and birth of our Japanese sales company and he has since headed the Medical Division.

Although our absorption of the BOC operation eased the competitive problems, it did not help the cash flow, and the overdraft limit was still holding back expansion. Beside our own production, a dozen or so difficult BOC projects had to be finished. These were not always profitable, in spite of the 'cost-to-complete' calculations we had insisted on using, alongside the book value of work in progress, when valuing the BOC assets. We sought further finance from ICFC, which agreed to provide a new loan. As part of the bargain, we made a one for two rights issue of shares at par (£1 a share) to existing shareholders, to raise £34,000 (which would be £233,000 at early 2000 prices). The modest number of low-priced rights-issue shares not taken up were sold to senior employees who had arrived since the previous opportunity in 1967.

With the difficulties frequently experienced in the magnet and cryogenic equipment business, Oxford Instruments sought further stabilization and growth through diversification. In 1974 the Company acquired Newport Instruments. Ian Boswell, its Chairman and MD, had studied physics in Oxford before the war, and knew Nicholas Kurti and other physicists still in the Clarendon Laboratory. Unable to join the armed forces for health reasons, he looked for other ways to help the war effort. He had been left an orphan at an early age, inheriting the family business, Tickfords Coachbuilders, in Newport Pagnell. The Company made various vehicle bodies for the war effort and Ian himself worked with the outfit that cracked the German Enigma code at the top secret intelligence establishment at Bletchley Park in Buckinghamshire. In 1944 Ian opened an Electronics Division at Tickfords, which became a separate company, Newport Instruments, in 1955. In 1957, wanting to concentrate on the scientific-instruments side, he sold the coachbuilding business to Aston Martin. The electronics company started developing pulse transformers under Douglas Bruce, a leading expert in the field. His group pioneered the first core memories for computers in the UK. Their early work was on the first version of ERNIE, the computer that chooses prize-winners for the lottery type of National Savings known as Premium Bonds. As other interests developed, this activity became a separate Components Division.

Meanwhile, the end of the war had put a stop to much of the secret development work at RRE at Malvern. A few RRE scientists were looking for useful

projects to occupy their skilled technicians until a full peacetime research programme could be established. After consulting the physicists at the Clarendon
Laboratory, who had, themselves, participated in secret research during the
war, they decided to make laboratory electromagnets. Nicholas Kurti sketched
out preliminary designs and specifications for one or two useful sizes, which
were developed and manufactured at RRE. These were in some demand by universities, and proved popular for both research and teaching. By the mid-1950s,
with renewed internal claims on the RRE workshop's time, the magnet production had become a nuisance. Nicholas Kurti suggested that it should pass the
business on to Ian Boswell at Newport Instruments, and this became the core of
an Instruments Division of the Company.

By the time Oxford Instruments came on the scene in 1974, Newport had
extended the range and had added magnetometers for measuring magnetic
fields. Another development was an analytical instrument for measuring the
water or oil content of a solid sample. Incorporating a permanent magnet, with
a low but very uniform magnetic field, this instrument was based on NMR,
which Oxford Instruments was also using for equipment for high-resolution
analysis of molecules in solution. This division interested Oxford Instruments;
and the Components Division, which had also grown over the years, was, by
then, quite substantial, with a turnover of about £250,000.

Ian Boswell now wanted to retire. He was virtually the sole proprietor and
was concerned for the future of his company. He wanted to hand on responsibility for his staff to a company he trusted, and settled upon Oxford Instruments. He and Martin had known each other since 1956 when Martin was
working in the Clarendon, and our company had since bought various magnetometers and other items from Newport. All parties wanted a rapid agreement
and it was soon reached, again largely for an exchange of shares. Newport was
a little old-fashioned and not as profitable as Oxford Instruments, but it had
more experience in electronic production, with spare capacity which we needed
badly, and some of its products complemented ours. The total turnover was
only about half Oxford's, and it was growing rather slowly, but it had a strong
asset base and a sound balance sheet, built up over many years, which would
increase the borrowing capacity of the enlarged Group.

When Newport Instruments joined Oxford Instruments it actually had three
divisions; the Instruments Division making the complementary magnets and
other instruments; the Components Division, making specialized wound components such as tiny transformers; and a third division, really a separate company, making caravans—a surprising mixture. Coventry Steel Caravans (CSC)
was a relic of Ian Boswell's early association with coachbuilding, and made
expensive one-off units for mobile exhibitions, libraries, health-screening units,
and even mobile operating theatres. Although it clearly did not fit easily with
Oxford Instruments' products, the caravan business was making profits and did
not demand a lot of investment in new equipment like our other operations.
This division had not been pushed with any energy for some time, and the rather

dilapidated rented space could accommodate a much larger turnover. The Group Board, now enlarged with representatives from BOC and Ian Boswell, decided to invest in a drive for more sales, both in the UK and overseas, with a view to a possible sale in a few years' time.

John Lawrence, then Chief Accountant of the Group, took on the Managing Directorship of Newport for a couple of years, and, among other actions, started to push for more sales of these specialized mobile units. This policy bore fruit. In 1975 CSC won a large order from Ferranti worth £220,000 (which would be well over £1 million at the beginning of 2000). This was for secure air-conditioned mobile vans for housing electronic equipment in the desert. In 1976 came more export orders, this time for special vans for Middle Eastern princes. The first was for a mobile 'hunting lodge'; then came an order for a pair of units, one to be a mobile 'court house' and the other luxurious living accommodation. These were to replace the large old desert tents for a prince when he travelled about his domain delivering justice—the equivalent of a judge going 'on circuit'.

With good margins, little competition, prepayments, low working capital needs, and an absence of the sort of difficult technical problems endemic in the magnet and cryogenic operations, CSC might sound like the perfect business to be in. But it was alien to Oxford Instruments, and was taking up a lot of management time. In 1977 John negotiated the profitable sale of the business to a much larger caravan company that lacked a facility for making these top-of-the-range units.

The Newport acquisition gave the Group more stability, and Barclays Bank in Newport Pagnell High Street, which had served that company for a long time, did not wish to lose its customer. It offered the whole Group terms that the Midland Bank in Oxford, which had come to our rescue some years earlier, would not match. Although now recovering fast, we had not been the easiest of customers over some of those years, and we understood its reluctance to better its existing terms. But as a result the Company transferred its business to Barclays branch in Newport Pagnell. As the new century began the Group was still using Barclays as its main clearing bank.

The factory at Newport Pagnell, now part of Milton Keynes, lay an hour's drive from Oxford along a slow road. It was hard to integrate the businesses at this distance. Newport was an older company than ours, and had active unions and traditional attitudes—there were actually three canteens for different levels of staff, those in the upper ones jealously guarding their status. It was hard to graft onto Newport the Oxford Instruments' philosophy and culture with its informality and widely shared objectives. It was necessary to transfer the magnet production to Oxford to join the resistive magnets there, but few of the staff were willing to make the move. The other instruments and the electronic components continued to be made at Newport. The original plan was for the whole of the Electronics Division in Oxford to go to Newport, but key Oxford employees were unwilling to relocate in that direction, so production was subcontracted to Newport instead.

At the time of the merger the Components Division was renting two outstations where groups of nimble-fingered women wound tiny doughnut-shaped coils of wire by hand onto minute rings of ferrite (magnetic core material). Also adding to the productive capacity were women outworkers who liked to wind these coils on a piecework basis, in their homes, while looking after their small children. The components made by the division were mainly pulse transformers, and were made for relatively low-volume products. There was little competition from the Far East, where companies are rarely geared to handle small 'runs'. These pulse transformers were mostly sold to go in computers; in 1974 the computer industry was flourishing, but it is notoriously subject to fluctuations and a dangerously large part of the output of the Division went to one company, Computers International. Although the Components Division performed well in its early years under Oxford Instruments, it was a very different sort of business from instruments and systems, and not easy to manage alongside the other Group operations; it tended to get neglected and starved of funds for the development of new products. When the sales dropped alarmingly in the recession of the early 1980s, it was formed into a stand-alone company with a view to a possible sale.

6 Medilog

How Oxford Instruments led the world into ambulatory monitoring

GERONTE. It seems to me you are locating them in the wrong place: the heart is on the left and the liver is on the right.

SGANARELLE. Yes, in the old days that was so, but we changed it all, and now we practise medicine using a completely new method.

(Molière, *Le Médecin malgré lui*, 1667))

In the early 1970s there was one other major new departure from the original type of business—into the medical electronics field. The story goes back into the 1960s, when Dr Frank Stott, in the Oxford-based bioengineering unit of the Medical Research Council (MRC), conceived the idea of a miniature physiological data recorder. He started developing an instrument, but found he needed a small and flexible electronics company to produce early prototypes for the doctors who were testing the idea, and to work on the necessary modifications. Oxford Medical Devices (OMD), the very small company he turned to in 1969, had an association with Oxford Instruments, and was occupying the old Middle Way stables. Its tiny staff, at first led by Peter Styles, worked with Frank Stott on small battery-operated tape recorders for a series of diagnostic trials. These involved the monitoring of patients' heartbeats over long periods of normal activity.

The results of these trials were very exciting, and OMD soon recognized the major diagnostic value of the instrument, even in its early prototype version, and foresaw a huge potential market. But it knew rapid further development was needed to produce a reliable product; and a substantial manufacturing and marketing effort would be required to serve that market. OMD was in no position to undertake this, so it offered to pass the business onto Oxford Instruments, which, in the depressed market for research equipment in 1971, was looking for new products. Barrie Marson was able to negotiate the purchase of the modest assets at valuation, with agreed royalties to both OMD and the MRC; this way, if the instrument proved to be a failure, Oxford Instruments would not suffer much. Barrie half expected the development to be overtaken rapidly by competition from the many electronics firms looking for exciting new products. The Company could see possible applications for the recorder relevant to its other products, including longer-term monitoring of experiments in a laboratory. It would have more knowledge and control in this area than in the unfamiliar medical market.

So the tape-recorder development and manufacture were transferred to Osney Mead. John Swadling came with the business and continued to work with Dr Stott, making one or two more batches of the recorders and a few more playbacks to the same early design before it could be developed much further (see Fig. 6.1).

'Ambulatory monitoring' was in its infancy. One company in America provided instruments to record physiological data onto magnetic tape, which was named 'Holter' monitoring after its initiator. This system used a reel-to-reel recorder that could be carried from place to place, but could not possibly be called 'ambulatory'—there was no way it could be worn by a patient as he went about his normal daily life. Frank Stott's new idea was to use the recently developed *cassette* tapes in a tiny recorder, geared down so far that a standard two-hour audio-tape would last for twenty-four hours, while electrical signals were magnetically imprinted on it by multiple recording heads. Fine wires hidden under clothing fed the signals from the electrodes, stuck to suitable parts of the body, to a tiny recorder, which could be worn in a pocket or on a belt. At the end of the recording time the tape was removed and played back at high speed on an electronic unit with an oscilloscope to show the rows of traces, one for each recording head, made by the signals from the heart. A doctor could then

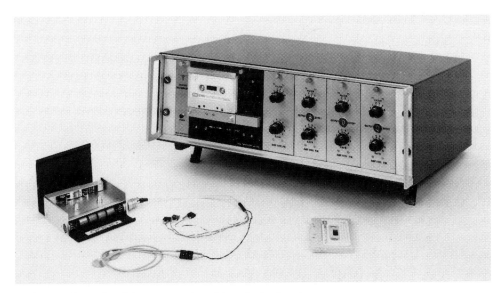

Fig. 6.1. An early miniature recorder with a playback unit

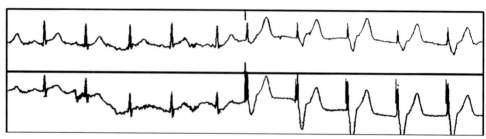

Fig. 6.2. Two channels of ECG traces from a Medilog recording

pick out interesting episodes of heartbeat and run through them slowly, and a printout could be obtained by using a chart recorder (see Fig. 6.2).

Although the early OMD instruments were far from perfect, the doctors involved in the round-the-clock trials were very enthusiastic and demanded more. These recordings had shown up features that had been missed in the brief standard electrocardiogram (ECG) recordings onto paper strip, taken in a hospital, with the patient linked to a static machine. They had also demonstrated the heart activity while the patient went about his *normal* daily life, away from the general apprehension often felt in a hospital atmosphere, but subject to the sudden stresses experienced in driving, or teaching, or having a family row. Clearly such an instrument could be used to record other small signals over many hours, and indeed some of the earliest trials used modified instruments for recording electroencephalograms (EEGs), the electrical signals from the brain. These are much weaker than heartbeat signals, and more amplification is needed to provide a reasonable recording. Two very early OMD prototype EEG recorders had been taken to the Antarctic by Dr Paterson, where he made interesting recordings of the brain signals associated with sleep patterns under twenty-four-hour daylight conditions (Paterson 1975). Fortunately he was an electronics wizard and had been able to keep one of the prototypes going in a remote research station by cannibalizing the other.

Interest in the recorders grew. Early in 1972 Barry McKinnon moved from his position as Oxford Instruments' Development Manager to run a small team to develop the new product further. As the Company was short of orders, the miniature analogue tape recorder (MATR), as it was called at first, had already been announced in a newsletter, with a photograph showing it linked to a CFC. Our newsletter readers were mostly physicists, not too interested in heartbeats, but a new way of recording the results of experiments might be attractive.

There was still much work to be done to provide a robust and reliable instrument, especially for the medical work, where the small size was very important. The dimensions were determined by the size of a standard audio cassette; all the components needed to be as small as possible and had to be fitted together rather like a Chinese puzzle. These included recording heads for four channels; four small high-performance batteries to run the tiny motor over twenty-four hours; a gearbox to keep the normally two-hour tape turning smoothly at twelve revolutions per minute over twenty-four hours; and a small printed circuit board with its components for controlling the operation and providing a time clock. It was the Company's first task involving tiny moving parts and electronic miniaturization, and it took a long time to get it right. This was well before the era of microprocessors, which have since made size reduction so much easier.

The launch of the MATR had brought an immediate, if small, market for which it was not really ready. The Company needed the sales, and the income from them, but it needed to sell instruments that would work well, and go on working, and the development task was quite challenging. Then the playback units needed improvements, including data reduction facilities, to make them usable by busy doctors who did not have the time to watch through vistas of four-track oscilloscope traces. Even when speeded up sixty times a whole tape took twenty minutes to play back.

The EEG brain recorders proved very successful at diagnosing the difference between hysteria and real epileptic fits, especially in children. The Company started to work with Dr Greg Stores of the Park Hospital in Oxford to develop an instrument and a playback system suitable for this purpose. In a minor form of epilepsy, known as 'petit mal', a sufferer may just appear to be daydreaming during an episode, and teachers may punish such a child for inattention. Diagnosis used to be almost impossible, as people cannot be attached to a static EEG machine for long periods, but with the new recorders any 'spike-and-wave' patterns, typical of epilepsy, could be spotted immediately as the tape was played back (see Fig. 6.3).

The development team soon had to limit the scope of their work. Early on they had started programmes on noise-level monitoring systems and on deep-sea measurements with the recorders in pressure-resistant cases. But these had

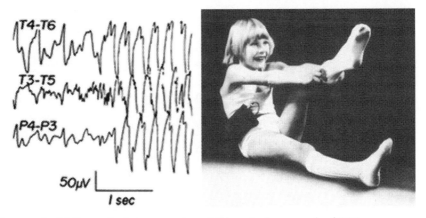

Fig. 6.3. An EEG recorder being worn by a child, and a few seconds of EEG traces on three channels showing the development of a mild epileptic seizure

to be abandoned as demand grew for the medical recorders. Enquiries came in for the monitoring of all sorts of other things such as vehicle performance, meteorological data, vibration, seismic activity, and sea-water temperatures near nuclear power stations. But, apart from our ignorance of the markets, there was no time to consider these ideas; heart monitoring became by far the largest application. The big advantage of Medilog, as it soon came to be called, was that adequate long-term heart data could be obtained, relatively cheaply, without sending the subject to hospital. Once 'ambulatory heart scans' became established, doctors started demanding a twenty-four-hour scan before a decision was made on the need for a heart pacer to correct irregularities. As other physiological data could be monitored on one or two of the channels at the same time as heartbeat, the instruments could be used for wider research. Medilog recorders played a part in monitoring the responses of fighter pilots, athletes, Himalayan mountaineers, and even racehorses (see Fig. 6.4). Later they were worn by adventurers on trans-global expeditions and by astronauts in space.

As more and more research doctors published papers on their results using our recorders, interest intensified. Along with a growing number of orders we

Fig. 6.4. A young woman wearing an ECG recorder while playing tennis

received requests for overseas agencies. At the same time the National Research and Development Corporation (NRDC), which held the rights to the MRC work, was pestered with applications from companies that wanted to manufacture similar tape recorders under licence. In order to keep our right to a sole licence we had to satisfy NRDC that we would be able to meet US demands for the product. It was a chicken and egg situation. Oxford Instruments was a small company with very limited resources; it needed to make sales rapidly, to establish this exciting product in the marketplace, to satisfy NRDC, and to keep the cash flowing in from the sales. At a time of restricted credit this cash was urgently needed to develop better and more reliable instruments that would also be cheaper and easier to manufacture and to use. This is a perennial dilemma for a small company making 'state-of-the-art' equipment. As Barry McKinnon put it early in 1973, 'Our basic problem is that we have launched a poorly developed product which, even so, has vast potential.'

The recorders already incorporated improvements for the doctor, and were easier to make, and there would soon be a new ECG data analysis system in place of the earlier playback units. Selling was not difficult. One only had to leave a system with a cardiologist for a few days and it became indispensable to him; in fact we had more orders than we could easily cope with. Early in 1973 Barrie Marson wrote, 'The Medilog situation grows more alarming as its market potential is realized.' This product was becoming something of a cuckoo in the nest and eating up our meagre cash resources. The rest of the Group was feeling the pinch in its funds for R & D.

What was to be done? The Board considered three possibilities: selling the medical product and concentrating on the smaller market for non-medical applications; licensing it to other companies; or getting in outside finance to force a rapid expansion with a complete redesign for larger-scale manufacture. Everyone wanted to keep this innovative new product—our largest single opportunity to date—and not to sell out our position in the most promising area, the medical market. But outside funding would mean a separately financed company for Medilog; this would curtail our freedom of action and would certainly be hard to negotiate.

Over the next few years Barrie and the Medilog managers talked to several companies about a sale, licensing, or a joint venture, without receiving any proposals to match the potential of the product. As the cash flow suffered further, there was even a suggestion that the other major section with great potential, the magnet production operation, should be sold to a persistent suitor to provide funds for the rapid exploitation of this rare opportunity. We feared that new competition from some large rich company in the medical field would erode our technical lead, and, with much greater marketing resources, would push us out of this new market we were establishing. There was already competition from relatively small companies, but they were well behind us technically, offering larger and heavier recorders that ran for more-limited periods.

In January 1974 the Group split formally into two divisions, for Magnets and Cryogenics on one side, and for Electronics on the other. Beside Medilog the latter included the chart recorder, and power supplies and other units for magnet systems. Following the takeover of Newport Instruments, it was able to relieve production pressure by using the spare capacity in Newport Pagnell for subcontracted manufacture of recorders and other instruments. Pressure on space in Oxford had become serious, as the Electronics Division was then still housed at Osney Mead and all operations there were cramped.

The equipment needed more development. Early in 1975 a Medilog recorder was selling for £300 (equivalent to around £1,500 at the beginning of 2000). In a memorandum Barry McKinnon outlined a project for a new design that would enable it to be sold for an attractive £200. A new analysis system would be more difficult; the current system cost £1,000 in hardware alone, and moving to a more desirable mini-computer-based system would push that up to £2,000 to £3,000—the ubiquitous low-priced PC was still far in the future. Unfortunately, the magnet business was then suffering from technical problems, and Barrie Marson, guarding the deteriorating cash flow, could find little money for this update of the Medilog equipment.

A year later the scene had changed dramatically. Early in 1976, with a rapidly improving financial situation, the Group Board finally turned down the latest inadequate offer to buy the product. It made a commitment to keep Medilog, and to provide enough funds for its proper development and expansion on a new site. Oxford Electronic Instruments became a separate company, and later that year moved to a rented factory and offices in Abingdon, some six miles from Oxford.

With much improved development, manufacture, and test facilities, sales could now be expanded. Pressure on public expenditure had cut back UK orders, but a drop in the value of the pound was enhancing export competitiveness. The Company began to study how to go about establishing Medilog as the market leader in the USA, the largest but most competitive market in the world for medical equipment. Two of our existing US distributors were becoming less interested in selling Medilog and more interested in setting up medical service bureaux that they called 'scan labs'. These would offer a service of twenty-four hour-heart scans and analyses for local doctors. The end product would be a written report on the patient with printouts of interesting heart episodes. This seemed the fastest way for the Company to penetrate the US market, so it started working with the new organizations, selling or hiring the equipment to the emerging scan labs. The Company set up a new subsidiary, Oxford Medilog Inc., run by staff from Abingdon, for the vital job of repairing and servicing our instruments in the USA. Scan labs proliferated. The larger ones in affluent areas needed two ECG analysis systems and some fifty recorders, and made an average of twenty patient tape recordings a day, five days a week. This was good business, both for the scan labs and for us. Soon some $50,000 worth of equipment was being shipped to the USA every month.

The equipment in the scan labs had to withstand a pounding for twenty-four hours a day, day after day. This was much harder than any previous usage by hospitals or for drug trials. New weaknesses came to light in our recorders. Before the advent of Medilog the Company had not made equipment with constantly moving parts. There were problems over friction, clearances, tolerances, lubrication, gearing, and so on. On the electronics side some bought-in components proved inadequate to the new demands on them. In America 'time is money'. As 'down time' rose we found these companies threatening to sue us for lost earnings. In Abingdon urgent searches for more reliable components, the tightening-up of manufacture and testing, and better quality control brought the relationships with the scan labs back from the brink.

Barry McKinnon had championed the whole development and marketing of Medilog, and had seen it through the many problems of its early days, caused mainly by the shortage of funds. But in the autumn of 1976, to our great regret, he was 'head hunted', and left the Company. The Medilog operation was the fastest-growing part of the Group and had become the most profitable. It needed an innovative electronics engineer, who was also capable of taking overall management charge, to carry it forward. Good managers were scarce, and several advertisements and 200 candidates later the right person, combining these skills, had not emerged. Meanwhile Barrie Marson was running the business from Oxford and Doug Reeve was managing production efficiently in Abingdon. In 1977, against Company philosophy, the top management role was split from the technical one. John Lawrence, the Group Finance Director, who had just concluded the satisfactory sale of the caravan company, took on the role of MD of the operation, while several bright young electronics engineers were recruited to keep the team in the forefront of the technology.

Antony Costley-White, a physicist who had been with the Group since 1969, was then running our German sales company very successfully. Soon after this he moved back to England to become Sales Director of Oxford Electronics, as it had then become. He was to follow John Lawrence a year or two later as MD of that company. Sadly, Antony died too early, in his fifties, in the spring of 1998. A complex person with many interests, he contributed a great deal to the Group between 1969 and 1984. He was an efficient manager, and, although he was sometimes frustrated, his bubbling sense of humour was never far from the surface. He is sorely missed by his colleagues and friends.

Sales of Medilog were booming. Two hundred recorders and ten complete systems were being shipped each month. So far we were succeeding in remaining at the forefront of this growing business, based on the system of diagnosis by cassette tape recording that Dr Stott had envisaged and we had pioneered commercially. Medilog was our most profitable electronic product, and by this time all other electronic production had been transferred to Newport Instruments or back to Osney Mead. The Company was again renamed, Oxford Medical Systems (OMS), later to become just Oxford Medical (OM). As time went on the ECG market slowly became more competitive, and margins began to shrink. Renewed efforts went into the development of ambulatory brain monitoring (EEG) systems, which had a much smaller market, but one in which we retained a virtual monopoly.

By the end of the decade OMS was outgrowing its first building in Abingdon, fitting out a second rented building for its R & D team, and planning to develop its own one-and-a-half-acre site nearby. From the original recorders and play-back units the Company had developed a family of ambulatory monitoring systems for health care. Early in 1979 it launched the Medilog 2 product, together with a sophisticated analyser. These smaller and lighter recorders incorporated new features and were easier to manufacture in quantity. Also that year, a better system was launched for twenty-four-hour EEG recordings, used mostly in the diagnosis of epilepsy and for assessing sleep disorders. Signals from the brain are much more difficult to analyse than the more pronounced and regular heart-beat traces, but the new unit for studying the tapes from our EEG recorders used a 'page-mode-display' system that enabled doctors to identify interesting episodes much faster than was previously possible. OMS later won a design award for this innovative equipment.

With the rapid advances in electronics, the 1980s were to see the development of 'smart' recorders with more analysis of heartbeat performed in the recorder itself. Later still there was a move to 'solid-state' recorders with no moving parts, where recording, storage of information, and analysis were all performed by integrated circuits or 'chips'.

7 Magnets for Modelling Molecules

The long road to our first significant business to come from superconductivity—in NMR

What an enormous revolution would be made in biology, if physics or chemistry could supply the physiologist with a means of making out the molecular structure of living tissues comparable to that which the spectroscope affords to the inquirer into the nature of the heavenly bodies. At the present moment the constituents of our own bodies are more remote from our ken than those of Syrius.

(T. H. Huxley, from his presidential address to the Royal Society, 1885)

NMR stands for nuclear magnetic resonance. The layman need not be daunted by the scientific name, as the basic concept can be understood quite easily. Resonance occurs when an object responds to the energy transmitted from an outside source and vibrates. Some objects respond to a single frequency and some to a wider range. Everyone is acquainted with resonance—our hearing depends on it. Our flexible eardrums behave like amplifiers and can vibrate to a considerable range of frequencies. These vibrations are transmitted through the little bones in the middle ear to the tiny sensory hair cells lining the cochlea of the inner ear. Here the hairs of each small region resonate to a very limited range of frequencies and the energy of vibration is converted into electrical signals fed to the hearing centres of the brain. The high-frequency vibrations, maybe from a violin string, excite the hairs near the base of the cochlea, while the lower-frequency notes, from a double bass or a lorry engine, travel to its apex. A considerable force can be built up by the transfer of energy from a source to a receptor. We all know that loud noises can hurt, and a high note emanating from the vocal cords of a powerful prima donna has been known to set up resonances in a thin glass tumbler strong enough for the transferred energy to shatter the glass. But I would be surprised if trumpets and shouting on their own could build up enough resonant energy to knock down the walls of Jericho (Josh. 6).

Every atom contains a nucleus, and many nuclei behave like microscopic magnets; they possess a 'magnetic moment', which makes them resonate in fluctuating external magnetic fields. The long list of elements susceptible to this nuclear resonance includes hydrogen, which occurs very widely in water and oils and also in living organisms. Other elements, in particular phosphorus, play important roles in the physiology of the body. This phenomenon of NMR can be used in a relatively simple way to measure the quantity of hydrogen in a sample, or, with a much more sophisticated instrument, to unravel the structures of large organic molecules such as enzymes and hormones or medicinally valuable plant products. Over the whole range it provides a very valuable analytical tool (see Box 7.1).

Box 7.1. The Basics of NMR

A simplified explanation for a complicated process will suffice here. Magnetic nuclei behave like small magnets. When placed in a magnetic field, they line up in the direction of the field like compass needles, but fundamental properties (the quantum theory) restrict them to certain particular orientations. If these nuclei are then bombarded with radio waves of the correct frequency, they will jump from one orientation to another, and in so doing will exchange energy with the applied radio waves. A strong pulse of these radio waves will stimulate the nuclei in the sample to act like little radio transmitters, broadcasting back their own radio message, which is loaded with all manner of detailed information about the molecules from which the signals come. This information includes the strength of the magnetic field at that nucleus, the concentration of nuclei, their position within a molecule, the nature of other atoms near them in the molecule, and detailed information about the molecular motion of the sample under study.

Of crucial importance among the elements that resonate in this way is hydrogen, whose nuclei (protons), when placed in a uniform magnetic field of 1 T, resonate at a frequency of 42.6 MHz (megahertz, or million cycles per second). In a field of 2 T they resonate at 85.2 MHz. Also important for the analysis of organic molecules are the nuclei of the ^{13}C isotope of carbon, which resonate at 10.7 MHz in a 1 T field, those of the ^{23}Na isotope of sodium at 11.3 MHz, and of the ^{31}P isotope of phosphorus at 17.2 MHz. These isotopes occur naturally, but in the case of ^{13}C it forms a very small percentage compared with the common form of carbon, ^{14}C.

A simple low-resolution NMR spectrometer formed the basis of the Newport Analyser (see Fig. 7.1). This could be 'tuned' to measure the water content of materials such as grain or cement, or to measure the oil in rapeseed, or the fat content of meat, or milk powder. This non-destructive measurement system for hydrogen-containing samples is valuable, but far more important is the role of NMR in the structural analysis of large organic molecules. Advances in this technique have been made possible by the evolution of high-field superconducting magnet systems and very sensitive spectrometers with increasingly powerful computers. Oxford Instruments has played a significant part in the development of this powerful analytical tool.

Sample in tube

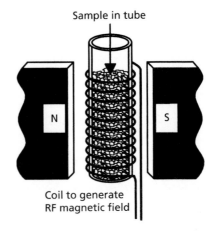

N   S

Fig. 7.1. The arrangement of principal
and RF magnetic fields in the Newport
Analyser, which was then based on   Coil to generate
simple continuous-wave NMR using a   RF magnetic field
permanent magnet

A decade before the Company acquired Newport Instruments we had become involved in developing superconducting magnets for NMR research at Oxford University. This led to a successful symbiotic relationship with the University over many years from which both parties benefited. The Company would stretch its technology to develop a magnet system to satisfy a newly demanding specification—ahead of any NMR magnet in the world. When we delivered this system, the Oxford scientists would achieve new insights into large molecules of importance in understanding body function, and so of importance to medicine and in the development of new drugs. On publishing their results they would receive acclaim, sometimes accompanied by another research grant for yet better equipment for further steps in this important work. Oxford Instruments, given credit in these reports, would receive a spate of enquiries and orders for similar equipment as well as a newly challenging request from the research group in Oxford. This virtuous spiral continued over many years.

The phenomenon of NMR was first predicted before the war and in 1945 scientists at both Harvard and Stanford universities succeeded in demonstrating these resonances. In Oxford Dr Bernard Rollin, a physicist, started making a spectrometer immediately after the first publication of these results, in 1946; the first paper on his work was published in *Nature* that same year. As we have seen, the essential starting point for this work is a stable and uniform magnetic field; all the early experiments were carried out using either a permanent magnet or between the pole pieces of an electromagnet with an iron yoke (see Chapter 1). The maximum practical magnetic field produced by these types of magnet is 1.4 T and 2.3 T respectively (corresponding to resonance frequencies for hydrogen nuclei of about 60 MHz and 100 MHz). Higher-field iron-free solenoids, such as those in the Clarendon Laboratory in the 1950s, were not of much use for NMR, as their fields could not be made uniform or stable enough for high-resolution work. By the early 1960s NMR, using permanent or electromagnets, had been established as a powerful method for studying molecular structure, although it was not then possible to study molecules containing more than a few hundred atoms (see Box 7.2).

Box 7.2. The 'Chemical Shift' in the Structural Analysis of Molecules

The reason why NMR is such a valuable analytical tool is that the magnetic environment of any particular nucleus in a molecule is affected by the tiny magnetic fields produced by electrons circulating around orbits in nearby atoms—the 'chemical shift'—and by other lesser influences. These lead to small, sometimes minute modifications to the frequency at which the nucleus resonates. Very sensitive high-resolution NMR spectrometers are needed to detect and measure these marginal differences. In a simple low resolution NMR system the spectrum produced by the resonance of the protons present shows a single wide resonance peak from which the amount of hydrogen in water or in oil can be calculated. Using a stable and homogeneous high-field magnet with a more sensitive spectrometer the spectrum becomes 'dispersed' or spread out, and the influences of neighbouring atoms may be seen in hundreds of different lines for hydrogen or phosphorus, many split into 'multiplets'. The higher the resolving power of the spectrometer the larger the molecules that can be analysed by chemical shift NMR. For the magnet-makers, the ultimate challenge is a high magnetic field that is uniform over the specimen to one part in 10^{10}—one part in ten thousand million.

Dr Rex Richards (now Sir Rex) had started building NMR spectrometers in Oxford in 1948, still a time of austerity in British laboratories. In the early 1950s he built an electromagnet, using locally made iron castings for the yoke, with sheet steel pole pieces ground by the nearby car-body plant, Pressed Steel. He wound the electrical coils for magnetizing the iron by hand, and he obtained the electronic components he needed for the system by carefully unsoldering them from redundant military radar equipment. Although used successfully to study many heavy nuclei with large chemical shifts, this system was not sensitive enough for studying protons (hydrogen nuclei). He then had a permanent magnet built by the company Mullards; its field was lower than that of the electromagnet, but it was more homogeneous and stable, and it *could* be used for work on protons. It later formed the prototype for a commercial spectrometer made by the company Perkin Elemer. In 1955, on the strength of his published work in Oxford, Rex Richards was offered a research fellowship at Harvard, where he continued these studies.

All scientists engaged in research using NMR would like to work at higher magnetic fields, in which susceptible nuclei resonate at higher frequencies, giving larger chemical shifts and so a better chance of sorting out small differences in the positions of atoms in a molecule. But the higher field must, of course, be extremely uniform and stable in order to register these small differences. The stability needed in the system can be compared to that in a powerful telescope where the tube must be kept extremely still, as a tiny knock may send the field of vision careering billions of miles away from the star under observation.

In 1964 Rex Richards, now back in Oxford, heard of a small company located in a slaughterhouse in North Oxford that was having some success making high-field superconducting magnets. He came to see us. We were already making what *we* called 'high-homogeneity' magnets, with fields that did not vary by more than one part in 100,000 (10^5) over a 5 millimetre spherical volume. But for high-resolution NMR work we would have to do very much better than this. Martin thought we might be able to meet the minimum specifications then suggested, given time for the development, and the Company got to work on the project.

Superconducting magnets have other advantages over resistive ones as well as the much higher fields possible. The most important is the stability of the field over time, which comes from the way they can be run in 'persistent mode', as in the old demonstration of energy storage (see Box 2.2). After DC electric current from a power supply has been fed into all the coils in an NMR magnet, and the current in the superconducting compensating coils has been 'trimmed' for maximum uniformity of field, and all the superconducting switches have been closed, the current goes on flowing round the circuits indefinitely with the magnetic field 'locked in' and the power supply removed. This gives a superb long-term stability of field. Some NMR magnets have been running like this for at least ten years, the only energy input being that involved in refrigeration for the very low temperature needed for this nearest thing to perpetual motion.

The first experimental magnet, delivered in 1966, was promising enough for Rex Richards to talk to the Company about a second magnet for which the Department would develop a new electronic system. Beside the magnet, a complete spectrometer includes equipment for the generation and control of the radio frequency energy, and the detection and recording of the radio signals from the resulting resonances. The end product of an NMR analysis was then an oscilloscope spectrum, with its printout, from which the skilled user could read much about the structure of the molecule under study and the positions and numbers of its hundreds or thousands of atoms. (See Box 7.3.)

By this time papers emanating from the work in Oxford University, which gave Oxford Instruments credit for the magnet systems, had led to orders for similar equipment from other universities. One company making NMR systems was already offering a superconducting magnet spectrometer, and others were considering them for making more powerful instruments. Our big breakthrough towards commercial sales was to come with the third NMR magnet system we made for Oxford University.

In 1970 Rex Richards (see Fig. 7.2) moved into the Biochemistry Department

Box 7.3. Early NMR Magnets for Oxford University and how Spectra were Acquired

7.3.1 The first two magnets

The first experimental magnet was wound from niobium zirconium. It produced a field of 5.5 T uniform to one part in a million (10^6) over a 10 mm spherical volume. The team in the laboratory improved this by adding room-temperature compensating coils. This homogeneity was still not adequate for separating close resonance peaks of protons, but it was used with some success for other elements.

The second NMR magnet system for Oxford, delivered in 1969, was made from single-filament niobium-titanium wire and reached a maximum field of 8 T. Its performance covered a wide range of resonance frequencies, but the bore was too small to achieve a homogeneity good enough for proton resonance. This limited the work possible in the system, but it was used for many years, mainly for biological work on phosphorus.

7.3.2 Frequency sweep and Fourier Transform

The older frequency-sweep method for acquiring and recording resonance was rather slow. It can be compared with sweeping the frequency band scale on a radio with the tuning needle when looking for a particular programme. The closer together the frequencies of the signals from the broadcasting stations, the more sensitive and selective the receiver needed for separating, say, an opera singer in Milan from a newsreader in Moscow and from the particular programme one wants to hear. Many interesting organic molecules occur sparsely or in a dilute form, and each pulse of RF energy may evoke only a minute response. The signals have to be built up slowly and charted from many repeated frequency sweeps.

The spectrometer developed in the University for use with our second magnet featured an early form of Fourier Transform (FT) sampling for obtaining and recording the resonances. FT techniques, made possible by the rapid advances in computers and software, were soon to revolutionize the speed of NMR research and routine testing. In FT sampling, the specimen is subjected to repeated bursts of RF energy covering the whole of the relevant frequency range at once, and the position and sizes of the resultant resonances are sorted out, built up, and charted by a computer.

to lead a new interdisciplinary team, the Oxford Enzyme Group, which included twenty senior members of the academic staff from eight departments in the University. This Group was to work on different aspects of the structure and operating mechanisms of enzymes, which are very important catalysts in the body. Their primary target was the enzymes involved in the breakdown of glucose in cells to provide energy. Alongside crystallography, high-resolution NMR spectroscopy was to be the principal technique employed. It would be used for studying the nuclei of hydrogen and of isotopes of fluorine, carbon, and

Fig. 7.2. Sir Rex Richards

phosphorus—all important elements in the chemistry of living systems. The Science Research Council (SRC) provided one of the largest grants it had ever given for a new NMR spectrometer—£371,000 (equivalent to nearly £3.5 million at the beginning of 2000). This was to provide the best possible performance achievable at that time. Bruker Physique of Karlsruhe, who made good electromagnet-based NMR spectrometers, was to provide the electronic part of the spectrometer, and Oxford Instruments the superconducting magnet system. This was based on our No. 2 magnet, but with a larger bore and a lower field.

For some years Varian, in California, a leading analytical-instrument manufacturer, had been selling NMR spectrometers incorporating a superconducting magnet and rated at 220 MHz. These were expensive, and I believe the only ones in the UK were bought by ICI. Bruker, its main competitor, wanted to enter the market for higher-field instruments, but had no experience with this type of magnet and no good source of supply. As the Oxford spectrometer project progressed, Professor Gunther Laukien, the Geschäftsführer (MD), sought a closer relationship with Oxford Instruments and pressed us to 'come under the Bruker flag', as he wanted his own supply of these magnets. In 1970 Oxford Instruments was going through a difficult period when our finances were at a low ebb, but we did not want to enter negotiations that might lead to the loss of our independence.

Dr Peter Hanley was in charge of the new magnet contract for Oxford University, which was to feature several innovations, including a more efficient and convenient cryostat. Rex Richards monitored progress throughout the project. By the summer of 1971, when the magnet had met its demanding specifications, he came to watch the acceptance tests, and, afterwards, repaired to the canteen with Martin and our NMR team. Martin commented,

'We never have time to do the fine tuning on these systems—we always have to ship them out as soon as they meet specification. If we had longer and the funds to cover the extra work I believe we could make quite an improvement in the homogeneity.'

'How long would it take?' asked Rex.

'Probably three months.'

'And how much extra would it cost?'

'Around fifteen hundred.'

'Done!' said this most unusual customer.

So the magnet delivered to the Enzyme Group in the autumn of 1971 had an exceptionally uniform field, which provided a resolution about ten times better than that guaranteed (see Box 7.4 and Fig. 7.3). The whole spectrometer, with this magnet as its crucial core, performed brilliantly. Sir Rex tells me this magnet has, for the past few years, been in Illinois, USA, and has continued to work perfectly after some thirty years of service.

Box 7.4. The 1971 Spectrometer for the Enzyme Group

Rated at 270 MHz (for protons), the magnet for this system has a field of 6.4 T, which, when fully compensated, is uniform to five parts in 10^{10} (ten thousand million) over a 5 mm spherical volume. Bruker's equipment for this multinuclear spectrometer included an advanced FT facility. With sample spinning to homogenize the signal, the system soon demonstrated separate lines on the spectrum where the resonance frequencies for protons in slightly different magnetic environments were less than 0.1 Hz apart. (See Fig. 7.3.)

Some months after its installation Martin walked into the laboratory when this system was in use. Excited by the results he was achieving, Rex exclaimed, 'The structure of Penicillin took many years to sort out before it could be synthesized—with this system we could have got there in a week!' I am sure this was an exaggeration, but it is certain that this advanced instrument was opening up a new world of rapid and detailed analysis for highly complex and important molecules.

Fig. 7.3. Barry McKinnon testing the 270 MHz NMR magnet system for the spectrometer for the Enzyme Group—magnet and spectrum inset

During the months of work in cooperation with Bruker, Professor Laukien had continued to press for a closer relationship. Oxford Instruments did not want to be taken over, and any joint venture looked difficult to negotiate from the standpoint of the smaller company. But the new magnet system would form the prototype for a standard commercial product that Barry McKinnon, then still Development Manager, had been busy formulating. It was now offered on an original equipment manufacturer (OEM) basis to the few companies making NMR spectrometers. Japan Electro Optics (JEOL) bought one or two for trials, but Varian was then making its own superconducting magnets. Bruker decided to go elsewhere, and bought the small US company Magnion, which had been among the very first to sell superconducting magnets in the early 1960s.

With no substantial customers in sight, Oxford Instruments discussed a possible joint venture with the UK subsidiary of Perkin Elmer, a US company making NMR spectrometers with lower-field magnets. We even considered 'going it alone' to develop complete spectrometers, with advice from our friends in Oxford University, and our own experienced NMR engineers, including Alan Simpson and Mike Biltcliffe. But we hesitated at the thought of head-on competition with the two leaders. For a couple of years Oxford Instruments sold very few of these magnets to other companies, although some continued to be produced for research scientists who could make their own electronic spectrometer systems.

In 1973, with the development of a higher-field NMR magnet well advanced at Oxford Instruments, Bruker came back to us. It was now willing to negotiate a deal for the supply of three or four types of magnet. This three-year OEM contract, with tight prices and substantial quantity discounts, was not the best deal in the world for the Company; but it enabled us to plan ahead and build relatively large numbers of standard magnet systems. It was the beginning of a real product business, for which we were very grateful to Bruker. The new magnet was made from multifilamentary niobium titanium, which, with the joints problem finally solved, could now be used for NMR as well as for other types of magnet. The use of this superconductor raised the magnetic-field limit to more than 8 T to give a frequency rating of 340 to 360 MHz (see Fig. 7.4). Bruker soon adapted a spectrometer system for the new magnet, and the first commercial 360 MHz system was delivered to Stanford University in California in 1974. This system was unobtainable elsewhere, and the standard price was about £30,000

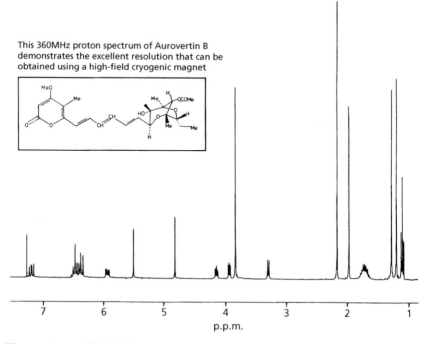

This 360MHz proton spectrum of Aurovertin B demonstrates the excellent resolution that can be obtained using a high-field cryogenic magnet

p.p.m.

Fig. 7.4. A 360 MHz NMR proton spectrum

(equivalent to nearly £170,000 at the turn of the century); this was almost double the price of the previous top systems, as the new ones were much more difficult to make.

Any fixed-price contract signed in September 1973 soon suffered severe erosion in its margins. In the following two years the UK Retail Price Index (RPI) increased by 48 per cent, and over the three years to September 1976 by nearly 70 per cent. Fortunately, spectrometers containing our more expensive magnets were in demand, and the 'learning curve', common with difficult new products, enabled us to make up for some of this price erosion by more efficient manufacture.

Through 1974, 1975, and into 1976 the Company worked on this bulk order from Bruker to the mix of models requested from time to time. As the second year of the contract closed, no further magnets were called off from the incomplete bulk order. With the raging inflation of the mid-1970s we were content to let this fixed-price contract die early, but we needed a market for our geared-up magnet production unit. We thought again about a joint venture with Perkin Elemer, which had been buying one or two more magnets, but it was uninterested. No other spectrometer manufacturer had ordered *any* magnet system from us for at least eighteen months. We were back at the 1971 situation, and our old customers appeared to be either developing or using their own magnets. Was our promising market slipping right away?

As in 1971, we considered making our own complete spectrometers in order to keep our NMR magnet production going. If we were to take this route, we would have two trump cards. The first was our relationship with the world-leading NMR research scientists at Oxford University. The Company would later be able to license the sophisticated electronic system the Enzyme Group was developing. This was for a new spectrometer it was planning, which would use a yet higher field magnet from Oxford Instruments when the development was complete. The second trump card was our reputation in magnet technology—no other company had yet matched our 360 MHz magnets after more than three years, and we were now working on an even higher-field system.

This new development was making rather slow progress. At that stage we had no dedicated R & D team spending all its time on a project. The work went on alongside production, as R & D had always done in the Company. This system has some advantages, as the engineers doing the development are also in charge of contracts and have close ties to customers. This makes them acutely aware of manufacturing problems, but development tends to take second place to production; with scheduled delivery dates and sales budgets demanding attention, the development timetable slips. But by 1976 the Company felt confident enough to accept an order from the Oxford Enzyme Group for a 10.5 T, 450 MHz-rated magnet, the inner windings to be made from multifilamentary niobium tin. The joints problem was at last almost solved (see Box 7.5).

The new magnet for Oxford University would take some time to complete, and from 1976, with this worrying dearth of orders for standard magnets, the

Box 7.5. Problems in the Development of Niobium-Tin Magnets for NMR

For other types of magnet the Company had, for some time, been using a multifilamentary form of niobium tin. A small magnet, made from this expensive and not widely available material, could be fitted inside an outer magnet made from standard multifilamentary niobium-titanium wire. These 'duplex' magnets could give fields of up to 15 T. But for NMR work the joints between sections of the wire had for long defied attempts to make them adequately superconducting for the necessary time stability. This problem, tackled in cooperation with RAL, was almost solved. But there were other difficulties; there was the sheer size of the outer magnet, and there would be serious magnetic stresses at such high fields, which might affect the field uniformity. As described in Chapter 5, the superconducting niobium tin layer on the tiny filaments was actually formed by a long reaction at a high temperature *after* the magnet had been wound. Some of the other components and materials would have to withstand these temperatures—as high as 700 °C or 800 °C—and would also have to operate in liquid helium at minus 269 °C. This development was the most difficult to date.

Company veered towards the risky route of making its own complete spectrometers in competition with its erstwhile customers. Later, we even had to consider the merits of selling the NMR magnet operation to one of them. But in the end the technical ability of the team was to save the day and lead to a profitable business that has continued to be world number one in high-field NMR magnet systems. Through those uneasy years the whole Company seemed to be dominated by this struggle to stay ahead technically, to keep *some* market for these NMR magnets and to remain independent in the face of strong pressures to sell.

Our designers worked out plans for two complete high-field spectrometers and a low-cost system rated at 200 MHz, where there seemed to be an unoccupied niche for fast routine NMR analysis. These products would take two years to develop, which would cost some £200,000 (equivalent to more than £800,000 at the beginning of 2000). More funds would then be needed for marketing, back-up facilities in the USA and Europe, demonstration units, and new staff such as software engineers and applications chemists. It would be a major undertaking. But the government agencies concerned wanted a British spectrometer for satisfying the growing number of requests for NMR equipment, and the NRDC offered to pay half the technical development costs under a royalty agreement. So, in spite of reservations about the competition in the market, the Board gave the project the green light.

While these new systems were being developed, priority was to be given to the elements of the work that would benefit sales to other spectrometer manufacturers should the OEM magnet business re-emerge. The most important matters were the completion of the 450 MHz magnet for Oxford University and improvements to the cryostats. If we were to succeed in this major programme of developing, manufacturing, and marketing our own spectrometers, and success was by no means certain, the whole of the magnet business would need to change its mode of operation and become product rather than project orientated. R & D in other areas would have to be curtailed and the Company would be concentrating on making well-tested systems rather than difficult innovative one-off equipment.

Would the total market for superconducting NMR spectrometers grow enough to allow a reasonable share for Oxford Instruments? Use of NMR analysis techniques was growing, but, of the 2,000 or so centres in the world with such facilities, most used low-field instruments with resistive magnets. These spectrometers were used in chemical and biochemical research as well as in routine testing in industry. A lot of this work could achieve better results using high-field superconducting magnets, but many chemists, unlike our physicist customers, put up a strong resistance to the very idea of using liquid helium. Somehow we had to convince these scientists that spectrometers based on superconducting magnets were reliable and stable instruments, and that liquid helium was now widely available and could be trouble free.

With these problems in mind Ian Herbert and Tim Cook planned to make the cryostats for NMR magnets more 'user-friendly' and more efficient. They aimed for a system so economical on liquid helium that a three-monthly refill under a service contract could replace the bothersome weekly top-up chore, often the task of the scientists themselves. With no power input needed for an NMR magnet once it is fully charged, these systems could become more convenient and cheaper to run than the electromagnets so many laboratories were using, as well as providing better resolution and long-term stability. In fact they would be more like permanent magnets with very high fields.

During 1977 the NMR team built a two-thirds scale model of the 450 MHz magnet to test some of the new features that would be needed in the final system. Peter Hanley, who was leading the development, became more confident in the outcome. In the October 1977 newsletter, the Company announced, 'Now we are going up to 450 MHz.' Perhaps it was this that stirred up our old customers; perhaps they had discovered we were developing our own spectrometers. For whatever reason, over the next two or three months we were contacted by JEOL, Bruker—again seeking to buy the Company—Varian, Hitachi, and by an American newcomer to the market, the Nicolet Instrument Corporation. Orders started to flow again, including some from Bruker. Our NMR future looked brighter, and Barrie Marson 'saw ahead the emergence of a major business for us'. However, if the OEM market opened up securely again, there would be new question marks over our own spectrometer development; it is always uncomfortable and often suicidal to compete with one's main customers.

The Nicolet Corporation, a major manufacturer of other spectroscopic equipment, was a strong new player in the NMR field. It entered this market with low-field systems and soon started to take market share from other companies in the business. Nicolet wanted to compete over a wider range, and John Woodgate visited its plant in California to negotiate an OEM deal for magnet systems. He came back with the makings of a half-million-pound-a-year contract, but also with a pressing offer to *buy* Oxford Instruments, or its magnet production operation, for $1 million (equivalent to about £3.5 million at the beginning of 2000).

Professor Laukien, of Bruker, who had been trying to buy the Company on and off since 1970, now started to press harder. He argued that, if we did not throw in our lot with Bruker, with its greater resources it would soon catch us up technically and stop buying our magnets. Nicolet would be forced out of the market because of its lack of experience in NMR; Oxford Instruments would then have a go at developing its own spectrometer, and would fail, and would finish up with nothing. There was some logic in this view; both Bruker and Nicolet were making it very clear that they could not afford to rely, forever, on a small company in England for such a core component of their spectrometers.

What was the Company to do? If we remained independent we would have to accept that our customers might catch us up technically, and this newly active market for NMR magnets might well last only for a couple of years. The alternatives were to sell the NMR magnet operation to Nicolet, or to sell the whole of the original magnet company to Bruker. There was no clear-cut right course to take, and the dilemma was debated over and over again in an attempt to clarify the future outcome of the various possibilities. No one wanted to sell the original Oxford Instruments, the source of ideas and new products. But a sale would provide funds for our rapidly growing and profitable medical electronics company, which then had a clear lead over its rivals; and there were new ideas for industrial automation systems emerging at Newport and needing development funds. Our own spectrometer project, part funded by the Government, was going ahead, and much of the work would be useful for the OEM magnet business, but the market for complete spectrometers was fast becoming a frightening battleground.

Meanwhile, as the months went by, and the pressures continued, and the debate went on and on, we were shipping a lot of magnets to Nicolet and others. Our margins were improving and the new magnet development was progressing faster. There was still no change in the cloud over our NMR business *after* 1980, but the offers to buy the Company were looking less reasonable. After a while the Nicolet bid was virtually ruled out, but Bruker had to be considered carefully. It was a flourishing European company keen on R & D; some felt our young scientists and technicians might find a better future there than with an independent Oxford Instruments if Professor Laukien's predictions were to come true.

The negotiating team decided to consult the middle managers concerned. All discussion to date was supposed to have been strictly confidential, but we found that the staff knew all about the matter and had long since made up their minds. Even if salaries were likely to be much higher under Bruker, they preferred to work for a small British company they knew and in which they could exert some influence and talk to anyone from top to bottom.

Opinion started to move away from any deal when we heard that Bruker was looking into the logistics of setting up a magnet factory in nearby Abingdon. Would all our experienced NMR staff stay with us? Although everyone was emotionally opposed to a sale, we had to look objectively at the situation and consider our outside shareholders—it was still possible that the Company could lose *all* the value of its NMR capability. But by May 1978 the order book, and current profits, and the next two years' budgets were looking so good that the best price Bruker had suggested would not match them. Independence would pay off, whatever the outcome in the market after 1980. Further success would depend on keeping ahead in the technology; if we were to lose our lead, the future would be bleak.

So the Company turned down Professor Laukien's offer. Bruker did not expect this reverse; as soon as it realized we were serious and had *really* rejected its proposals, it came back to sign a contract for magnet systems. It still needed Oxford Instruments. John Woodgate soon negotiated a new two-year half-a-million-pounds-a-year deal for the supply of magnets, and resisted all demands for options over equity or other future commitments.

In May 1978, as the long debate on the right course for the Company was coming to an end, the advanced magnet for the Enzyme Group passed its tests with flying colours. It went to its specified field of 10.5 T for a frequency rating of 450 MHz, and beyond to 11 T for 470 MHz. As our press release soon proclaimed, it represented 'a quantum jump in technology' and the fully assembled system in the University was 'probably the most powerful and convenient NMR spectrometer in the world'. The magnet system won for Oxford Instruments its third Queen's Award—this time for the technology. And there was another award, won jointly with the RAL, which had assisted in the development of the joints in the multifilamentary NbSn. This was a coveted IR100 Award in the USA, where the magnet system was cited as one of the hundred most significant new technical products of 1978. The assessors were probably right. High-field NMR spectrometers played a vital role in the development of most of the new drugs that transformed medicine in the last two decades of the twentieth century.

The new low-loss cryostat, built for this NMR magnet, achieved its targets; by careful attention to every possible 'heat leak' into the system, the evaporation rate was reduced by an amazing factor of ten. It now had to be refilled with liquid helium only four times a year. The development of this cryostat cracked the cryogenic barrier and made superconducting magnets much less trouble to run. A while later, seeking a technical explanation for something I was writing, I asked Derek Shaw, the Sales Manager, 'what can you do with these new systems you couldn't do with the old ones?' His answer was simple: 'sell them!' From that time the worldwide sales of all types of superconducting NMR systems grew much faster than sales of resistive systems. By July 1978 three 470 MHz systems had been ordered, and by January 1979 the Company had upgraded the product to 500 MHz (see Fig. 7.5 and Plate 5). From then on the sales took off.

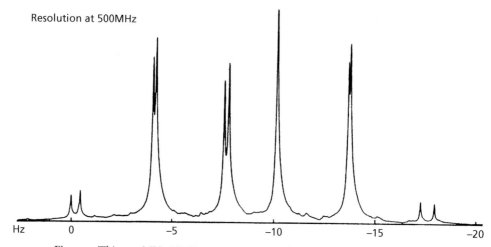

Resolution at 500MHz

Hz   0   −5   −10   −15   −20

Fig. 7.5. This 500 MHz NMR proton spectrum shows resolution of peaks down to 0.08 Hz
Source: Department of Biochemistry, University of Oxford

The May 1978 consensus agreement to keep the magnet company independent was undoubtedly the right decision. Over the next two years sales of NMR magnet systems increased almost tenfold. By 1980, instead of the feared famine in NMR orders, a five-year contract was in place with Bruker, and orders continued to come from Nicolet and JEOL; and from 1985 orders came also from Varian. In the 1990s Varian was to become the Company's main customer for NMR magnets. But at the end of the 1970s, the main problem was the Company's capacity to manufacture the magnets fast enough.

As the fortunes of the magnet company rose, those of the medical electronics company faltered for a while. Growing competition cut into its profits until newly developed products restored its position. This see-saw in fortunes proved the value of encompassing a spread of products and markets within the Group, and reinforced our philosophy of separating activities with different markets into semi-autonomous subsidiaries, or divisions, once they reached a viable size.

The decision to keep our independence turned out to be even more important in the longer run. As early as September 1978 we were in discussions with two or three international medical companies over supplying much larger magnets. These were for a new experimental technique, which used NMR instead of X-rays for body scanning. This further development of the Company's expertise in NMR magnets was to be our principal engine of growth in the 1980s.

8 Where is the Company Going?

The turnaround in the Company's situation in the late 1970s

'Would you tell me please, which way I ought to go from here?'
'That depends a good deal on where you want to get to', said the cat.

(Lewis Carroll, *Alice's Adventures in Wonderland* (1865))

Here I must track back to January 1975, the month when Barrie Marson made his last *cri de cœur* over threats of catastrophe from the negative cash flow.

There had been two general elections in 1974. Early in the year the Labour Party had taken over the government from the Conservatives, but without an overall majority; then in a second election in the autumn they had won a majority of only three. This weak government had to propitiate the powerful unions, and had expensively settled the outstanding miners' strike that had been the downfall of Mr Heath's Tory administration. The newly elected Prime Minister, Harold Wilson, had set in place a 'social contract' to try to limit excessive pay settlements, but wage inflation still roared ahead; it went up to an average of 24 per cent in 1974 and reached 33 per cent by mid-1975.

In order to generate more income to cover large increases in planned public expenditure, the new Government raised taxes, and was particularly hard on industry. National Insurance contributions and corporation tax went up and, for the first time, part of the latter had to be paid in advance, creating a one-off cash-flow problem for many companies. With inflation still rising, industrial production started to fall; the *Financial Times* index of the shares of industrial companies fell by 50 per cent in six months. Later, in this difficult period, the employers' contributions to National Insurance were increased again, to raise a further billion pounds a year (equivalent to over £4 billion at the beginning of 2000). Unemployment went on climbing throughout these years, companies of all sizes got into difficulties, and insolvency accountants had a busy time. Among many small companies in trouble was Thor Cryogenics, which was sold by the receiver to another British company. (Its new parent was itself taken over a year or two later by a US company not interested in cryogenics, and Thor's managers were able to buy their company back.)

Barrie's gloom in the early months of 1975 came from another wave of super-conductor problems as well as from these increasing financial burdens. Expenditure had to be controlled as tightly as ever. The Cryomagnetic Systems Division, then managed by Roger Wheatley, had to hold back acceptance of new orders until several magnets, delayed by the current technical problems, had been delivered and paid for, and the cash flow was healthy enough to buy the necessary materials. The Company was still making some profit; in the year to March 1975 it managed £81,000 on a turnover of £1.6 million. This was only a 5 per cent return on sales, but the Board still congratulated Barrie warmly for trimming the ship and riding out such a stormy sea.

By the end of 1975 the cash situation in the Company was improving fast (see Fig. 8.1). The Government had brought in measures to relieve industry from paying tax on unreal profits due to inflation, and the general climate was a little better. In Oxford Instruments the technical problems, both in magnets and in Medilog, had been largely resolved, the high-risk hybrid magnet for the Clarendon Laboratory had finally been accepted, and the German sales company was at last making a good profit. Both the bank and ICFC were now happy to consider financing expansion with further loans. The old cash starvation had ended, and new equity partners would not now be needed for the medical operation. Early in 1976 the Board was able to approve finance for new plans to

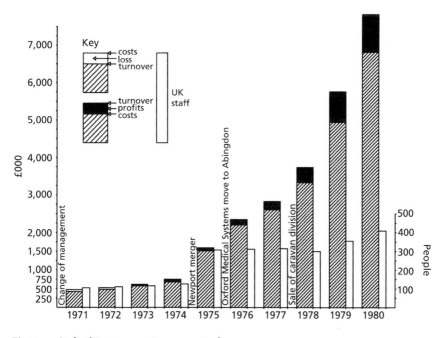

Fig. 8.1. Oxford Instruments' progress in the 1970s

satisfy the explosive growth of the Medilog market. In Newport, the expansion of the caravan company, with its substantial prepayments, was largely self-financing, and, when it was finally sold in 1977, more resources were released for investment in the science-based businesses.

The accounts to March 1976 showed pre-tax profits up by 80 per cent to £145,000 on a turnover of £2.3 million. The Group directors at that time were Martin, the Chairman; Barrie, the MD; John Lawrence the Finance Director; Dr John Gardener of BOC; Ken Fuller, an accountant, representing Ian Boswell's investment company, and myself. I was also Company Secretary, a statutory post I had held from the beginning. In a small private company this legal position is not difficult for an unqualified person to manage. Tasks include the keeping of the statutory books and other records, sending out notices of meetings, writing minutes of Board and Shareholders' meetings, sending Annual Returns to Companies House, and the occasional issuing of shares. For any serious negotiations, or vetting of proposed agreements, contracts, or leases, one can always consult the company's solicitor.

Free from immediate money worries, the Board started to concentrate on the future, and on defining more formalized objectives and structure. Barrie wrote a stimulating paper entitled 'Where is the Company Going?' With his experience of a much larger and older company, Kent Instruments, he looked forward to a board structure more like that in a public company, with a majority of executive directors and fewer shareholders' representatives. The structure of Oxford Instruments had grown up with its history. Martin and I, the 'family' founders, still performed some executive functions, and the objectives of the other directors and shareholders varied quite a lot. Most were not seeking straight financial goals—in fact, their objectives were generally rather remote from the aims of institutional shareholders in a public company of the late twentieth century. The only shareholder in that category, ICFC, would have liked dividends on ordinary shares, but was not pressing. None of the shareholders wanted a quick sale of the Company for quick gains. BOC wanted to keep a window on the technology, and Ian Boswell took a very long-term view of his investment.

Martin and I were wrapped up in the Company we had founded, and saw it as a growing organic body, with its committed and hardworking employees, its exciting technical developments, its scientist customers, its growing reputation—and its occasional lapses. We wanted the Company to be willing to take risks when new opportunities arose, and we wanted profits to go towards building up the Group for the long term—it was not yet time to pay dividends.

We also wanted to share more of the ownership of Oxford Instruments with the employees, which, we felt, would be equitable as well as providing an incentive for the staff. It was not a common view at the time; dividends and capital gains were for investors and 'wages were for workers'. We had already succeeded in getting some shares into the hands of twenty or so senior employees in the two rights issues in the past, but employee option arrangements at that time could bring severe tax penalties. We suggested instead that the Board should give a percentage of the shares to the Pension Fund and they agreed to keep the idea in mind, but saw no strong incentive reasons for doing so. It was to be 1979 before these ideas progressed; the Liberal Members of Parliament, who were then keeping the Labour Government in power through the 'Lib.–Lab.' pact, forced a scheme through the House for tax-effective employee share ownership. It took a little longer for the Board to agree what to do, but finally, in 1980, a decision was made. The Company was to set up a group-wide scheme for all employees, financed from profits under the new regulations; to introduce a less tax-effective share-incentive scheme for some twenty key senior executives; and to provide the Pension Fund with a block of shares. This was later to give the fund a very healthy boost; at the end of the century it was worth well over £50 million.

Back in 1976 the Government was again in financial difficulties. That October the pound fell to $1.55 and the bank rate, then the basis of all interest rates in the UK, went up to 15 per cent. Dependent on the International Monetary Fund for vital loans, the Government had to cut public spending, and UK scientific research inevitably saw its funding slashed. But with a low pound, our competitive position overseas was improving. The profit for the year to March 1977, which Barrie had feared would be badly affected by the UK's economic situation, was actually up by nearly 50 per cent at £216,000, on a turnover up 20 per cent at £2.8 million. A feature of the last few years of the 1970s was the size of the sums by which our results exceeded budgets. The management had got used to low expectations in those years of underfunding.

Barrie was developing an effective executive team, which started to spend two days a year on strategy meetings away from the distractions of the Company. Change was in the air. Every part of the Group seemed to be growing and on the move; planning or building new factories; buying expensive computers and other new equipment to aid design and manufacture; and developing advanced products. The payroll was going up by about fifty a year, and senior people were moving from one job to another within the Group. Hardly any of them left; people liked working for a small, rapidly growing, science-based firm. In spite of the increasing size and changes in structure and organization, the Company retained its informality and friendliness—its long-established culture.

In most parts of the Group engineers were starting to develop sophisticated new products that made use of the latest advances in microprocessors. This sort of development needs bright young innovative scientists and engineers, who have skills that were not in existence when the previous generation got its training. Finding the right people is always difficult and was then probably our main limiting factor on future growth. Some of the people recruited then were to become good managers as well as being design, development, production, or sales engineers. The medical company, located in Abingdon, was growing fast and brought in several new senior recruits. Among them was Julian Morris, a computer and bioengineering expert, who became Technical Manager of the heart-monitoring activity. Paul Brankin and Alan Simpson moved from other jobs to work on advanced brain-monitoring equipment in a newly separated EEG section.

In the still overcrowded Osney Mead premises Tim Cook, who had arrived from Thor Cryogenics a couple of years earlier, was joined by Bill Proctor from the same source. Both were experienced magnet designers who were to become senior managers—Tim in several successive positions and Bill largely in the NMR operation, which he managed successfully over a long period. In 1978 Ian McDougall, who had been involved in making superconductors at IMI, joined the Company, strengthening our technology with his deep knowledge of metallurgy and physics.

Overseas, Roger Wheatley was in charge of Oxford Instruments North America, and, early in 1978, was joined by Jack Frost to manage the embryo Oxford Medilog, the US subsidiary of Oxford Medical. This was later moved to Tampa, Florida, where it eventually grew to some fifty people engaged in marketing, sales, and service functions. Later, Jack was to head the whole of Oxford Medical for many years.

In his Chairman's Review with the 1977 Group Accounts Martin wrote:

Since its foundation the Company has operated out of somewhat awkward sites and with equipment which has often been less than fully adequate for the job. Although this is a common state for new small companies, the problems have been particularly acute for our Group because of the sophistication of the products and the high growth rate. Success has relied heavily on the resourcefulness, flexibility and hard work of the staff . . .

The moves this year towards concentrating and expanding our activities in the three main areas of scientific, medical and industrial instrumentation at Oxford, Abingdon and Newport Pagnell, combined with improved accommodation, more sophisticated equipment and a generally better working environment, represent a milestone in the history of the Group. They are made possible by the Group's growing financial strength.

The first new factory to be built since 1971 was for Newport Instruments, at Milton Keynes. This was made necessary by the sale of the caravan company, which would shortly need all the old rented premises. The Instrument Division at Newport had been making medical and other electronic units for the rest of the Group. After the magnet section had been moved to Oxford, only one substantial product of its own had remained, the low resolution NMR instrument, the Newport Analyser. This was an industrial instrument, and Barrie Marson, then working on Newport's forward strategy, wanted to turn the Division into an industrial monitoring and control systems company. Barrie had a lot of experience in digital electronics, and was a consultant for Carlo Gavazzi, an Italian company in this field. He was sure digital data acquisition would soon become important in process industries as the power of microprocessors developed. Through his influence Newport started making and installing systems, as subcontractors to Carlo Gavazzi. The equipment was a microprocessor-based 'distributed intelligent data acquisition and control system', known as DIAC.

The first DIAC system to be made at Newport was ordered in 1977. It was a remote monitoring and command system for controlling the cold terminal at Tilbury Docks—then the largest refrigerated container terminal in the world. The story goes that the management at Tilbury decided to automate the system only after a Rolls-Royce car got put in the wrong place in its container and was deep frozen. In 1978 Newport obtained a two-year agreement to make and sell DIAC under licence. The work involved a lot of one-off systems design and software writing. The sites needing this type of equipment were mainly oil refineries or industrial process plants for companies such as ICI. A Systems Division was set up, and John Lee, who had managed the DIAC contracts for Carlo Govazzi, joined the Newport Instruments' staff as its General Manager.

The new factory at Milton Keynes was completed at the end of 1978, and the remaining divisions moved in over the Christmas break. All the high-profile politicians and other dignitaries approached were busy, so the Company held a low-key opening party a month or two later. Then in May 1979 Martin received a mysterious message from Conservative Central Office. Would it be possible for Margaret Thatcher (now Baroness Thatcher) to open the new building? The election campaign had started and a go-ahead science-based company was just the background she needed. When they heard it had already been opened, they still wanted a visit to go ahead. It all had to be kept a dark secret, but, at Martin's remonstrations that *some* of the staff would have to be consulted, they agreed to the MD, John Lee, being let into the plan 'in the strictest confidence'. A few days later the *Milton Keynes Express* got on the line to John asking for the date of Mrs Thatcher's visit. So much for the security of Central Office.

The Instrument section at Newport was then manufacturing a new device for Oxford Medical along with Medilog. This was Unibed, a 'stand-alone' mobile bedside monitor for patients who needed individual intensive care. Developed by doctors in Rotterdam, this clever unit, which could display several physiological parameters at the touch of a panel, was very advanced for its time. Oxford Medical had a contract to tidy up the design and provide units for Rotterdam, and an option to make and sell these Unibeds under a royalty agreement. When Mrs Thatcher toured the new factory (see Fig. 8.2) Martin invited her to be 'wired up' with electrodes on her wrists, and to see her ECG traces on the monitor of a Unibed. She was game for anything, and happy for reporters to see 'what good heart' she was in. But she wanted to know why Denis Thatcher's heart produced a much larger ECG signal on the monitor. Of course it was because his electrodes had been placed on his chest, nearer his heart; but a voice in the background could just be heard to murmur 'all politicians have thick skins'.

Mrs Thatcher used the occasion to point out to the press that, if she got to No. 10, she would be the first Prime Minister to hold a science degree. She surprised everyone by her grasp of this complicated medical equipment; in fact, she quite tactfully took over the explanation from a rather hesitant technician—she had done her homework in the car on the way from London. Unfortunately,

Fig. 8.2. Margaret Thatcher talking to employees at Newport Instruments in 1979

scientific research in this country did not benefit from Margaret Thatcher's years in office. There was so much emphasis on applied research in universities, and on industrial sponsorship, that many first-class basic research proposals failed to win funding.

Martin has often spoken against this neglect of fundamental research, which may be curiosity driven, but may also result in an Aladdin's cave of new knowledge from which who knows what prizes may come in the future. As he points out, anyone casting their eyes over the work going on in Oxford University in the late 1950s, looking for groups doing research with commercial potential, would probably have put the high-magnetic-field team in the Clarendon Laboratory at the bottom of the list. That work was the starting point for Oxford Instruments, and the taxes paid and generated by the Company over the years would, in total, be enough to finance, say, a new university or a hospital. The economic ramifications of one successful company can be very wide.

At Abingdon, OMS saw its sales and profits dip in 1978 because of increasing competition in the USA, and a six-month delay in the launch of a new product. It always seems difficult to get developments, finished on schedule, and products properly launched, with full documentation, at the right moment. This is a perennial problem, and not only for small companies. It is hard to quantify the unknown elements in order to develop a realistic time budget. Oxford Instruments has since developed better-constructed project control programmes, which certainly help; but many developments are still not finished on time. With the eventual launch of Medilog 2 and the new microprocessor-based ECG analysis system, sales recovered and surged ahead again, and OMS got back its technical lead. But following this hiccup the building on its own new site had to wait for a year or two.

At Osney Mead in Oxford the management was preoccupied for nearly six months with the big debate over the future of NMR magnets, and indeed the whole of the original magnet company. After the decision to remain independent, made in May 1978, the huge increase in orders from the spectrometer manufacturers, and the urgent need to keep ahead in the technology, forced the Company to concentrate on NMR magnets at the expense of almost everything else. There was not the space nor the development staff to advance equally hard in the other magnet and cryogenic areas, and the Company started to slow down its traditional project-type business. This had to be controlled at a workable volume by raising prices and by curtailing the 'hairier' projects.

The designers became more cautious and conservative. Competitors, using cheaper materials and less prudent design criteria, started to undercut our prices by quite large margins—and they usually succeeded in completing these 'risky' projects. There was now another competitor, Cryogenic Consultants, in the business as well as Thor Cryogenics. Some directors and managers became worried that the trend away from challenging contracts had gone too far. It is from these semi-development projects that future products come. Our non-standard magnet systems were too expensive, and the engineers started to look again at the Company's design philosophy to see if they could safely cut costs by being more innovative and adventurous. Companies should never be so confident in their own expertise that they cannot learn from others.

At the time of maximum pressure the dilution refrigerator came under threat. These sophisticated systems needed renewed development to catch up with current top performance, but there was no spare effort available, and there were not many orders to justify the work. Although a French Government laboratory in Grenoble was achieving excellent results, the only real *commercial* competition was from SHE, a company in California that was advertising lower temperatures than our machines could achieve. The Ultra Low Temperature Section was then managed by Nick Kerley, and its reprieve came from a wave of new orders for units that could provide high cooling power, rather than the lowest temperatures (see Fig. 8.3). This was an area where we still had the lead and where little extra development was needed. Soon after this a scientist from the prestigious Grenoble laboratory became our consultant, and helped improve our heat-exchanger design. Before long the Company also started to recover its lead in commercial refrigerators for the lowest achievable temperatures.

Fig. 8.3. Ian Duff assembles the heat exchangers for a high cooling-power dilution refrigerator

Osney Mead had become badly overcrowded. With the surge in NMR orders, the congestion was affecting production capacity and profits. There was a long search for vacant premises, and, to Barrie's dismay, further delays while the outside directors argued over the relative merits of leasing, or borrowing to build freehold. Finally the Board agreed to the redevelopment of the back end of the existing Osney Mead site. The new factory was actually built over the top of the old boathouse while work went on underneath (see Fig. 8.4). When the new machine shop area was ready, the machine tools, some, by now, computer-controlled giants, were moved over to their new positions during a weekend. The old boat house, home for fourteen years, looked like a small shed below the new roof, and was demolished in a day.

While the new 20,000-square-foot design office and factory were being built, several departments had to move out to temporary rented buildings, and some of these had to be kept on until the next building operation in the early 1980s. The delays in starting the factory had meant a slowing-down of our expansion

OXFORD INSTRUMENTS ROADS TO GROWTH OVER 40 YEARS

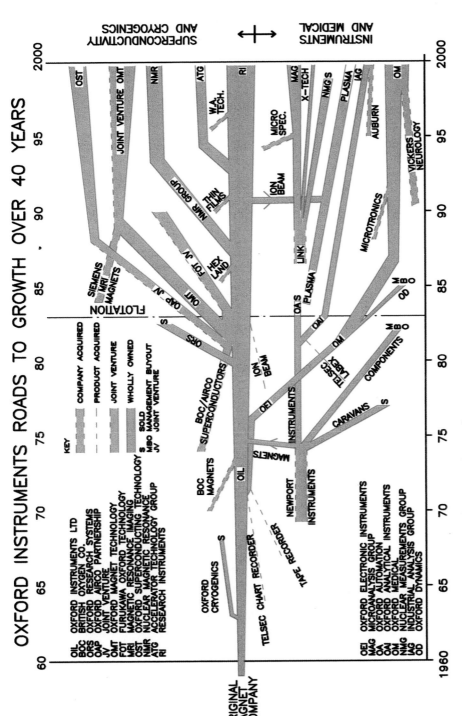

1. Oxford Instruments' roads to growth over forty years

2. 'The Small Company' in 1968: About half the employees were in the picture

Back row: Pierre Perrot Kevin Smith Jos Schouten Bill Proctor Lester Sideropoulos Terry Martin Mike Russell Della Mabey John Benson Barrie Lowe Allyson Reed Simon Bennett

Sixth row: Jill Shepherd Sam Klaidman Paul Barrett Elaine Outlaw Tracy Cuthbert Kate Nayler Damian Stone Masahiko Nishiyama Norioh Saito John Francis Roger Humm Vince Kempson Ron Jones John Field Brian Price Alan Goodbrand

Fifth row: Peter LoDico La-Verne Tyler Seung Hong Jack Frost Kevin Hill Rick Seymour David Hemsley Keith Murray Ian Cotton Jim Steward Peter Roberts Brad Boyer Jim Hutchins Ian Radley Kevin Hole

Fourth row: Karen Taylor Ian McDougall Peter Penfold Peter Statham Michel Hascoet Dietmar Fischer Patrice Perrin Pierre Schneider Richard Clark Scott Reiman Tom McNulty Kevin Timms Mike Carstairs John Pilcher Paul Noonan Howard Dunham

Third row: Tony Ford Mike Dadswell Laure Chatelain Tony Sayers Colin Cleaton Thomas Marx Jeremy Smith Victor Regoczy Robin Higgons Andrew Searle Dave Andrews Anthony Williams Andrew Mackintosh Nick Wilkins Paul Brankin

Second row: Gernot Amm John Comfort Larry Darken John Lewis-Crosby Hugh Hardiman Andrew Davies Peter Walton Doug Gilbert Dietmar Budzylek Alan Street Nick Kerley Martin Townsend Andy Baker John Gordon Herbert Albrecht Craig Sawyers

Front row: Glenn Epstein Jennifer Dandridge Alistair Smith John Hearn Jim Worth John Woodgate Kumar Bhattacharrya Peter Williams Martin Wood Martin Lamaison Jiro Kitaura Mike Brady Pauline Hobday Judy Voght Ken Bailey

3. The International Management Conference in 1995: Some 7 per cent of the employees participated

The Garden Shed—our first work-shop photographed in its decline many years later

The boathouse on Osney Mead with extensions and additions—in 1969

The Osney Mead site with the 1971 factory on the right and the 1982 MRI factory on the left

The 1997 Tubney Woods factory—then one of the fourteen production sites in the Group

4. Progression in factories

Our first superconducting magnet—1962

Our first 10T magnet—1967

500MHz NMR magnet—1979

Assembly of a split-pair magnet—1983

Small-bore horizontal NMR magnet—1985

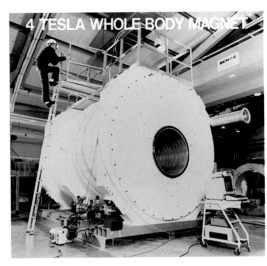

A 4T magnet for invivo spectroscopy—1994

The 'Kloe' detector magnet for Frascati in Italy—1997

5. Progression in superconducting magnets

High field superconducting magnet for biochemical research.

Examples of superconducting materials shown in various stages of the wire production process.

Assembly of an ultra-low temperature cryostat.

Installation at Oak Ridge National Laboratory, USA of an ultra-low temperature refrigerator.
(Courtesy: Oak Ridge National Laboratory operated by Union Carbide for the United States Department of Energy)

Testing heat exchangers for an ultra-low temperature refrigerator.

Fitting insulation to a cryostat.

Automatic testing of printed circuit boards for Medilog equipment.

Medilog EEG recording and display system.

Medilog recorder in use.

6. Pictures from inside the back cover of the prospectus of 1983

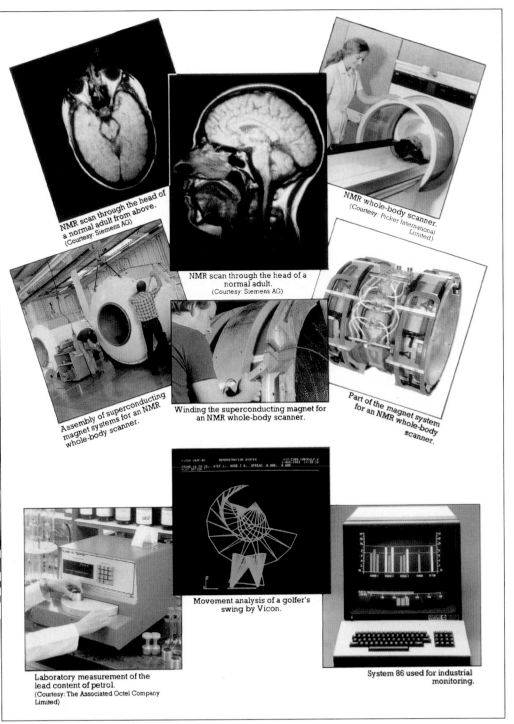

NMR scan through the head of a normal adult from above.
(Courtesy: Siemens AG)

NMR scan through the head of a normal adult.
(Courtesy: Siemens AG)

NMR whole-body scanner.
(Courtesy: Picker International Limited)

Assembly of superconducting magnet systems for an NMR whole-body scanner.

Winding the superconducting magnet for an NMR whole-body scanner.

Part of the magnet system for an NMR whole-body scanner.

Movement analysis of a golfer's swing by Vicon.

Laboratory measurement of the lead content of petrol.
(Courtesy: The Associated Octel Company Limited)

System 86 used for industrial monitoring.

7. Pictures from inside the front cover of the prospectus of 1983

8. The production of superconducting wire

Above: OST's factory in Carteret, New Jersey, USA
Below: magnified cross sections of a few examples of multifilament wire and cable products

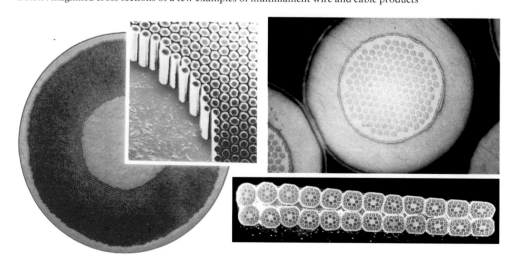

Fig. 8.4. The new factory at Osney Mead being built over the top of the old boathouse while work went on inside

and proved expensive. Barrie estimated the Group would have made an extra quarter of a million pounds profit had we been able to use these new facilities six months earlier.

In 1978 Martin had written:

We are at something of a crossroads as we emerge, finally and completely, from the early stages where survival was the real motivating force. Profits and the technical muscle of the Company have become solidly based, and this has become apparent to everyone from our bankers to the workshop staff. With the survival motivation gone and profits rising, everyone is asking 'where do we go from here?'

The question was how fast we could, or wanted to, expand. The profits were good. In 1978 they were up 86 per cent on 1977 at £402,000 on sales of £3.7 million. In 1979 they were even better, doubling to £813,000 on £5.8 million turnover (equivalent to more than £2.5 million profit on £18 million turnover at the beginning of 2000)—a return on sales of 14 per cent. The 'headcount' had risen to 355. Although well above expectations, these profits were clearly not enough on their own to enable us to take full advantage of our opportunities. We needed more funds to maintain our pre-eminence in NMR and to work on the exciting new developments stemming from it; and there were other promising ideas in Abingdon and in Newport. John Woodgate, then MD of the magnet company, expressed the dilemma: 'The danger is that if we do not capitalise on our present position we shall damage our long-term prospects.' There was not really any thought of slowing down; the debate was on how best to finance our expansion. We had a much improved borrowing capacity, but at 15 per cent interest rates how much was it possible or wise to borrow? Should we seek more equity capital through a 'private placing' of shares with one or two City institutions? Were we ready to contemplate a public flotation, which could make us vulnerable to a hostile takeover? Would it be better to negotiate an 'arranged marriage' with another group with plenty of resources to back our future developments?

Between 1978 and 1980 both the political and economic situations were unstable. For the time being we decided to borrow more. In those two years our borrowing on secured debenture rose from £45,000 to £571,000 and bank overdrafts from £117,000 to £568,000. At least we were now in a *position* to borrow, but the high interest rates made the exercise painful. The decision on where to go for the permanent funds we needed was put off until the Board had a clearer view of future trends.

9 Making the Human Body Transparent

How Oxford Instruments' expertise in NMR magnets led to exciting new opportunities in the 1980s

I think it is rather important to bear in mind that the really big leaps forward that occur are often due to engineering.

(Sir Rex Richards from the Wellcome Witness Seminar, 'Making the Human Body Transparent' (1996))

Grasp opportunity by the forelock.

(A proverb)

In the 1970s fresh ideas in a few universities led to two new developments in NMR. One idea was to form the basis of a substantial new industry in medical diagnostics; the other was to remain as an interesting tool for advanced research in a few well-endowed laboratories. For some time we did not know whether one or the other or neither could become a successful business for Oxford Instruments, but the Company invested substantially in both these opportunities.

As already described (in Chapter 7), NMR can provide a wide range of information from the simple counting of hydrogen nuclei (protons), to the modelling of molecular structures using 'chemical shift' spectroscopy. The Group was already supplying equipment for both extremes of the technology. At one end of the range was the Newport Analyser, for finding how much water or oil there was in a material by counting hydrogen nuclei in a weighed sample in a test tube; at the other end were the magnet systems for high-resolution NMR spectroscopy.

In the early 1970s the British company EMI launched the first really new type of X-ray system for many years. Invented by Godfrey Hounsfield (now Sir Godfrey), the new technique brought a major advance in medical diagnostics and won for him a Nobel Prize. The technique was 'X-ray computerized axial tomography', generally known as CT or CAT scanning. The word 'tomography' comes from the Greek *tomos*, a section of a book or a slice, and the Greek ending -*graphos*, which means something written or recorded. This well describes the drawing of cross-section slices of the body, by using a computer to integrate information from a rapid series of X-ray shots taken all round the segment. The new tomography gave much more information than older photographic X-ray plates.

I have been told that these X-ray tomographs were not the first images of cross sections of the human body. Apparently the Russian doctor who looked after the St Petersburg prison in the late nineteenth century was very interested in anatomy. If a prisoner who had no relatives died in the deep winter, this doctor had the body placed on the ice on the nearby River Neva. When it was frozen, he sawed a series of cross sections, which he then drew and coloured for the instruction of students. But X-ray CAT scanning without doubt produced the first cross sections of *living* bodies.

In 1973 Dr Peter Mansfield (now Sir Peter) at Nottingham University, and Professor Paul Lauterbur in New York, independently published their work, which showed that NMR could be used to produce images of simple heterogeneous objects. Earlier that year Dr Damadian had filed a patent in the USA proposing, without much detail, that it would be possible to detect cancerous tissues using the differential relaxation times of the NMR responses. This patent was not published until 1974. The idea grew that it might be possible to produce NMR images of living organisms using proton NMR, and these might be comparable to X-ray CT images. Living plants and animals contain a large percentage of water, which varies considerably from tissue to tissue. If a measurement, similar to that made by a Newport Analyser, could be made at many discrete points across a section of a solid object, an image of the varying water densities might be built up by a computer. Such an image might be compared to an old-fashioned newspaper picture made up of thousands of small dots, making some areas darker and some lighter; unique information might be obtained, especially on the soft tissues that contain so much water.

Several scientists started working on the concept of making computerized NMR images of sections of the body; it was a desirable goal, as repeated exposures to X-rays can be damaging to body cells. To begin with they worked on tomatoes or lemons, fruit with some internal structure, then on small animals. The Nottingham team used an existing NMR system to show that the principle could be made to work on a small scale. Peter Mansfield called it 'spin mapping' or 'spin imaging', and his team produced images of a finger as well as other objects; in the USA Paul Lauterbur's team demonstrated similar early images. In Aberdeen the group led by Professor John Mallard, who had for long been looking for a use for magnetism in medicine, produced an image of a live mouse as early as 1974.

The UK was to play a major role in the early development of scanning by NMR as well as by X-rays. Much of this work was funded by the NRDC, and resulted in patents that have brought substantial royalties to this country.

Aberdeen and Nottingham remained leaders in the exploration of this new use for NMR, and development started a little later in the EMI Central Research Laboratories, and at GEC Medical. Oxford Instruments worked closely with all these organizations to develop the magnets they needed for these exciting programmes.

The second new NMR development had its roots in Oxford University's Biochemistry Department, with our old friends in the Enzyme Group under Rex Richards. The high-specification NMR magnet we delivered to them in 1971 formed the core of a powerful NMR spectrometer, which was used for work with phosphorus—a very important element in metabolism, crucially involved in energy conversion processes in the body. George Radda (now Sir George), Dr David Gadian, and their research students were working on enzyme regulation, using muscle extract. (NMR studies are almost always carried out using samples in solution.) This team was hoping to extrapolate their results to try and work out what was happening in *living* bodies. As George reported at the Wellcome Witness seminar held in 1996, 'the solution studies didn't really tell us how things behaved *in vivo*. And a very bright graduate student of mine, Steve Busby, suggested ". . . why don't we look at the metabolites in a *living* organ like a piece of muscle, then we can immediately tell what is happening". Everybody pooh-poohed the idea, but nevertheless they went away and did the experiment' (Wellcome Trust 1998: 23). To the surprise of most, they obtained good phosphorus spectra. They immediately recognized that they might be able to study detailed chemical processes or 'metabolic pathways' in living tissue. *In vivo* spectroscopy had been born.

Excited by the results of this first *in vivo* tissue experiment at the end of 1973, the team soon realized they needed a larger bore magnet for working on muscles and other organs. At the 1996 Witness Seminar, George Radda described how he then went about getting a new magnet for this work:

Rex took us to Oxford Instruments and asked them if they could design what was then called a wide-bore magnet, 11-cm bore, at 4.2 Tesla, and of course they could do it they said and it is going to cost us £25,000, which in 1975 wasn't a trivial amount of money . . . So I went to the British Heart Foundation [BHF], and said, 'Look, I think I could have a beating heart inside a magnet of that sort and find out the biochemistry of the heart during a heart attack.' And they sent Sir John McMichael down, who was Professor of Cardiology . . . with [Professor] Peter Sleight [from Oxford] and we had this 3-mm tube with a little mouse heart beating in it and we put it in an old spectrometer and we watched it for about 20 minutes and saw the signal building up. Then we said to Sir John, 'Now we'll turn the oxygen off, that's a heart attack, and you can see those signals go away', and he got so excited that he went back to the BHF and said, 'We must support this, its going to be tremendous', and so that's how our first wide-bore magnet was delivered in 1976.   (Wellcome Trust 1998: 22–3)

These two ways of looking inside living creatures, although both based on NMR, are substantially different. The first essentially measures the number of protons at each of many locations in the cross section being mapped for water density, the second, based on chemical shift spectroscopy, provides a means of studying the biochemistry within a single type of living tissue such as muscle or liver. They also differ in their magnet requirements. As NMR imaging depends primarily on the relative densities of protons across a section, and they have a strong resonance response, *some* sort of image can be built up slowly in a field as low as 300 gauss (0.03 T). A good basic homogeneity is needed and the field has to be modified across the subject by additional coils to produce a field gradient. For *in vivo* spectroscopy a very high field and maximum homogeneity are essential for obtaining the necessary chemical shift spectra, and the volume from which the signals are obtained must be small so that they are confined to one type of tissue.

In 1975 Barrie Marson, and Ian Richards of Newport Instruments, joined in tentative discussions at Nottingham with Professor Raymond Andrew, who had first worked with NMR in 1947, and with Peter Mansfield and Bill Moore, who were much involved in spin imaging. NRDC was interested in the possibility of Oxford Instruments or Newport, who were making the NMR analyser, developing some sort of an instrument in the future to exploit the techniques being pioneered by the Nottingham groups. At that time NMR images could be built up only very slowly. The new technique needed a lot more research, as well as development at the industrial level, before it would stand a chance of competing with the X-ray CT scanners, which several large companies were beginning to sell. It was too early for any decisions, but the Company registered a strong interest in the new possibility.

At Aberdeen University John Mallard and Jim Hutchinson were developing a different technique, which they called 'spin warp imaging', for collecting the information from the different points and reconstructing it into images. Early in 1977 Oxford Instruments delivered to this team a strange-looking cage-like resistive magnet consisting of two larger and two smaller horizontal coils—an approximation to a homogeneous 'Helmholtz' arrangement. Between the two larger central coils was a 30-centimetre deep gap. This was just large enough for one of our technicians to try it out, uncomfortably balanced between head and heels, in the space that was later to take a stretcher (see Fig. 9.1). The static field of this magnet was only 300 gauss, but, with a ninety-minute build-up time, the system produced surprisingly good images. The Aberdeen Medical Physics Department was probably the first group in the world to produce useful results from real patients.

The Company developed a more powerful type of resistive magnet, which had four or six vertically mounted coils with a horizontal bore for the patient. The first was to go to Peter Mansfield at Nottingham (see Fig. 9.2). As usual, the development took longer than expected. With promises to deliver before the end of 1977, the last adjustments were made and the heavy magnet was loaded onto a large hired van on Christmas Eve. The van kept breaking down, but, after an

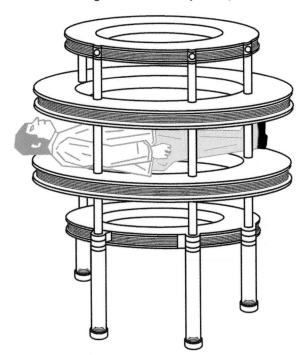

Fig. 9.1. Prototype vertical-bore resistive magnet for NMR imaging at Aberdeen University—being tried out for size

epic drive in a growing snowstorm, Tony Groves finally succeeded in finding someone sober enough to receive it into the department at Nottingham. Peter Mansfield recollects, 'I was at a party in Nottingham at the time, and I had to run around and muster a group of people to help to hump the heavy magnet off the truck.' In this magnet he soon demonstrated abdominal 'line scan images'— on himself. This magnet, painted bright yellow, now reposes in the Science Museum in South Kensington. It was originally red, but the Nottingham team found the colour too threatening. Brilliant ideas were developed into innovative techniques in this magnet, and all who benefit from the diagnostic power of magnetic scanning owe a debt of gratitude to Peter Mansfield.

During 1978 and 1979 several companies in the diagnostic industry ordered resistive magnets similar to those at Aberdeen or Nottingham. Oxford Instruments had been deeply preoccupied by the debate on its future, and it was not until September 1978 that we began to recognize the possibility of a substantial business in magnets for body scanning. Interesting images from the early research were then just beginning to appear in scientific and medical journals. Barrie wrote,

There is a major upsurge of interest among some very large international companies in the potential use of NMR instead of X-rays for whole body scanning. We are currently engaged in discussions with many companies, and it is still not clear whether the technique itself will be successful . . . The situation may, in the future, be similar to that in NMR Spectroscopy, of our being vital suppliers to a few very large companies. In order to survive in such a situation we need to maintain a substantial lead compared with those customers.

The doubts over the technique revolved around whether induced currents in the surface of the body, due to the high-frequency RF pulses needed for proton resonance in high-field magnets, might distort the response. Six months later Barrie wrote again, 'We are unlikely to know for another one or two years whether this is a scientific flop, a business which is much too large for Oxford, or a unique opportunity.'

The EMI Research Laboratories, where Godfrey Hounsfield had developed X-ray CT scanning, had started working seriously on NMR imaging in 1976. First they used a 0.1 T electromagnet, for which they developed field-gradient and RF systems, and imaging electronics. NMR differs from X-ray CT scanning in that information comes from several different signals. On top of the basic resonances of the nuclei, information can be obtained from the time taken for them

Fig. 9.2. Ian McDougall with the first horizontal bore resistive whole-body NMR imaging magnet, for Nottingham University

to absorb and then to release the RF energy. These times, known as the relaxation times, depend on the characteristics of the particular tissue, mainly its density and viscosity, and they affect the image. The system can be tuned to obtain the information or contrast required. In this, timing and sequences are all important. In the early machines, unlike the early work with CT scanners, it usually took some months for the users to learn how to get a reasonable image. Dr Ian Young, the engineer developing the system at EMI, told the Wellcome Witness Seminar of an experiment with their first prototype: 'We got a bit bored one day so we thought we would put ourselves in the machine, see what happened and get some head images. And that was fine, except that the head image looked something like a dislocated lop-eared rabbit after somebody had put a cleaver through its brain' (Wellcome Trust 1998: 42).

EMI sought support from the Department of Health and Social Security (DHSS) for the development of a serious NMR imaging machine in which the techniques could be evaluated clinically. In 1978 it received a contract to build a whole-body scanner. Soon afterwards it approached Oxford Instruments for a one-metre inner diameter superconducting magnet with a field of 0.3 T. This was a much larger magnet than the Company had built before, but the potential opportunity was becoming very tempting. So we took the plunge and accepted an order for its design and development, although we knew this would cost a good deal more than the price we had quoted.

EMI was jumping straight in to work with a whole-body superconducting magnet, but most early imaging, and *in vivo* spectroscopy trials in higher fields, were performed on animals, using relatively small bore superconducting magnets. San Francisco University ordered an NMR magnet with a 20-centimetre diameter horizontal bore to give a field of 1.5 T. This was bigger than any NMR magnets made in the past, and Peter Hanley and Ian Herbert worked on its development alongside that of the whole-body magnet for EMI. Delivered in the summer of 1979, the San Francisco magnet was the prototype for hundreds of similar magnets used in research and in the development of imaging techniques.

By this time there was worldwide interest in this new type of scanning, but in Oxford Instruments there was anxiety. One memorandum reads, 'There is a race on between the major companies for a supposed three billion dollar market of which Oxford Instruments produces an essential element. We are in danger of repeating our uncomfortable situation in NMR spectroscopy. We are likely to become essential suppliers of a vital part to several larger companies who may want to kill us or buy us.' Our deep experience of superconducting magnets for NMR had made us virtually the only immediate supplier in the world. We had no illusions that a small company in England could remain the *sole* supplier of a vital element for an important medical system, likely to be marketed, at the end of the day, by a handful of large and powerful international companies. We hoped, by remaining ahead technically, to retain a slice of this industry. We had to hold our own, to avoid being bullied, and to keep our nerve. As the decade closed the development team successfully tested the first superconducting whole-body magnet (see Fig. 9.3).

In the meantime George Radda, in the Oxford Biochemistry Department, had been making progress on *in vivo* spectroscopy. He believed this unique way of looking into the chemistry and metabolism of the living body would provide many new insights into physiology and metabolic diseases. In 1976, with the new wide-bore NMR magnet, he started experiments on animal kidneys. One possible outcome of this work was a way of determining whether a donated human kidney was in a fit state to be transplanted into a patient, or had been short of oxygen for too long to be viable. Using chemical shift spectroscopy, he could show the proportions of the different phosphorus-containing molecules present. Too much of one and too little of another spelt morbidity. He started talking to NRDC and to Oxford Instruments about the possibility of developing a special spectrometer for this particular task.

Fig. 9.3. John Woodgate with the first superconducting whole-body magnet for NMR imaging— holding an NMR spectroscopy magnet

At that time the Company was already developing its own high-resolution NMR spectrometer with financial assistance from NRDC. This work was well underway when, in 1978, the OEM market for our NMR magnets opened up again, and the Company decided not to launch its own complete spectrometers in competition with its customers. Instead, with the agreement of NRDC, the development was redirected towards an *in vivo* spectrometer, where there seemed to be no competition. Estimates of the cost of a kidney-assessment machine rose, and it became clear that few hospitals would be able to afford an instrument dedicated to this task alone. The team, now under Peter Hanley, aimed at more general instruments, mostly for research. The early machines would use magnets with a 20-centimetre diameter clear access, which would be large enough to study blood circulation in the limbs, and muscle activity, as well as small organs like kidneys, or little animals. The magnetic field had to be quite high to get good chemical shift spectra in phosphorus, and was fixed at 1.89 T.

The Company committed funds for developing the first two spectrometers for *in vivo* spectroscopy. These would be complete instruments, with a specialized end-user market—up to this time we had sold only the NMR *magnet* systems. The team now called the technique topical magnetic resonance (TMR). The word 'topical' is derived from the Greek *topos* for a place or spot. In contrast to NMR scanning, where the signals come from a wide segment of the subject, TMR focuses on a small volume in a single type of tissue, using field-gradient techniques. George Radda soon ordered the first of these systems, again with funding from the British Heart Foundation. This was for further studies on ischemia, the 'die-back' of heart muscle starved of oxygen during a heart attack. The second system would be used for in-house developments, and as a demonstration and applications facility for potential customers. A third system was soon ordered by a Californian hospital.

The first tests in the new system were promising, and it was used at an open day in 1980, to show the changing phosphorus metabolite spectra in a student's arm as the blood flow changed. When he first put his arm into the magnet he moved his hand and fingers a little, and the spectrum showed normal metabolism. Then a tourniquet was tightened on his upper arm, cutting off the blood flow, while he kept his muscles working hard by clenching and unclenching his fist. The spectrum now demonstrated some non-aerobic metabolism, with a marked change in the proportions of the phosphorus compounds. This mechanism enables muscles to go on working after their oxygen supply has been used up. The spectrum took a little time to return to normal after the tourniquet was released and the hand and arm had relaxed (see Fig. 9.4).

One or two research scientists brought their experiments to the factory to try out their ideas in this first machine. Professor John Griffiths of the St George's Hospital Medical School, who had earlier worked with the Enzyme Group, was one. He asked to use the newly developed spectrometer for preliminary work on cancers, and he produced the very first NMR spectra of tumours right there in the Company's laboratory.

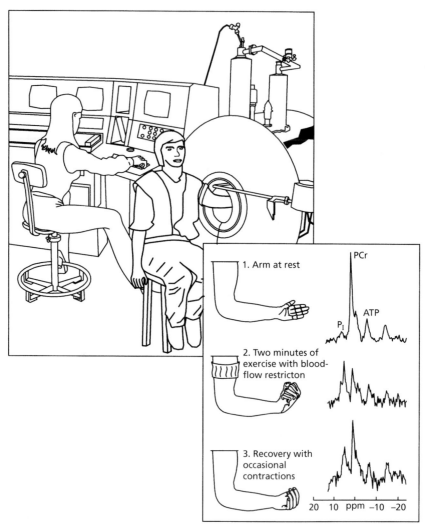

Fig. 9.4. Demonstration of *in vivo* spectroscopy, showing the changes in the phosphorus spectrum when the oxygen supply to the muscles is temporarily restricted. The spectra show the peaks for the three main forms of phosphorus in metabolism: P_I is inorganic phosphorus, PCr is phosphocreatine, ATP is adenosine triphosphate, which has three peaks.

The scientific work using *in vivo* spectroscopy might be exciting, but the business future was still very uncertain. Barrie called the project 'a highly speculative technology venture. If a flop it could be expensive; if a success it could be too big for us.' The Company started to think about finding a 'big brother' to share the development costs and the risks, and to help in the US medical market, where sales and service are so expensive. Oxford Research Systems (ORS), became a separate subsidiary in which NRDC had an investment, but no other suitable partner was found. In March 1980 the TMR spectrometer was announced in the USA, and generated a lot of interest. Some thought it more important than the still-experimental NMR imaging. But the world was moving into recession following a second 'oil price shock' and few research groups could find grants quickly for a big and expensive new system, which had no track record in research.

ORS delivered two 20-centimetre systems in 1980, and by 1981 there were three more orders for these systems costing £180,000 each. One system was to be used by Professor Osmund Reynolds, at University College Hospital in London. This was for studying the biochemistry of the brains of babies who needed intensive care, to give information on the damage caused by shortage of oxygen at birth. Their results were later to show that *permanent* brain damage may not occur until two or three hours after birth when the baby has been starved of oxygen.

Several groups were trying to raise the £300,000 needed for a whole-body spectrometer. ORS decided to go ahead and make a complete large system for George Radda in Oxford, although only part of the funding had been secured. The Board eventually agreed to subsidize the system to the tune of £100,000. When completed, this instrument was set up in a special building near the John Radcliffe Hospital in Oxford (see Fig. 9.5).

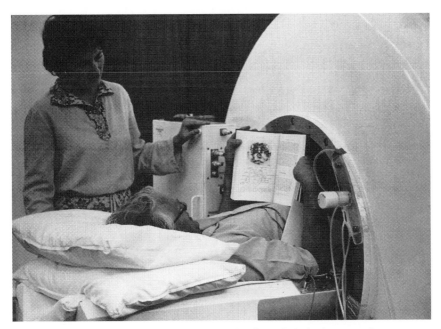

Fig. 9.5. Martin Wood acting as guinea pig in the first whole-body system for *in vivo* spectroscopy—he was found to be in good shape

A few interesting results were coming from the smaller systems already installed. George Radda's first clinical diagnosis was on a patient who had always felt tired. He had rather painful muscles, and could never hold down a job for long. The TMR spectrometer showed he was not just lazy, but had a rare disease of the muscles known as McArdle's syndrome, which was due to a missing enzyme. The abnormal muscle metabolism showed up as abnormal patterns of the phosphorus metabolites. This case was widely reported and raised the profile of TMR, but, as yet, no widespread vital diagnostic need had been identified for the technique. By the end of 1981, in contrast to a growing avalanche of orders for imaging magnets, ORS had only seven orders for smaller systems and two for whole-body ones. They were still confident that this remarkable way of watching the metabolism of the body in action would prove its worth, in time.

On the other hand, NMR scanning had caught the imagination of the medical world and of the public. TMR gives esoteric spectra from which the initiated can interpret what is going on in the biochemistry of a muscle or a kidney. NMR imaging provides recognizable pictures of the inside of the body. Its advantages over X-ray CT scanning were becoming clearer; top of the list was the absence of damaging X-rays—high magnetic fields were, and are still today, thought to be harmless. Then, with NMR, images could be produced in any plane of the body simply by selecting the right data from the computer's memory store; the angles needed could be keyed in and the images examined long after the patient had gone home. X-ray CT scanning was limited to cross-sections of the body. Differentiation between soft tissues was normally better with NMR, which was particularly valuable for the brain, where white and grey matter could be clearly distinguished—an early success was the first definite image of the lesions in the brain due to multiple sclerosis. Heavy bone structures can reduce the clarity of X-ray images of soft tissues inside them, but magnetic fields pass through bones without any distortion; so magnetic scanning is particularly useful for looking at structures within the knee joint and the pelvis, or tumours inside bones or in the brain (see Fig. 9.6).

Soon NMR imaging was hailed as the greatest advance in medical diagnostics since X-rays. By 1980 most of the research laboratories of companies selling X-ray CT scanners were planning or developing prototype NMR imaging systems. One or two were reluctant to move fast, having made large investments in the development of the X-ray scanners some six or seven years earlier. Technicare, a subsidiary of Johnson & Johnson, a large US medical supplies company, hoped to be the first to offer a complete scanner, but full acceptance by the medical profession was still some way off.

Our first two commercial whole-body superconducting magnets were delivered in 1980. One went to EMI and Ian Young installed it in the Hammersmith Hospital where he worked with Professor Graeme Bydder on the clinical trials. Although the first, this magnet and its spectrometer system eventually per-

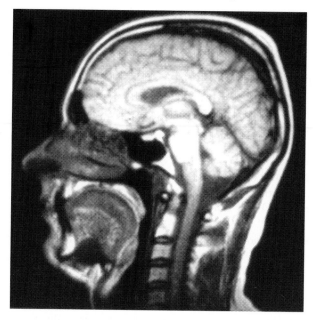

Fig. 9.6. Early saggital MRI scan of the brain

formed very well and it was used for a lot of trail-blazing research—it was still in use twenty years later. The other went to Professor Leon Kaufman at the Imaging Laboratory of the University of California at San Francisco. The work of the team there, originally financed by Pfizer, a large drugs company, was later taken over by Diasonics, a young US company, founded by Dr Al Waxman, which specialized in diagnostic imaging. Diasonics was to buy many of our magnet systems, and to offer one of the first complete high-field NMR scanners. Most of the original imaging systems used resistive magnets, which needed some 25 kW of power, and cooling water to remove the heat generated. In these systems variations in the field are bound to develop owing to small changes in the flow of the cooling water. Although initially more expensive, and involving the use of liquid helium, superconducting magnets have clear advantages. Apart from the higher magnetic field strength and better homogeneity, there is the benefit of using the magnet in the persistent mode, which provides a very stable field. These early superconducting magnets for scanners had a field rating of 0.35 T, a lot higher than the resistive magnets available at that time.

All the potential manufacturers of magnetic imaging systems came to visit Oxford Instruments, including at least five Japanese companies. In the year to March 1980, the turnover of the whole Group was only about £8 million (equivalent to about £20 million at the beginning of 2000). Not surprisingly, our potential customers were dubious about our capacity to manufacture, commission, and service these magnet systems, which they needed before they could develop their own instruments. Apart from the Japanese, there was a race on between some twelve companies, many of them working with, or financing, university research groups studying imaging techniques.

As early as October 1980 there was a symposium on NMR imaging—at Nashville, Tennessee. The participants were radiologists, physicists, and people from the companies involved. Derek Shaw represented the Company's imaging and spectroscopy interests, and gave one of the papers; but most academic researchers were muzzled by their commercial paymasters. Derek described secretive little cliques in corners studying and comparing private prints of images. A few slides of animal or limb scans were shown, probably all obtained using resistive magnets, and none was of very good quality. The electronic signal generation and resonance mapping systems, and the computer image reconstruction technology, still had a long way to go as well as the magnet development.

In the light of future events, Derek's debriefing, following the Nashville meeting, is very interesting, and points out the difficulty of making decisions about new products in a market pushed ahead by the technology. As well as reporting on the papers and formal discussions, he noted the widely held opinions of those present, gleaned from private conversations. The first conclusion was that NMR imaging would have a fight ahead if it hoped to compete with X-ray scanning; one must remember that many participants were radiologists who had only recently got used to CT scanning. As far as clinical doctors could see from the poor early images, NMR scanning would have few advantages over X-ray CT, and considerable disadvantages, including cost and the unfamiliar technology of magnetic fields.

The second general conclusion of the medical people at Nashville raised the expectations of ORS—its newly launched system for *in vivo* spectroscopy was thought much more interesting than NMR imaging systems, and more likely to be important in clinical diagnosis. The third pointer was towards cheaper and simpler imaging systems. A recession had begun and the cost of an NMR scanner, especially one using a superconducting magnet, was expected to be prohibitive. There was a young US Company called Fonar, run by Dr Damadian, who may have been the first to work on one aspect of NMR imaging. Fonar was already offering a complete NMR scanner, the first on the market, based on permanent magnets, and simple data collection and image reconstruction techniques. Many at the meeting felt this low-cost system would prove the right way ahead. *All* these predictions were to be proved wrong.

Oxford Instruments shipped a third whole-body superconducting magnet system early in 1981, and eight more were on order. These included a higher-field 1.5 T magnet needed by the US Company General Electric (GE) for experimental work on imaging and *in vivo* spectroscopy. We were still the only company to have delivered large superconducting magnet systems, but competition was now on the horizon from Intermagnetics General Corporation (IGC), a small US company, which manufactured superconducting wire and smaller high-field magnets. Beside IGC, at least two of the big companies in the race for complete systems had some in-house experience of superconductor technology, and we expected them to develop their own magnets in time.

In October 1981 there was a second conference on NMR imaging and spectroscopy at Winston-Salem in North Carolina. In the year since the previous meeting the technology had made a great leap forward. Peter Mansfield had developed a very fast 'echo planar' imaging system; he showed rapid 'snapshot' images—almost a film—of a heart beating, which caused great excitement. The British General Electric Company (GEC) was supporting this work in Nottingham as well as work at the Hammersmith Hospital. GEC had recently acquired a US medical equipment subsidiary, Picker International, which was preparing to manufacture NMR imaging systems using Oxford magnets. It was aiming to be in business by the end of 1982.

At Winston-Salem the conclusions from the previous year's meeting were obviously far adrift. It had now become clear that NMR imaging *would* rapidly become an important diagnostic tool—there were new predictions that the market would grow to 300 units by 1985 and 1,000 by 1990. But *in vivo* spectroscopy no longer held the same interest. One company was indeed aiming for a high-field system that would serve a dual purpose. These developments were bad news for a struggling ORS in Oxford. As far as cheap permanent magnet systems were concerned, the long time needed for the development of an image, and their poor quality, were counting against them; and they weighed about 100 tons, which made it necessary to reinforce laboratory floors.

By this time hospitals trying out pre-market NMR machines were scanning more patients. The public is frightened of anything with 'nuclear' in its name. The bomb springs to mind, and radiation damage and genetic abnormalities. X-rays were familiar. Elementary science teaching in schools is often poor, and most people seem ignorant of the basic fact that every atom in their bodies contains a nucleus! After a long debate on an acceptable name for the new diagnostic procedure the 'nuclear' was dropped in favour of magnetic resonance imaging or MRI. The instruments became known as magnetic or MRI Scanners.

High magnetic fields were passed as safe. In the UK, the National Radiological Protection Board (NRPB) had looked at the problem. Among other enquiries it had examined our test laboratory staff, some of whom had worked in close contact with high fields since the mid-1960s. In the scanners, apart from claustrophobia, and the noise, which seems hard to eliminate, the only effect patients noticed was occasional transitory psychedelic colour patterns when the magnetic field was changed; but this left no aftermath. The NRPB suggested a limit of 2 T until there was any new information on the subject.

The main danger seemed to be from metal objects drawn into the powerful fields in the magnets. Hospital staff knew about electricity and gas and X-rays, but high magnetic fields were new to them. You could not *see* anything in the air in the centre of a magnet, or within four or five metres, where there was a considerable 'stray field'. Before the installation of metal detectors, like the ones found in airports, the main dangers were said to come from the medical professors or the hospital maintenance staff. Both these groups assumed they had an established right to go anywhere they had work to do. Metal pens and keys would leap out of the professor's pocket into the magnet, and gas cylinders would disappear from passing trolleys. There were few serious accidents; people soon learned about magnetic fields. But damage there was, at first, to the magnetic recording strips on credit cards, which could be wiped clean. Before warnings went up, quite a few important people found themselves embarrassed. One wag called it 'magnetic endocarditis'.

There is an amusing story, which could have turned out to be far from funny. In George Radda's first TMR spectroscopy system in Oxford the superconducting magnet was, of course, kept running in persistent mode, day and night, month after month. One day the cleaner came to him saying the magnet had grabbed her floor polisher, and would he please come and retrieve it for her. In this high-field magnet it was held by very strong forces. The field had to be run down before this large foreign body could be removed. The system then had to be restored to field and homogeneity. Neither the instrument nor the floor polisher suffered any permanent ill effects (see Fig. 9.7).

In the early years Oxford Instruments received orders for trial magnets from almost every serious entrant to the MRI race, large or small. Soon there was some consolidation in the field; small companies and development groups were taken over by the larger and more serious competitors. We continued to supply many resistive magnets as demand for the high-field superconducting systems grew. Professor Mallard, in cooperation with Aberdeen University, set up a company to make good low-cost scanning systems using improved versions of the first resistive magnet we had supplied to him in 1977. But he failed to get adequate financial backing for his company, and, after a few years, the technology was sold to a Japanese company, which made some 150 scanners from the same basic design. This is one of the many sad cases of British developments going overseas.

Customers often fail to realize how easily magnetic fields can be disturbed by electrical currents in nearby conductors. We supplied a vertical axis magnet to the Aberdeen Company to be installed in an old building belonging to a private clinic in Geneva. Soon we got complaints that it was functioning irregularly. One of our engineers found it in perfect order, but, as soon as he left, the problem started again. A careful study with sensitive magnetometers found sudden surges in the background magnetic field. These were eventually traced to the Geneva tram system; this was run on DC electric current from a huge ring main, which acted as a vast circular coil. When two or three trams happened to start moving at the same time the magnetic field produced by the pulses of electricity in the ring main were enough to upset the delicate homogeneity of the magnetic field in the instrument. More magnetic shielding was necessary.

Fig. 9.7. George Radda's first *in vivo* spectroscopy system, which had captured a floor polisher in its strong magnetic field

By 1982, with confidence in the new technique growing fast, Oxford Instruments was negotiating larger and longer-term contracts with several firms. Our strategy was to work with as many companies as possible and retain at least a proportion of their requirements when the inevitable day came when they found a second source or developed their own successful magnets. As with NMR spectroscopy in the past, customers would have preferred to have the sole rights to our valuable technology, and our limited production capacity, to help them ahead in the race, and some of them wanted to buy the Company.

So, how had we been keeping up with this rapid expansion in demand for magnets? In 1980 the brand new factory on Osney Mead was already full to the doors, and the development team continued to work on imaging magnets in an old rented factory four miles away at Wootton. This building provided space and the necessary height but little in the way of offices. It was not a very good advertisement when we were trying to convince large companies that we would be able to meet their future demands for a key component of an important medical system. Fortunately experience in the technology of superconducting magnets was in short supply, and Oxford Instruments was undoubtedly the world leader. Companies wanting these specialized magnets in a hurry almost *had* to come to us. This situation would not last, and we were determined to satisfy the market as well as we possibly could to establish our position. For our longer-term future we also needed to keep at least two years ahead of any one else in the technology.

There was a spare plot of land of nearly an acre next door to the Osney Mead factory—an overgrown wilderness frequented by foxes. We had tried to buy it before, but now our need was urgent, so we made an offer well above the market value, and reached rapid agreement; soon we had planning permission for a 20,000-square-foot factory. In 1980 NMR imaging was still unproven as a valuable clinical technique, so we paused before putting actual building work in hand. But the site and plans for the new factory were enough to convince our customers that we were in the business of making imaging magnets for the long term; by the time they wanted them in any quantity we would have the capacity to satisfy their needs for a while. Very soon the plans had to be activated; the question mark over the technique had gone. NMR imaging was proving more and more valuable as the radiographers and doctors lucky enough to have early trial machines learned how to use them to best effect.

The growth in our market was impressive. In the year to March 1981 the Company received about £1 million of orders for imaging-magnet systems and delivered seven resistive magnets and our first two 0.35 T superconducting systems. In the year to March 1982 we received orders amounting to nearly £5 million, and delivered eleven resistive magnets, twenty 0.35 T superconducting ones, and two high-field 1.5 T magnets. Back in 1980, early predictions of sales had been £5 million a year by 1985, which Barrie Marson then considered would be all we could produce without major changes. If demand was to prove higher than this, the Company might have to consider licensing its technology. In the event, there *were* to be major changes before reaching 1985—the capacity would be increased by an injection of new funds, there would be more new buildings, and longer-term sales agreements would be in place. In the year to March 1985 Oxford Instruments' operations, and part-owned associates, would together succeed in shipping over £50 million worth of imaging magnets.

As the orders and predictions for the future market grew, it was clear that production would need to be organized on a scale not seen before in the Group. In 1981 Nick Randall, who had a good track record in production management in a larger company, came to manage the newly designated NMR Imaging Division. At first he was working in the Wootton factory, but, from the beginning of 1982, in the new purpose-built Osney Mead premises (see Plate 4). Delivery times were still longer than we would have liked, and our customers pressed hard. Magnet production was not free from the old perennial problems of inconsistent superconductor, and quite a few other headaches emerged. In this new market with frequently changing demands, production sometimes ran ahead of the proper development of some new feature. The Company had to balance the need to make time for 'debugging', in order to maintain quality, with the need to respond to the strong pressures for quick delivery. It was also vital to keep abreast of new ideas on scanning techniques, and to go on developing the magnets needed for them.

There was some internal questioning about the volume of magnets the Company would be able to manufacture. Licensing the technology would be one method of keeping a share of the market, and in 1982 Oxford Instruments signed an important contract with the German company, Siemens AG. It would make some magnets for its own systems to our designs and, in parallel, would buy a proportion of its requirements from the Company. In another contract, the technology of the Aberdeen type of resistive magnet was sold to the Japanese company Asahi, which was then taking over the Scottish imaging technology.

Nick Randall installed new computerized systems for production and resource planning. Documentation improved, but, as the diagnostic systems were developing rapidly, there still had to be frequent technical changes in the magnets. It took the Imaging Division several years to achieve anything like smooth large-scale production. In 1982 the Division was expanding fast, but still falling behind the swelling order book. At that time Oxford Instruments probably had some 90 per cent of the available market, and the market itself was growing fast. Unless we wanted to hand over most of it to the developing competition, something would have to be done about capacity.

The Company started to consider manufacture in the USA, the source of the majority of our orders. US companies were always anxious about our ability to service our systems there. BOC had retained its shareholding in the Group, and David Craig, one of its very top managers, had recently become its man on our Board. He is an American and, knowing the situation in the USA as well as the Company's dilemma, he could see a possible solution. Airco, a substantial American gas company, was BOC's largest subsidiary. It owned Airco Superconductors, which had an underused 100,000-square-foot factory in Carteret, New Jersey, which was making superconducting materials. These were supplied mainly to US national research laboratories for high-energy physics machines, but this market was in decline at the time, and the operation was losing money. BOC was already looking for positive ways of solving this problem, and David Craig suggested a joint venture for making MRI magnets there, as well as continuing with the existing superconductor business.

The Oxford Board was doubtful about a joint venture with a huge and inevitably slow-moving partner; any new operation would have to be flexible and quick to react. Fortunately David Craig understood this, and within a few months agreement was reached. The Oxford Airco Partnership (OAP) was announced in July 1982. At first the new operation was only to assemble, test, deliver, and service the systems; later it would go on to full manufacture. Soon the Group's other magnet divisions started to use superconductor from Carteret, and this facility has proved of huge value over the years. Dr Eric Gregory had been in charge of the wire operation, and, for the time being, remained Chief Executive Officer (CEO) of the Partnership. In the US market our credibility rating went up, and the new magnet operation got off to a good start.

In September 1982 the Oxford imaging operation became a separate Group subsidiary, Oxford Magnet Technology (OMT). With the large order backlog, staff numbers were escalating. The potential was by now clear, but the funding of further growth might well need an equity injection into this fast-growing business.

10 What of the Rest of the Group?

How the other operations had been faring during the dash for magnetic imaging

The universe is change; life is what thinking makes it.

(Marcus Aurelius (AD 121–180))

Innovation . . . consists not in invention and scientific break-throughs, but in a day to day commercial endeavour to introduce improvements of one kind or another with a view to profit.

(Alec Cairncross, *The British Economy since 1945* (1992))

The rapid expansion of the imaging magnet operation would soon be in danger of unbalancing the Group, but in 1980 the future of MRI as a diagnostic technique was still in doubt. Customers were buying these magnets for their own early work on imaging systems in order to be in the running should a large market emerge for this new type of scanner. The whole business was still very speculative, and, in the new two-year plan our executives were then drafting, reliable growth was still scheduled to come from the older operations.

The objectives in the 1977 plan had almost all been exceeded. Because of the queries over MRI, imaging magnets did not feature strongly in the new 1980 plan, but the executives were still confident of growth from *in vivo* spectroscopy, then attracting wide interest, and forecast to have a substantial market. The team making spectroscopy systems was then still part of the original magnet company, but it was soon to be separated as Oxford Research Systems (ORS). It was looking for partnerships to help in serving the big expected market in the USA. Real black clouds were not to gather over ORS until late in 1982 when the US company GE succeeded in performing both imaging and spectroscopy in the same high-field magnet. This had previously been thought impossible. GE started talking of a future dual-purpose product, and potential customers for ORS spectrometers delayed their purchasing plans.

The original magnet company was still known as Oxford Instruments, although this now caused some confusion with the parent company, which had been renamed The Oxford Instruments Group. (It had previously been called Romeyns Investments for an obscure historical reason.) Following the rapid strides at the end of the 1970s, there were flourishing sales in magnet systems for NMR spectroscopy, for the analysis of large organic molecules. In 1980 this Company celebrated its new Queen's Award for Technology earned through these magnets. Queen's Award presentations were always the occasion for a big party, and her representative in Oxfordshire, Sir Ashley Ponsonby, the Lord Lieutenant, came to represent the Queen. To everyone's delight, he made the presentation to John Woodgate, the MD, wearing his full uniform including sword and spurs.

The old demonstration of the storage of energy in a superconducting magnet was on display at this celebration, and, among other sideshows, there was a newer party trick. This presented the magnetic levitation of lead—a Type 1 superconductor, identified back in 1911 by Professor Kammerlingh Onnes. The demonstration took place in a 'Cryorama', a special cryostat developed by a French professor. The flat insulated vessel was covered by a transparent plastic dome under which one could actually *see* the liquid helium. Normally kept in a metal cryostat, this liquid boils at minus 269 °C, and, if poured from its container, evaporates amid a cloud of frozen air before it reaches the ground. In order to demonstrate this magnetic levitation, a small permanent magnet was placed in the liquid helium in the Cryorama, and a lead disk was dropped onto it. At first nothing happened, but suddenly the disk shot up away from the magnet and remained floating above it. When cold enough the lead had become superconducting, and currents induced in it from the permanent magnet then continued to circulate, creating another magnetic field in opposition to that of the magnet. Enough magnetic force was generated to raise the lead and keep it floating in the air for a short time until it warmed up, lost its superconductivity and dropped back again onto the magnet. This demonstration was always popular and soon after was shown to the Duke of Kent on a visit to the Company (see Fig. 10.1), and later to the Queen Mother at a Royal Society Soirée.

While Oxford Instruments had been concentrating on the blossoming NMR magnet work over the previous two years, there had been a deliberate cutback in the traditional project-based magnet and cryogenic activity at Osney Mead. There had been a little opposition to the curtailment of this work, but by 1980 the other groups were recovering some of their lost position in leading-edge technology for the very highest fields and the very lowest temperatures. By the early 1980s these groups were tackling some extremely difficult projects, now at more realistic prices than in the past. An example of the kind of equipment they were then making for specific experiments was a 'neutron bottle'. Tim Cook described this system in the Group's internal newsletter of November 1981. Tim has a flair for simple elucidation of complex subjects, and I can do no better than to reproduce his colloquial description (see Box 10.1)

Fig. 10.1. The Duke of Kent watching Martin Wood demonstrate magnetic levitation in a Cryorama

As the 1980s began John Lee was managing all three divisions of Newport Instruments in the new factory at Milton Keynes. The Components Division had contracted into the factory from the outstations, and was having a difficult time. Following successful pilot shipments of components, this group had expectations of large orders from the Post Office for its new 'System X' telephone equipment. It became frustrated as the development of the system dropped further and further behind schedule. The computer industry, on which it depended for too many of its sales, was itself suffering from the recession of the early 1980s. The output of the Division halved over six months, and it drifted into losses, with the miseries of a short working week and redundancies. The Components Division had never fitted well in a Group concentrating on products and systems, and in 1981 its future was much debated. That year it was turned into a 'stand-alone' company for a possible sale, and a buyer was sought. One agreed deal collapsed on the very day Barrie Marson went to London to sign the contract—the would-be purchasing company had itself been taken over earlier the same day.

In April 1982 Newport Components was finally sold to its management under John Lee. With him went John Baxter, Peter MacWaters, Geoffrey Bolton, and the production team. This was the beginning of a success story; the operation

Box 10.1. . . . and a bottle of neutrons please

<u>....AND A BOTTLE OF NEUTRONS PLEASE.</u>

Neutrons are small (33 million to the inch) and light 30,000,000,000,000,000,000,000,000,000,000 to the ounce). They are present in every substance (except hydrogen) and are the 'flames' in a nuclear reactor. Scientists are interested in neutrons because they are the 'glue' that holds the nucleii of atoms together and so an understanding of neutrons is essential if we are ever to understand the fundamental properties of matter.

The trouble with neutrons is that they move very quickly (5500mph for an average one) so you don't get much time to look at one, and it's hard to collect a bunch of them together. Another problem is that because of their high speed they go through things, so it's hard to contain them in a box. Neutrons can be slowed down by passing them through a tank of cold liquid and to date by using deuterium (a special sort of liquid hydrogen), they have been slowed down to about 300mph. However this is still rather fast and what is needed is neutrons going at about 1m/sec (2mph). Now to get neutrons this slow means cooling them using liquid helium to a temperature of 0.5K (272.7°C) i.e. half a degree above absolute zero. A bonus in cooling neutrons this far is that, although they can go through the walls of a steel tube on the way in, once they get cold they can't get out again because they haven't enough energy to go through the steel (although they can still go through thin aluminium or titanium). So you can fill a tube up with ultra cold neutrons, then open a door in the end and let the whole bunch out to do experiments on.

In 1978 scientists from I.L.L. in Grenoble approached Oxford Instruments and asked if we thought such a 'neutron bottle' could actually be made. (The bottle itself is 10ft. long and 4" diameter, all at 0,5K) After a design study by Nick Kerley and others in the fridge group at Osney we thought we could and the project was ordered. The only problem remaining was to build it and get it going. The position at October 1981 is that the refrigerator has been built and tested, without the bottle connected and reached 0.37K and now the bottle itself is being fitted ready for a system test. So if you are visiting Osney Mead and wonder what the odd shaped thing in the fridge area is (sorry Murph) its only a Neutron Bottle.

TIM COOK

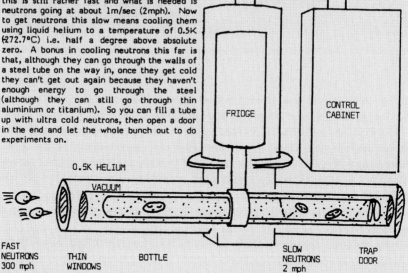

STOP PRESS
The bottle reached 0.47K and we put liquid helium into it on Friday 7th November.

fared much better on its own. The original Newport Components became the core company of the Newport Technology Group. As the twenty-first century began there were still many faces there familiar to old-timers at Oxford Instruments. Until his tragic death in 1998, Antony Costley-White had, for some years, been the Chairman, and John Laurie has been the the Group MD for more than a decade. The components themselves have changed a great deal, thanks to the innovations of John Baxter, the Development Director. In a fund-raising exercise in the early 1990s the wheel came round full circle when Ian Boswell, founder of Newport Instruments, once again became a shareholder. Sadly, he died early in 1997.

There has been speculation as to whether Oxford Instruments should have kept some shareholding in this management buyout, which was made under pretty soft terms, but the executives at Newport are adamant that it would not have worked to the benefit of the newly launched company. It is probable that an over-large majority of the equity was allowed to go into the hands of its rather conservative first Chairman and MD, which may have stunted the Company's growth in the late 1980s, until he retired and sold his shares.

I must go back to 1980. Following its work on DIAC, Newport's Systems Division launched its own product for monitoring and controlling equipment and plant in hazardous or remote positions on a site. The development of this 'System 86' had been delayed by the chronic shortage of electronic and software engineers, but in 1980 and 1981 it was still at the forefront of the technology. These systems were very flexible, with a variable number of microprocessor-based remote stations linked to a central computer by one data highway. Most installations were in oil-refining or chemical complexes, but the system was also installed in a new commuter jet, where it was used for high-speed data scanning and processing during flight trials; and it was also used in a special ship for controlling cable laying under the sea. Bass Breweries commissioned a system for monitoring and controlling thirteen large fermentation vessels from a console 200 metres away. The accurate control of the temperature in the vats reduced fermentation time by two days—a considerable saving.

John Lee's departure to manage the newly independent Newport Components left a gap at the top in Milton Keynes. The Systems Division was, by then, the only remaining part of the old Newport Instruments, and in 1982 the name was changed to Oxford Automation (OA). While the Company searched for a new MD, Barrie Marson stepped in to take responsibility for this operation—along with his many other commitments. The Division seemed in danger of drifting; OA needed a new leader, conversant with digital electronics, who would propel the successful System 86 team to fresh heights and revitalize the whole business. No candidate who was 100 per cent suitable came forward through word of mouth or advertising. The Technical Manager, John Thomas, stood in as General Manager while the hunt went on, and Barrie tried to make time for strategic planning. But the clock went on ticking, and other companies were catching up. In retrospect, the Company would have done better to take on a candidate with a good management record who was 80 per cent suitable. It was a case of the best being the enemy of the good.

In 1980 the Newport Analyser had been the last remaining product in the Instruments Division at Milton Keynes. In 1981 it was taken into a new company, Oxford Analytical Instruments (OAI), formed to provide analytical tools for industry, largely for the expanding quality-control market. In several areas instruments based on physical principles could replace 'wet chemistry', giving a quicker, less messy, and often more accurate analysis. The Newport Analyser had received a big boost with its acceptance in the USA as a standard tool for determining the hydrogen content of aviation fuel. This gives information on just how much fuel is needed to reach a plane's destination in safety, and the Analyser could replace the slow and messy old calorific method.

The new OAI immediately acquired more products from Telsec Instruments, the source of the old flat-bed recorder that had helped maintain turnover during a recession nearly a decade earlier. These instruments were 'radioisotope-excited X-ray fluorescence analysers'—known as Lab-X. They are portable instruments, used mainly for metallurgical analysis and ore assay. Common applications are in measuring the quantity of sulphur in oil or coal, and the lead content of petrol. The Lab-X instruments needed updating, with microprocessor technology and with new software to extend their range and make them more user-friendly, but they were to become a very successful addition to the Group's range of products. In the new Company, the teams responsible for Lab-X and the Newport Analyser settled down in the old rented Telsec premises on the eastern edge of Oxford. The team coming from Newport included Ray Millward, Reg Wright, Ray Bailey, and Doug Brown, while some twenty ex-Telsec staff remained in the building to continue producing the Lab-X products. OAI soon established an applications laboratory and engaged a chemist to extend the range of processes in which these instruments could be useful. Potential customers could come to learn about these rapid modern methods of analysis.

At OMS in Abingdon, the new and unique brain-monitoring equipment was selling well. One interesting application was its use by a British army climbing team in the Himalayas to study the effects of reduced oxygen on brain activity at high altitudes. In the much larger heart-monitoring market, OMS was suffering from fierce competition, mostly from fairly small companies. In the USA alone there were now twenty competitors, and profit margins were under pressure, but Jack Frost's team was increasing Oxford Medilog's share of a sluggish market hit by recession. In 1981 Antony Costley-White, who had been Marketing Director of OMS, became the MD, and joined the Group Board.

OMS had developed several sideline products, and in 1981 the Board decided that two of them should be spun off on their own into a new subsidiary company where they could receive the undivided attention of a small team. This company was at first called Oxford Medical Computers, but this was later changed to Oxford Dynamics (OD). Julian Morris moved from OMS to manage the new Company with Paul Bradstock as Marketing Manager. The Unibed intensive-care monitor, which had been demonstrated on Mrs Thatcher, was one of these products.

Acquired partly developed, under a licence from a Rotterdam hospital, Unibed looked like a winner. It contained clever software and could monitor many clinical indicators such as ECG, blood pressure, and respiration, and it could process and store all this information for retrieval by the medical team at the touch of a panel. Doctors could question it at the bedside to find what had been happening to the patient, and could select areas for more detailed study. It was even capable of controlling drug and fluid infusions and of guiding an attendant through complex procedures. Unibed could virtually turn a single bed in any ward into an intensive-care unit. It was an attractive system, but, apart from more units for the Rotterdam Hospital, orders came only for single trial units or for batches for a few appreciative hospitals. Unibed was, in fact, a system ahead of its time. Although it saved doctors' and nurses' time, and possibly patients' lives, the many clever features made it expensive, and it could not compete with more basic systems. Over the next two years the Company let the operation run down gently, and later sold its product rights.

The other product in OD was Vicon, developed by Julian Morris in collaboration with research groups at the universities of Dundee, Glasgow, and Oxford. This was a three-dimensional movement analyser. It used up to seven video cameras to receive signals from small reflective markers stuck to defined points on the body and on limbs. These signals were processed and analysed by a computer to give objective information on orthopaedic problems. The system was useful for following up progress after an orthopaedic operation, as well as for initial diagnosis. There was a lot of interest after its launch in 1981, and an early system went to the well-known Boston Children's Hospital in the USA for studying patients with cerebral palsy. Beside its clinical value, the system could be used to analyse movements in sport, such as an athlete's technique, or a golfer's swing. Jo Durie's tennis serve was analysed by a mobile system, and the army used another, not for sport but for analysing the movements needing to be made in a tank cockpit. Doctors are notoriously conservative, and take a long time to accept a new way of assessing patients, especially when it means spending money in a recession. It took many months for sales to pick up, but Vicon would eventually prove to be a success.

Marketing and selling are key functions in any industry, but have not always been among the main strengths of British companies. The subsidiaries in the Oxford Instruments Group were no exception to this. With unique products in the early days, keeping up with the orders had been the main problem. But with growing maturity and more competition, a lot of attention was, by this time, being given to the sales side. 'Marketing' includes researching the potential markets, advising on which are the right products to develop, what the timescale should be, where to launch them, and finally presenting them to possible customers through exhibitions, explanatory publications and advertising. Selling is the art of persuading customers to buy; in particular, to buy one's own products rather than a competitor's. In this area the companies in the Group made steady progress with occasional leaps forward. Most had professional marketing managers and staff by 1980, and overseas there was a wide variety of outlets, including our own companies or offices, agents, and distributors. Our products were in several different markets; some, like Lab-X, had originated outside the Group and the old agents had been retained.

In 1981 Paul Winson started to oversee marketing for most of the Group, slowly bringing the various operations into line to present a common Group image. Our earliest logo had been a stylized red ox head—we used to say it represented the horns of a dilemma on which we got impaled from time to time. In the early 1970s, a logo of the superimposed letters O and I was introduced, which, to me, looked like a hexagonal nut. This remained with us until the mid-1980s when the Group appropriated the word 'oxford', with the second O looking a little like a large magnet (see Fig. 10.2). Oxford University is known and respected throughout the world, and this name has always been important for the Group. Soon all our companies were to incorporate 'Oxford' into their names, and 'Oxford blue' became the standard colour. Besides bringing more uniformity to the Group, this gave a sense of its strength and breadth, which was especially helpful for the credibility of the smaller companies in their sales drives.

Fig. 10.2. Succession of Oxford Instruments' logos

At the beginning of the 1980s there were few internationally known companies linked by their names with this city and county. In the 1980s and 1990s a new climate developed for scientific entrepreneurship. Some new firms spun off from other companies or were formed by independent people, but quite a few came out of Oxford University. Beside the companies in the Oxford Instruments Group, several other science-based or technology companies, mostly of recent origin, began to use 'Oxford' in their names. Among the electricians, bakers, and computer traders starting with 'Oxford', the 1999 edition of the local phone book listed: Oxford Applied Research and Development Ltd.; Oxford

Asymmetry plc; Oxford Bio-Medica plc; Oxford Bio Sciences Ltd.; Oxfordshire BiotechNet Ltd.; Oxford Cryosystems Ltd.; Oxford Diversity Ltd.; Oxford Glycosciences plc; Oxford Intelligent Machines Ltd.; Oxford Medical Knowledge Company Ltd.; Oxford Lasers Ltd.; Oxford Metrics Ltd.; Oxford Molecular Group plc; Oxford Optronix Ltd.; Oxford Orthopaedics Ltd.; Oxford Scientific Films Ltd.; Oxford Semiconductor Ltd.; and Oxford Technology Venture Capital Trust plc. Things have changed a lot since 1960.

11 Strategies for the Future

The new political era and the positioning of Oxford Instruments for its next phase

For now sits expectation in the air

(Shakespeare, *Henry V (1600)*)

The Conservatives, under the leadership of Margaret Thatcher, had come to power in May 1979. This had heralded what was probably the most far-reaching change in political and economic philosophy since the war, which in many ways appealed to industry. Companies were required to look after themselves and not expect subsidies from the state; taxes on industry were to be reduced, and with them tax concessions, administrative complications, rules, and regulations; enterprise was to be stimulated and encouraged. Most personal taxes were cut first: the top rate of income tax soon went down from 83 per cent to 60 per cent, and after a few years to 40 per cent. But capital gains tax was then raised from 30 per cent to a top rate of 40 per cent. This was done to bring income and capital taxes into line, but it had one negative effect; there was no longer much reason for shareholders to want companies to invest for growth rather than distributing high dividends. Soon there was pressure on company boards to maximize profits in the short term.

The new government also wanted 'sound money'; inflation had to be defeated to provide a stable business environment; 'monetarism' was the way to do this—if the money supply was adequately controlled, so they said, inflation would automatically fall. Interest rates remained high, and this policy was pursued ruthlessly right into the deep recession at the beginning of the 1980s. Inflation was a hard dragon to slay, and industry suffered badly from this unflinching attempt.

The recession, which started in 1980, was precipitated by the second oil price shock late in 1979. OPEC, the cartel of oil-producing countries, raised prices sharply, provoking widespread repercussions among developed countries. By this time the UK had its own North Sea oil supply, and the pound was already strong, so the country remained free from the type of currency crisis suffered after the previous oil price shock in 1973. In fact, the crisis for exporters was the over-strong pound, which reached nearly $2.50 in 1980. It was not only exporters who suffered; imports from the newly emergent producer countries with low wage economies competed with UK products in our home market. Interest rates, currency strength, the money supply, and inflation seemed to be stuck in a vicious circle. In spite of lower costs for imported raw materials and finished goods, inflation roared ahead, preventing a reduction in the high interest rates. These high rates were pushing up the pound and sucking in overseas investment. These extra funds were boosting the money supply, which was supposed to be the underlying cause of the inflation, and so on, and so on. It was rather like the old circular song 'There's a hole in my bucket, dear Liza dear Liza', where the solution to the last of a chain of problems is frustrated by the original problem.

As the recession bit deeper and deeper, large reputable companies dipped into losses and there were many bankruptcies. Manufacturing output fell by 16 per cent, and the 'dole queue' went up to two million, inexorably pushing up the public spending needed for unemployment benefit, and cutting the tax take from salaries and pay packets.

In his Chairman's Review for Oxford Instruments' March 1980 Accounts, Martin talked of 'a year of increasing difficulties . . . High inflation, high interest rates, erratic exchange rates, unreliable supplies and massive bureaucracy all make our jobs more difficult—as if the technical difficulties of development and manufacture were not enough.' The Group results that year show the turnover up by 36 per cent to £7.8 million, and profit up by a lesser 24 per cent to just over £1 million, a measure of the squeeze on our margins.

After a while the Government began to see the need to control other factors as well as the money supply. Interest rates slowly came down, and with them the exchange rate of the pound. By October 1982 it was down at $1.70 and in March 1983 it bottomed at $1.49 before rising gently again. The economy steadied, and was soon to enter a phase of sustained expansion. The mid-years of the 1980s were to be a particularly good period for entrepreneurial high-technology companies, and good for UK industry in general.

By 1982 the worst of the recession was over and the Government was doing its best to encourage enterprise. For Oxford Instruments it was a year of discussion and planning for the future. The 1980 three-year plan had again been overtaken by events. It was now clear that magnetic imaging *was* going to be a success, but there were still many question marks over the Company's eventual role in this exciting new industry. During 1982 the order book for MRI magnets grew to almost the same level as the turnover of the whole Group in the previous year. The staff of Oxford Magnet Technology trebled in the year. This lopsided growth caused strains and stresses throughout the Group, as Barrie and other senior executives perforce concentrated on this voracious cuckoo in the nest that was devouring all our resources. The management became stretched and the smaller activities somewhat neglected. It was still hard to find good managers with a technical background—the Group was then looking for at least three for high positions. Important discussions on the future of two or three sub-sidiaries kept getting postponed. Borrowing soared and, with interest rates still rather high, overheads were cut back again.

The Board also was changing. David Craig was then the BOC representative, and mediated in any OAP problems. Sir Rex Richards had been closely involved with the Company's NMR activities since the 1960s, and had recently finished his five-year term as Vice Chancellor of Oxford University. He had also been involved in industry, most recently as a non-executive director of IBM's UK subsidiary. In 1982 he joined the Group Board as a non-executive director. Ken Fuller was approaching retirement and, with the Company's increasing strength, Ian Boswell saw no further need for representation on the Board. Roger Cotterell, who, as manager of ICFC's Reading office, had kept a watching brief on the Company, no longer attended Board meetings.

The Company was now involved in more overseas matters, with its own sales companies, the OAP and large new contracts with scanner companies. These included legal matters beyond my knowledge, and for several years I had been appealing for a properly qualified Company Secretary to take my place. In the autumn of 1982 the Group recruited Tim Eustace for the job, and I was able to give up my executive tasks, which included writing the minutes of Board meetings. But the years between 1980 and 1982 had kept me busy, as there had been several changes in shareholdings and other corporate matters that came under the responsibilities of the Company Secretary.

Some who are fascinated by the growth and development of a company, its technology and its management, are thoroughly bored by discussions of its share structure. But ownership is of vital importance to the behaviour and future of a company. The shareholders, however quiescent they may appear, can ultimately control its management, objectives, investment policy, and forward strategy. It is rare for the executive directors of a company to be ousted by the shareholders, but their views may still have wide-reaching repercussions. They may influence whether a board seeks to make a quick profit and pay high dividends, or to build up its technology and assets for the longer term. In a private company, the shareholdings reflect its history; the philosophy and aspirations of its founders; the wisdom of its early legal advisers; its skills in negotiating agreements with financial backers; and the results of acquisitions in exchange for shares.

In the formative years of a company, when everyone seems to be preoccupied with product problems, the cash flow, or next year's profits, there can be a danger of giving away too much, too easily, in vital corporate negotiations. As the company's success becomes more apparent, a 'slice of the action' may be demanded as a precondition for obtaining a large contract. But it can be a mistake for customers or suppliers to be shareholders; either may later be in a position to force unpalatable actions. Joint-venture deals are preferable. Oxford Instruments had not done badly over the years. In the 1960s and 1970s we had parted with blocks of shares to ICFC and BOC, and in the Newport Instruments acquisition; but we had resisted demands, from spectrometer manufacturers and others, to buy the Company or to obtain a shareholding. Our concern, ever since the 'family' had ceded majority control in the early 1970s, had been to keep the larger outside shareholders in balance and to spread the shares more widely among the employees.

When, in 1980, the Board finally agreed to a Share Participation Scheme for the employees, it was necessary to reorganize the capital of the Company. A few old £1 shares had changed hands at £10 to £15 over the previous few years, but with the new scheme, a junior employee would, perhaps, be entitled to 2.7 shares from the Group's profits. To make the shares more manageable, the old £1 shares were subdivided into 10p shares. There was an accounting reserve that had grown from profits that had never been distributed as dividends, and that formed part of the 'shareholders' funds' in the balance sheet. This was not, of course, a pot of money, but, like funds from investors, had long since been spent to promote the growth of the Company. This reserve was now turned into new 10p shares, and existing shareholders were given nine new shares to add to each 10p share they already owned. The underlying value was the same, but the numbers had grown to a hundred times the old share numbers. Employees who had bought 100 shares, back in 1967, for £167.10s., now found they had 10,000 shares. Instead of 2.7 shares, the junior employee would have 270 shares earmarked for him or her in the new Share Participation Scheme Trust, a much better-sounding nest egg for the future.

Until 'capitalized' in this way, the reserve from past profits can, in theory, be distributed as dividends, so the move also strengthened the balance sheet in the view of bankers, and reflected a maturer company. The Pension Fund had, by now, been given the resources to buy 5 per cent of the shares in the Group, and about 2.5 per cent had gone to some twenty senior managers in an incentive scheme.

In 1980 David Craig was not yet a director, and the BOC representative on the Board was an accountant. BOC had been showing little interest in the Company, and when it received an offer for its shareholding at what everyone felt was a very high price, it was interested. Another offer was made to Ian Boswell's private company, VF Investments. The offers came from our old friends at Bruker Physique. They had been interested in our magnet operation for many years, but we had resisted selling shares to Bruker in the past. Although buying some 20 or 25 per cent of the shares in the Group would not amount to control, it might *appear* that way to other customers in the marketplace, and would certainly give Bruker a good deal of influence.

One of the advantages in a *private* company is that its board may be permitted by its rules to invalidate any sale of shares by refusing to register a new shareholder, and no reason need be given for the rejection. The Company also had a complicated pre-emption procedure to be followed on the sale of shares from any of the larger holdings—the result of the 1973 negotiations with BOC. Our Articles of Association—our rulebook—now seemed a little ambiguous. BOC thought that, if it followed the pre-emption procedure first, it would then be free to sell any leftover shares to anyone it pleased. Our lawyers said the Board would still be able to refuse to accept any new member it did not want. BOC consulted four lawyers on the question and, with amused frankness, told us that two pronounced in its favour and two agreed with our lawyers. It was at this impasse that David Craig, a very senior BOC manager, stepped in, looked into the Company for the first time, liked our technology, and became its representative on our Board.

When both BOC and VF Investments were considering selling shares, up to 25 per cent of the Company might have been on offer. The Board decided the time had come to fix up a relationship with experienced advisers. The Company would need an organization well placed to find acceptable patient institutional investors, because, at the 'high' prices being discussed, we expected most of these shares to be left over after the pre-emption exercise. As we anticipated 'going public' one day, we also needed a merchant bank with experience as an 'issuing house' for launching companies on the stock market. A subcommittee of four directors interviewed several institutions on their own territories. We found one too aggressive for our style and another too inexperienced in some aspects of our potential needs. In the end the choice fell between Barings Brothers & Co. Ltd. (Barings) and Robert Fleming & Co. Ltd. (Flemings). We chose the latter, largely for reasons of personal rapport between John Lawrence, our Finance Director, and Strone McPherson of Flemings. Strone had already advised us on one or two matters including the Bruker offers, and Flemings agreed to place with City institutions all the shares left over from any impending share sales.

By the summer of 1981 BOC wanted to buy rather than sell shares, but VF Investments decided to dispose of half its shares at the Bruker offer price. The Company was developing fast, and the price of 50p for each 10p share no longer seemed so outrageously high. BOC and ICFC both wanted to take up the leftover shares, but we wanted a wider shareholding, rather than more added to the existing large blocks, over and above their entitlement. We persuaded all individual shareholders who could manage it to take up the offers—quite a few employees had significant holdings by then. In the end only 11 per cent of the shares offered remained unsold, and everyone agreed to most of these going to recently arrived managers, while Flemings was allocated a small block for one of its investment trusts.

With so many employee shareholders, both individually and through the Share Participation Scheme, Martin wanted to give them some guidance in understanding the Annual Accounts. In the newsletter in November 1981 appeared his graphic attempt at explaining where the money came from and where it went (see Box 11.1). I'm not sure it's as lucid as he hoped.

Box 11.1.

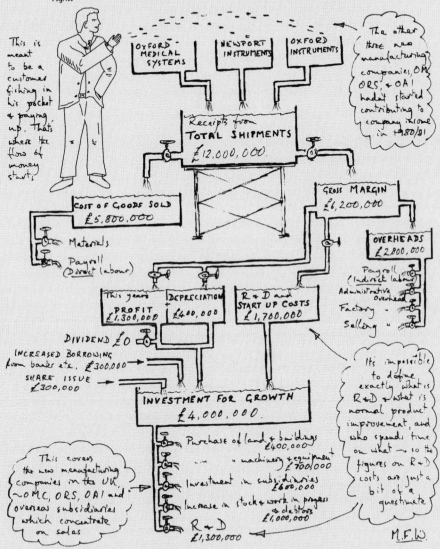

WHERE THE MONEY COMES FROM AND WHERE IT GOES

This isn't a re-hash of the accounts, it's more of a cash statement based on our figures for 1980/81 - an attempt to show how the money flows, from the pocket of our customers at the top to all the things we pay for in the company. To fit it all on one page and not make it too confusing lots of details are left out and many simplifications brought in - but the general picture is about right.

During the summer of 1982 several employee shareholders of long standing wanted to sell shares—mostly to help in buying houses. Flemings explained that if its investment trusts were to buy substantial numbers of shares there should be a proper valuation of the Company, which it undertook. It delayed the final figure until after the signing of the OAP agreement, on which it placed a high value for the rapid exploitation of the imaging-magnet business in the USA. The figure, when it came, was unbelievable—£3.50 a share; everyone was dumb-founded. Martin, severely embarrassed, wrote at once to Ian Boswell, whose company had sold shares for so much less within the previous year. Ian was surprised: 'I just don't understand how they could possibly reach such a figure'—but he was his own usual charming philosophical self, glad that his remaining shares were worth so much, and glad that many of the ones he had sold had gone to employees.

The City, the square mile of financial institutions in the East of London, was beginning to figure in a lot of Group thinking. Barrie and his executive team had, for some years, aimed at matching the growth rates and various financial ratios expected of a small but expanding public company. In the year to March 1981, in spite of severe conditions in much of British industry, the turnover had been up 53 per cent to nearly £12 million (equivalent to nearly £28 million at the beginning of 2000). Pre-tax profits had been up by 27 per cent to £1.3 million (before a big tax rebate). This growth was largely due to the success of NMR magnets and Medilog. In the year to March 1982, with a fast declining pound, sales were up by a further 47 per cent to £17.7 million and profits by 54 per cent to nearly £2 million, both way above the 1980 plan; and the profit was still coming from the older operations. In 1982, bowing to advice from Flemings, the Company paid a dividend on the ordinary shares for the first time ever—it was 1p a share.

Barrie preferred to work to simple financial objectives uncluttered by statements of good forward-looking practice, which he considered went without saying. Martin liked to spell out the Company's non-financial aims. He did not like the bald traditional objectives said to be needed for the City. In his Chairman's review with the 1982 accounts, which, in the event, was to be his last, he drew up a statement of aims that is really an affirmation of the Company's culture:

- For our customers we aim to supply the most advanced and reliable equipment.
- For our employees at all levels we aim to provide satisfying and fulfilling careers, a pleasant working environment, security, participation in ownership, and good financial rewards.
- For our shareholders, some return on their investment, an increase in its value and the satisfaction of being members of a successful and forward looking company.
- For the community at large we aim to provide job opportunities as older industries contract . . . We aim to play constructive roles in the communities in which we work and to contribute nationally by earning foreign currency through our exports.
- For science and technology we devote a lot of resources to research and development

and we collaborate with outside research organisations so that new ideas and discoveries can be developed into practical instruments.

A new plan was due in 1982. In the post-mortem on the 1980 plan and the review of subsequent events, there was satisfaction, but also some criticism. This included the Group's 'over-cautious' start on imaging magnets, in spite of the predicted market; the failure to go all out and exploit Newport's System 86 while it had a technical lead; and a management recruitment policy that was 'inconsistent and mean with salaries, and had left the Company far too thin from Board level downwards'. Expansion had been very rapid, and old habits of cutting costs to the bone die hard. A growing company needs to have enough good and experienced managers on board to fulfil its ambitions and catch its opportunities.

Martin was worried by what he saw as a growing aversion to risk-taking. Barrie's firm control of overheads, and cost discipline for his managers, had proved their worth in the growing profits. Now, with profits of nearly £2 million, Martin felt the Company needed to spend more to keep its lead in its technologies. Concerned that too many internal proposals for new advanced products were being refused funding, on the grounds that the shareholders would not want to risk our steadily rising profit line, he called an informal meeting of all the major shareholders. In a very fruitful discussion they proved to be more adventurous than the management was becoming; they were keen to support any properly assessed development projects that were potentially profitable, but risky. Martin sees this as an important turning—or returning—point. With new ideas lurking, and competitors always wanting to steal a march on us in our technologies, avoidance of risk is itself a danger for a science-based company such as Oxford Instruments.

The new strategic plan, built on the outcome of the previous two, aimed at turnover growth of 40 per cent a year, and profit growth of 50 per cent—very demanding figures. The overall message was that nothing must be done to impede the progress of the imaging magnets business—clearly the Group's biggest opportunity—but that it would be too risky to allocate *all* available resources to OMT. Developing other products and markets was necessary as an insurance policy and for longer-term balanced growth. The other operations were to have more resources—for more managers; for new product development; and for better marketing and sales. One part of the Group acted rapidly on this recommendation. In the summer of 1982 John Pilcher, then General Manager of the original magnet company, took on five young physicists. Among them were Alan Street, who was for some time head of the Accelerator Technology Group, before moving to be MD of the NMR Instruments in 1998, and Neil Killoran, who was to become the Technical Director of NMR Instruments.

The Company was short of managers right to the top. On and off for at least two years Martin had been urging Barrie to think about the succession. He needed to find a really promising younger manager he could train up to take some of the burden off him before he worked himself into the ground. What were the desirable qualities? An understanding of the physical sciences and engineering; familiarity with scientific research; first-hand contact with major areas of management; past involvement in corporate negotiations; a flair for business and finance; clear leadership and communication skills; commitment to the long haul and to hard work; a cool head in a crisis; and a track record in a senior management position in an international science-based company. It was a tall order. The executives at that time had many of these attributes but none had the width of experience needed. That summer Barrie told Martin that he could identify only three people he would be happy to see doing his job. They were all in secure, satisfying, and well-paid positions near the top of substantial science-based companies.

One hot August day I met an old friend, Barry McKinnon, who had managed the early development and launch of Medilog and was then working in the VG Group. In passing, he asked whether Oxford Instruments was, by any chance, on the lookout for any senior managers. There had been some disagreement between the Chairman and MD of VG, and the Deputy MD; the latter, who had a lot of support from colleagues, had left in consequence. Barry wrote down the name for me—Dr Peter Williams (see Fig. 11.1). It was one of the three names on Barrie Marson's list. Peter did not respond to the phone, and Barrie found he had gone to the USA, and feared he was receiving job offers in Silicon Valley at astronomical salaries. Barrie's secretary caught him on the phone later, just as he walked back into his house carrying suitcases. He had no intention of living in America.

After Peter had met the senior executives, Barrie offered him an appointment—there was as yet no mention of the succession. He was to concentrate, at first, on two of the smaller companies that were causing concern, ORS and OAI. This was expected to lead to Group Board membership in six months' time. Peter Williams started work in Oxford Instruments in September 1982. He was then 37 years old. After obtaining a Physics degree at Cambridge, he had taken a doctorate there, working on surface science and electron microscopy. He then spent five years as a lecturer in the Department of Chemical Engineering and Technology at Imperial College, London. His productive research there led to a consultancy with VG, which eventually led on to his appointment as Director of Research at VG Scientific. His career in that group included the management of successful growth at two other VG subsidiaries, and he became a Group Board member and finally Deputy MD. It would have been hard to find a closer fit with the list of desirable qualities and experience.

Peter's first six months in Oxford must have been a difficult time. Whatever the track record, it is always hard for a young new arrival from outside to join a long-standing team of senior people—especially when the succession is open. Barrie had been in office for nearly twelve years, and the other three main board

Fig. 11.1. Peter
Williams

executives had worked for the Group for forty years between them—none less
than ten. Added to this, Peter's immediate executive tasks turned out to be
something of a poisoned chalice. He would have to prove his good judgement in
recommending the future course to be taken over these two subsidiaries. In one
he was to face up to advising unpopular action, and in the other to risk asking
for more time and patience.

As General Manager of ORS, the *in vivo* spectroscopy company, Peter made
a thorough analysis of its problems. Investment was still needed for development and for marketing and service facilities; but it was hard to see how this
could be justified unless a market opened up for some imperative new diagnostic use. The American company GE had developed a prototype dual purpose
whole-body system, for both spectroscopy and imaging, using a high-field magnet. MRI scanners were expensive, and the idea of a dual-function product had
attracted a lot of interest when it had been foreshadowed a year earlier. This
was one reason why so few orders had come to ORS. The traditional NMR

spectrometer manufacturers such as Bruker and JEOL were also becoming involved in this *in vivo* spectroscopy at the lower end, competing with ORS's smaller systems. Peter could see no alternative to selling or closing this Company. In the end, Bruker made an offer for ORS. This was not enough to pay back the substantial investment in development, but it kept some of the team together and was accepted. The operation eventually moved to Abingdon.

Peter's other early charge, OAI, had been put together from the rumps of two other groups some eighteen months earlier, and had failed to find a new General Manager. OAI had never really settled down, had been disrupted by a recent move, and was making serious losses. It was also suffering badly from the recession as its main markets were in the oil, mining, and food industries, all seriously depressed; but these industries were capable of providing a substantial market in brighter times. Discussions were held with another company, Link Systems, which will appear again later in this story. Early in 1983 Link made an unexciting offer for OAI. Peter's report had made it clear that any further disturbance would lead to this Company's collapse; there was considerable investment there, and he felt the operation should be given at least another six months' grace. Much to his relief, at a key meeting with Martin and Barrie, the decision was taken to reprieve OAI and give it more time to prove itself. Getting rid of *both* Peter's first responsibilities, ORS and OAI, might have made any further career in the Group rather difficult.

These two companies are interesting in hindsight. *In vivo* spectroscopy did not take off as a stand- alone diagnostic technique. Dual-purpose instruments never became practical clinical systems; the lists of patients requiring MRI scans were always too long to make the necessary adjustments to the machines for occasional spectroscopy work. The introduction of a later method of localizing the NMR response made the technique easier. But it remained largely a research activity, for which the Group continued to make a number of very high-field whole-body magnets each year—some up to 6 T. Over the remaining years of the century imaging systems developed a long way. Higher-field 'functional' MRI gives a lot of information on physiology, and much more can be learned about the chemistry of the body in a modern MRI scanner. ORS would never have been commercially successful. On the other hand, OAI, with better management, eventually got over its problems. The new generations of microprocessor-based instruments it had been busy developing proved popular, and it has since contributed well to the profits of the Group. They got it right.

12 The Road to Flotation

Pressures and decisions leading to the stock-market debut

There is a tide in the affairs of men
Which taken at the flood leads on to fortune.

(Shakespeare, *Julius Caesar* (1599))

The order book for imaging magnets was growing at a startling rate. At first all manufacture had been at OMT, but by late 1982 OAP in New Jersey was in full production of one model. All the managers were under pressure; constant transatlantic liaison was needed to coordinate and monitor the successful transfer of the technology, and all the other aspects of the business.

At the beginning of 1983 the members of the main executive team under Barrie Marson were: John Lawrence, the Finance Director; John Woodgate, in overall charge of the burgeoning magnet operations; Antony Costley-White, responsible for the recession-afflicted medical companies; and Peter Williams, the 'new boy' wrestling with the disposal of ORS and nursing OAI along. It was difficult for Barrie to forge these over-busy, constantly travelling, managers into a team, and get them all together with time for an objective analytical look at the future of the Group as a whole.

The main questions needing answers were: how far the imaging-magnet companies should be set apart from the rest of the Group; how to finance their almost exponential growth; and whether all the magnet companies should be managed in one large superconductor-based division. The latter would leave a much smaller second division made up of the medical companies, OAI, and OA in Milton Keynes. Views differed on these questions and on the best management structure for the Group.

Soon it became very clear that, in order to exploit the Company's exceptional opportunity in magnetic imaging, it would have to make rapid decisions on how to finance its growth. A few months after OMT had moved into its new factory on Osney Mead it was already full; in New Jersey the large old factory was also filling up fast. The turnover in these magnet systems had been £1 million in 1981/2; it would finish the 1982/3 year at £7 million, and £20 million was being budgeted for 1983/4. This growth would need a lot more working capital for people and materials; it would need investment capital for machinery, equipment, and computers, and, of course, much more space for production, design, and development.

Early in 1983 OMT found a suitable site of eleven acres near the still-active old Swinford toll bridge over the River Thames at Eynsham, a few miles from Oxford. On it stood a large dilapidated camouflaged factory where the Wolsley Motor Car Company had made tanks during the Second World War. In February the Board gave the go-ahead to buy the site for £1 million. To build and equip the new 100,000 square feet of factory and office space OMT needed would cost a further £3.5 million. The necessary funds could have been obtained through loan finance, but this would have pushed borrowings rather high, and would have left all the financial risk on the shoulders of the existing shareholders.

The Board started to think seriously about an offering of shares on a stock market. Perhaps the time had come to share with the public the risks and the potential rewards of this unique opportunity. If the Company was to go this route, should it be in the UK or in the USA, where premiums were a lot higher? Should the whole Group be floated, or just the Imaging Division? Our corporate advisers, Flemings, counselled against splitting the Group. The imaging operation had no independent track record, and would be seen as a high-risk single product company. The whole Group had a ten-year uninterrupted record of rising turnover and profits, and a spread of markets and technologies. Flemings was, of course, keen for any flotation to be in the UK, and urged speed while the 'technology sector' was in fashion.

The Company still had a lot to learn about the City of London, but it knew even less about Wall Street in New York. We needed continuing advice in the USA, and in February 1983 Dr David Ellis was appointed a non-executive director. It was David who, as manager of the ICFC office in Reading, had supported the Company through the black days of 1970 and the difficult early years of that decade. By 1983 David was a Partner in an American venture-capital enterprise, investing in small science-based groups, mostly in the USA. He was to assist with US matters.

By March 1983 many questions on objectives, structure, and how to raise funds were still unresolved. Martin called a series of meetings to galvanize the executives, Board, and major shareholders into taking the necessary decisions. In all these meetings there were decisive votes in favour of a stock-market flotation for the whole Group on a 'full listing' in London. The Board passed the formal resolution on 30 March; at the same meeting Peter Williams, having served his six months' 'probation', was appointed an executive director of the Group. But a further three months passed before the management structure was finally agreed and implemented. Barrie Marson then became Executive Chairman; Martin stepped down to the position of Deputy Chairman; Peter Williams became Group MD; John Woodgate remained answerable for all the magnet companies, but was to spend most of his time guiding OAP in New Jersey. In the original magnet company, John Pilcher became MD, and in OMT, Nick Randall was promoted from General Manager to MD. Antony Costley-White retained his existing position leading OMS and OD. Top managers were still needed at OAI and OA.

The Company had been warned. The flotation process involves a small army of professional advisers and an ever-accelerating proliferation of documents, all requiring rigorous scrutiny, and culminating in the prospectus. Flemings was to be our 'issuing house', and, along with our own auditors, outside 'reporting accountants' would be necessary. Our own UK and USA lawyers would need to be assisted by specialist authorized legal experts, and this team would steer us through the stringent legal and stock-market rules for a public offer. All the Company's financial and business affairs for many years back had to be scrutinized; corporate relationships had to be analysed; the key executive team had to be assessed, and each director's credentials documented and checked. Later we would need stockbrokers for 'underwriting' the issue and for 'making a market' in our shares; we would require 'registrars' to deal with all the applications and the allotment of shares to thousands of new shareholders. For designing the prospectus and its newspaper version it would be necessary to use experienced City publicity agents. Finally we would have to employ City printers who could turn out and distribute 50,000 copies of the prospectus in three days. All this would cost a lot of management time and money.

News of the decision to float soon spread among the employees. Some had to be informed before the cohorts of accountants started crawling round every operation and asking questions. It was going to be an exciting time, and the employees were proud of their Company. Later, press reports were to comment on the excellence of the management of the Group, and its subsidiaries, and the positive spirit among the staff.

The heavy costs of a public flotation come out of the proceeds; it *had* to go through and it *had* to be a success. Barrie, John Lawrence, and Martin formed a flotation committee of 'three wise men'. At least one was needed at each general flotation meeting and at the weekly drafting meetings, all held at Flemings. At the end of June, Flemings sent a list of the sixty-seven documents to be assembled, and a timetable for the many meetings, essential actions, and press conferences through the countdown to 'Impact Day' on 7 October. That was when the press would receive the prospectus. Any delay in finishing that crucial document would mean missing our stock-market slot—and a lot of redrafting to match a new date weeks or months later. Some of the sixty-seven documents were to run to many drafts before they were acceptable, and would eventually form part of the prospectus. Others were certificates needed for legal backup; specialist lawyers' opinions on contracts; validation of factory ownership; disclaimers of any possible litigation, and so on. Then the accountants would need a budget from each operation to support the Group budget, on which the all-important profit forecast for 1984 would be based.

Drafts of the prospectus followed each other through our door at frequent intervals, delivered by courier—strict confidentiality had to be maintained. A lot of the descriptive part of this document was contained in the 'Chairman's Letter' to the issuing house, Robert Fleming & Co. This outlined the history of the Group, descriptions of its various businesses, and its objectives, strategy, and policies. The directors and principal managers would all feature there, along with the trading record, profit forecast, and future prospects. Every word had to be correct and no unverifiable clauses were allowed. By August the team was confident in the accuracy of the prospectus, but felt it had become very flat in the drafting process. It was handed to Martin and me as we left for a conference in Grenoble, in France, with instructions to try to inject into it some of the excitement and interest so palpable in the Company. While Martin was busy at the conference I spent a happy week poring over our product brochures for interesting applications, and playing with the text. I think it regained some sense of the vitality of the business. Martin introduced two pages on the science of magnets, superconductors, cryogenics, and NMR—our core technologies. One press report later welcomed this little physics lesson, a novelty in a prospectus.

Parties of stockbrokers' analysts and journalists had already been visiting the Company and its various operations, and there was much favourable press comment. 'Road shows' were later to be held in London, Edinburgh, and Geneva for investing institutions—the insurance companies, pension funds, investment trusts and unit trusts that own so much of British industry. By late September most documents were ready, and all concerned were receiving frequent lists of the diminishing number of outstanding points. On 27 September at the Annual General Meeting (AGM), the Company approved the March 1983 Accounts, which showed a turnover of £26 million and a profit of £2.7 million. Immediately afterwards the Company held an Extraordinary General Meeting (EGM—'extra ordinary' only in that it is intended for the transaction of business not permissible at the AGM). There, the members present converted the Group to a Public Limited Company, a plc, and adopted a new set of Articles of Association (the rulebook), necessary for the more regulated and visible status of a plc.

There would have to be another EGM before the flotation to reorganize the capital again. The 10p shares were by then worth about £10 each—awkwardly large for a public offering. Each was to be split into two 5p shares, and some of the profit reserve was again to be capitalized, turned into shares, and given to shareholders as a 'one-for-one' bonus issue. Of course this again would make no difference to the *underlying* value of the holdings. There had been a lot of movements of shares in the previous few months as recently arriving managers, and people who had helped the Company, were given the chance to buy shares at the low pre-flotation price from some shareholders who had done well from past opportunities. At this EGM new shares would also be created for the public offering. The Stock Exchange insists on about 25 per cent of the shares of a company becoming available so as to give an adequate number in circulation in the market.

On the day before 'impact day' all available directors went to the completion meeting at Flemings' offices. A series of meetings and rehearsals was going on that day, including the second EGM. The focus of the day was a long Board meeting, with fourteen advisers in attendance. It was at this meeting that I resigned my directorship. The Company was now well served by professional managers and experienced non-executive directors. I knew more about small struggling companies than larger ones, and I felt I would have little to offer in the new world of a listed public company. At this meeting various documents had to be approved, signed, and sealed, including service contracts for the executive directors and recent share allocations in the Share Participation Scheme. But the main business of the meeting was the approval of all the arrangements for the flotation.

At the beginning of the session the forty-nine documents, that needed to be signed or approved lay in order down the long table, propped up between document boxes. As each was dealt with it was removed in its box and lined up in strict order on the floor against one wall. These documents included copies of the prospectus—now signed by all the directors, who took legal responsibility for its contents; the accountants' report; the backing for the profit forecast; the certificates of title; a professional opinion on the adequacy of the Pension Fund, and many others. They were to be locked in a safe box where they would remain far into the future, unless some legal problem was later to emerge, or questions to be asked about the propriety of the flotation and the claims in the prospectus.

The meetings went on throughout the day in an atmosphere of suppressed excitement—of a long race nearly won. As we left Flemings, the bulk printing of the prospectus got under way. We had advance copies, and others would be given to stockbrokers and the press the next morning. This smart final version, which followed some seventeen or eighteen drafts, featured many small colour photographs inside the covers (see Plates 6 and 7), giving a vignette of the Group's products, which were described in the Chairman's Letter.

The prospectus is the key document presenting the Company to potential investors. One of the last stages had been the 'verification notes', sent to all internal directors. The lawyers had combed through the latest draft document for every statement they felt to be unsupported by clear evidence. The first question was, 'What is the basis for the statement that the business was founded in 1959?' Fortunately, having been Company Secretary from the beginning, I still had all the early Annual Accounts, including the very modest ones for 1959/60. I had taken them home for safe keeping away from those who like to clear out anything over seven years old—the statutory period for keeping most documents.

In its final form, the prospectus was forty-eight pages long, including the tear-off application form. In it, details of the share capital, indebtedness, directors and advisers, and the track record, were followed by pages and pages of accounts. But the flavour of the document was in the sixteen-page Chairman's Letter, which had taken so long to draft. The directors had insisted, against mild protests from advisers, on spelling out their continued commitment to substantial spending on R & D, which might sometimes be risky. They also stated their intention to continue to be rather mean on dividends, ploughing back most of the profits into investment for further growth. Since this letter encapsulates the Company's position and culture in 1983, I will include extracts from two of the twenty-two sections, as they may be of interest after the lapse of so many years (see Box 12.1).

The offer for sale was by tender. In the recent past there had been a move away from fixed-price offers. In one or two cases companies had floated on a prudent offer price which had been oversubscribed many times, and the price had shot up 20 or 30 per cent as soon as dealing started. Those to benefit had been the 'stags', who buy as many shares as possible in an underpriced new issue, always intending to sell them for a profit a few days later. When a company called Superdrug had floated early in 1983, the stags had made a profit of 54 per cent on the first day's trading. An article by Tony Jackson in the *Investors' Chronicle* (21 Oct. 1983) read, 'One has the feeling that before Superdrug reduced the fixed price system to absurdity, the City had lost sight of the purpose of an offer for sale—to raise money for the company.'

Tender offers introduce the forces of supply and demand into the flotation process. A minimum price is stated in the offer document, and applicants have to choose the maximum price at which they are willing to buy shares, and the number they want at up to that price. If the response is poor and most offers near the bottom figure, the company and its advisers set the 'striking price' low. If there is a lot of competition for a few shares, it will be pitched higher, eliminating all offers below that price. Oxford Instruments asked for a minimum price of £2.30 a share for the 4.8 million new shares being offered, and some being sold at the same time by a few shareholders.

At 7.45 a.m. on 7 October the die was cast. In a prearranged phone call, John Crosland, the leader of the Flemings team, and Barrie Marson, Chairman of the Company, agreed to proceed. This last-minute chance to back out is normal, in case some sudden event rocks the stock market, the company, or the country— an overnight crash on Wall Street; an unexpected writ for the company; or the outbreak of war. One or two companies on the point of flotation were to draw back after the stock-market crash in October 1987.

At 10.00 a.m. the press release was issued. At 11.00 a.m. individual 'in-depth' interviews started. At 12.00 midday the first copies of the prospectus arrived for journalists and at 12.30 Martin and the executive directors met the press for lunch and questions. The press conference was well attended by financial journalists, who asked some astute questions. Through earlier visits and press releases the media had some idea of our spread of businesses; magnetic imaging, the main reason for seeking finance through the market, had already caught the imagination of both press and public. Stockbrokers De Zoote and Bevan had underlined our key role in this new diagnostic revolution in a research report in September: 'NMR imaging only became feasible as a result of the pioneering advances made in magnet design by Oxford Instruments.'

The first press report to appear was on 7 October, impact day itself. The local evening paper, the *Oxford Mail*, a long-term supporter of the Company, had its home opposite our factory on Osney Mead, and had been given the chance to be first into print with the news. In a long article John Chipperfield applauded Oxford Instruments, its management, and its history, showing, side by side, photographs of the old garden shed and the huge new factory going up at Eynsham. The editorial was almost fulsome in its praise. There were to be many applications for shares from the Oxford region.

We were hopeful of a good press the next day, although one can never be sure in advance, and press reports do not guarantee success. The computer company Acorn had been launched about a week earlier and, in spite of good press comments, had barely been fully subscribed at the minimum tender price. But when the next day came all the principal papers carried stories of our flotation, and comment was basically enthusiastic. There were occasional questions about motivation, competition for imaging magnets in the future, and remarks about our low-dividend policy and the high minimum tender price. This was nearly twenty-four times the forecast profit per share after tax (the price/earnings ratio (P/E)). Here is a selection from press comments:

The directors and staff of Oxford Instruments represent the crème de la crème of the British scientific establishment and the fruit of their labours over the last 20 years or so is not to be cheaply plucked by a few passing investors now that it has come to the stock market.   (*The Financial Times*, Lex column, 8 Oct. 1983)

Dr Martin Wood, the unassuming founder of Oxford Instruments Group, marks the culmination of 25 years' work at the frontiers of magnetic technology when he brings the company to the stock market next week . . . Mr Marson said . . . substantial investment in product development was needed to meet the group's objectives in the longer term and this may sometimes take priority over the demands of short-term profitability.' [One point had been taken.]   (*The Times*, 8 Oct. 1983)

There will be no helicopters or yachts paid for out of the proceeds of next month's share issue. This point may cause a few qualms among potential shareholders in Oxford Instruments. The City understands greed and can predict the behaviour of those motivated by it. The greed of shareholders who control a company is the protection for the greed of the outside shareholders. [This view of the City was rather negative, but a month later the comments were more positive.] One product has caught the City's imagination. That's the superconducting magnet for the new breed of body scanner. Since Oxford has sold 107 out of the 110 such magnets in scanners in the world . . . the interest is understandable.   (*Investors' Chronicle*, 16 Sept. and 14 Oct. 1983)

It is not often that British companies establish themselves as world leaders in some branch of high technology. And when such a group decides to float its shares on the stock market it is time to celebrate.   (*Manchester Evening News*, 7 Oct. 1983)

Box 12.1.

CHAIRMAN'S LETTER

This is the text of a letter from G. B. Marson, Executive Chairman

The Directors,
Robert Fleming & Co. Limited   6th October, 1983

Dear Sirs,

I am writing to you to describe the background, strategy, operations and future prospects of The Oxford Instruments Group p.l.c. It may be helpful to your understanding of the Group's activities to draw your attention first of all to the section below headed "The Technology".

Harnessing science and technology

The Group's business is the development, manufacture and marketing of high technology products in the fields of scientific, medical and industrial equipment. Our success has come by harnessing certain advanced technologies and scientific discoveries to the development of new products for identified markets and building a profitable business around them.

In particular the Group has built a leading position in the supply of superconducting magnet systems for scientific and medical applications. It also produces a range of advanced instruments and systems for patient monitoring, biomechanics, materials analysis and for the monitoring and control of industrial processes. . . .

OBJECTIVES, STRATEGY AND POLICIES

Objectives

Our objectives for the Group's long term development are to achieve high profitability and growth while maintaining financial stability. We believe this to be to the benefit of our customers, our staff and our shareholders.

Growth

Our strategy for growth is to concentrate our activities on certain market sectors which we expect to show significant expansion and in which our expertise is relevant. These markets are currently identified as health care, industrial efficiency and automation, analysis of natural resources and defence.

Profitability

Our strategy for continued profitability is the efficient development and manufacture of technically superior equipment for specialised product areas within these markets. Through our strong scientific and technological capabilities we aim to achieve and maintain a significant technical lead over competing products. We have found that when combined with sound management and marketing this strategy has generally proved profitable once a new activity has been established.

Financial stability

We believe that our financial stability is best maintained by a realistic approach to expansion of our current product areas, coupled with diversification leading to a spread of business in different and non-interacting markets and geographical areas.

Staff

One of the key factors in achieving our objective is our emphasis on people. It is important to attract high quality technical staff with appropriate personal motivation. We believe we have achieved a good reputation as an employer by our emphasis on responsibility and career development, a stimulating environment, good employee benefits and opportunities for share ownership. Turnover in technical and managerial staff is extremely low.

New ideas

The origins of the Group and of many of its products are in university research. The benefits of this close relationship are mutual. Our university customers require technically difficult equipment for advanced experiments in science, medicine and technology. Detailed discussions help us to develop the equipment they need, while at the same time increasing our expertise and awareness of new ideas and the "state of the art". Our scientists and engineers participate in numerous conferences and contribute to scientific journals.

Product development and associated risk

From this scientific awareness many suggestions for new products arise and, after careful technical, financial and market appraisal, a small number are selected for development. In recent years research and development expenditure has been heavy. Of the 10 per cent. of turnover for the year ended March, 1983 spent on research and development, approximately half was written off as incurred and the remainder was spent on advanced customer projects and is recoverable when these are sold. This expenditure on innovative product development, which is necessary to transform ideas into marketable products, has been a cornerstone of our progress but can never be devoid of risk. In our experience, not every development project succeeds, but we believe that the overall level of risk in this context is substantially reduced by our policy of seeking a spread of technologies and market areas.

PROSPECTS

The major opportunity before us in the short and medium term lies in the growth of the market for NMR whole-body scanners. Although some of our larger customers are likely to manufacture part of their own magnet requirements, we believe we are well placed in technology, reputation and resources to continue to build a significant and growing business in superconducting magnets for this rapidly expanding market.

In order to spread the Group's activities over a range of products and markets in line with our strategy, we have programmes planned for the introduction of new or improved products in all our non-magnet businesses, several of which derive from our collaboration with universities and research establishments.

To meet our objectives in the longer term, substantial investment in product development is intended and necessary and may sometimes take priority over the demands of short-term profitability.

The Board has confidence that the Group's management and staff are well equipped to generate profitable growth from these opportunities.

Yours faithfully

Barrie Marson.

Barrie Marson

Executive Chairman

Everything you want in a share is offered by Oxford Instruments, except a good yield. A world leader in high technology, fat margins, an expanding market, fast-growing profits, weak competitors, and it's British . . . Oxford are a unique mix of boffins, brains and business sense . . . The greatest danger for Oxford is a market rating that demands too much even of their proven management. [This last comment on the high rating—the high P/E—was significant in hindsight.]   (*Daily Mail*, 8 Oct. 1983)

Nothing dreamy about Oxford. If the name of Oxford Instruments conjures up images of university spires, that is only appropriate for the company has been built on boffins. Its products are so far up the technology ladder that they can peer down on such mundane items as mini and even microcomputers with lofty disdain. This quality image alone should ensure that the company's stock market debut . . . is well received . . . there are no similar companies quoted on the London Stock Exchange.   (*Daily Telegraph*, 10 Oct. 1983)

By 10.00 a.m. on 13 October applications had to be delivered to the New Issues Department of Barclays Bank in Farringdon Street, London. The task ahead for the directors and their advisers was to decide on the right 'striking price' at which all the shares on offer would be sold. During the afternoon I went with several of the directors to the Barclays offices to watch the applications being opened and sorted into piles on trestle tables. As the piles spread from the tables along the floor and grew in height, we could see that the response had been tremendous. The offer was going to be oversubscribed. Counting numbers, at each price tendered, was going to take some time. So the whole party repaired to the Savoy Hotel, where, in the atmosphere of euphoria that follows success after so much labour, the Flemings team entertained the Oxford contingent to an excellent dinner.

Back in the Flemings offices, quite late at night, with the latest figures being phoned through from Barclays from time to time, the Board debated the sensible level for the striking price. At the minimum tender price, the offer was more than nine times oversubscribed, so it would have to be well above that. All could have been sold at £3.11, but if the price was fixed too high there would be disappointed financial institutions that would get no shares, and the Company needed to keep them happy for the future. A very high P/E rating would also make it hard to live up to expectations. The price was eventually agreed at £2.85, where it was still oversubscribed nearly seven times. The P/E ratio of over thirty, valuing the company at thirty times forecast net profits, was at a level clearly demanding a rapid growth in profits over the next few years. The Company would receive nearly £13 million after expenses and would be valued by the stock market at £126 million. Applications had to be scaled down, and 15,000 applicants got only 100 shares each. Company employees, applying on special pink forms, were allotted all the shares they asked for, and between them bought 65,000—a strong vote of confidence.

The actual debut on the stock exchange was on 19 October, when a party from Oxford, invited by our stockbrokers, watched the progress of the first day's trading from the gallery of the old exchange. From that height people trading round the hexagonal booths looked like bees round a honey pot—a lot seemed to converge on the cell with the Oxford Instruments' logo (see Fig. 12.1). We were told the market was in an uncertain mood, but our shares opened at £3.00. They dipped to £2.93 as the 'stags' unloaded their not-very-profitable allotments, but recovered to end the day at £3.03—a satisfactory premium of about 5 per cent.

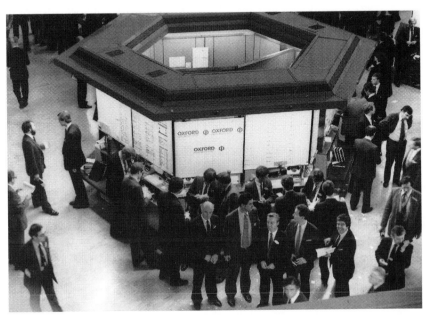

Fig. 12.1. The floor of the old stock exchange on the day of Oxford Instruments' debut

So Oxford Instruments entered a new phase as a public company. As Martin wrote in the newsletter that December, 'That was the year that was . . . We now have a market for our shares, a very big factory going up in Eynsham and are much more in the public eye. Best wishes for 1984.'

13 The New Public Company

The early years as a listed company brought successes and disappointments

> Prosperity is not without many fears and distastes; and adversity is not without comforts and hopes.
>
> (Francis Bacon, *Essay on Adversity* (1625))

After the flotation things would never be quite the same again. People ask whether it was a 'good thing' to go public. The main reason for it was, of course, the Company's need for a lot of money, fast, for the exploitation of the exceptional 'window of opportunity' in MRI magnets. The move also gave our patient long-term investors a proper market, and a sharing of the risks of MRI with a wider public. But a public listing certainly involves 'cons' as well as 'pros'; on several occasions when the City has put pressure on the Company for short-term performance at a difficult time, the directors have regretted the passing of our less visible private-company status.

City commentators are always looking for growth. Without it a share price may dwindle to some ten or twelve times annual earnings—a P/E of ten or twelve. At its launch Oxford Instruments' P/E had been around thirty. As financial journalists had commented at the time, this was a very demanding rating, and would need substantial continued growth over several years to justify. The profits would have to keep going up by something like 40 per cent a year to satisfy the new institutional investors. Public companies have to report progress and profits to their shareholders twice a year. (In the USA it is a daunting four times a year.) Stockbrokers' analysts issue periodic reports, on the companies they follow, which include estimates of sales and profits for the next year or two. The pressures to achieve these expectations can feel like a treadmill; no company grows smoothly by the same amount every six months, and occasional out-of-line results can lead to disproportionately large falls in the share price. What the directors *say* when announcing the results also weighs heavily, and it is not always easy to transmit exactly the right tone.

Profit performance and trends are far from the only influences on the share price, as we were soon to discover. For the first six months after the flotation, and before the 1984 results had been announced, the share price fluctuated between £3.33 and £2.60—that is by 25 per cent. Soon the Board resolved not to be oversensitive about the City's comments and valuations. Its prime objective was to build up a successful group for the long term, not to maintain the share price in the short term at all costs. But the hard-working employee shareholders were disconcerted when the shares went down for no apparent reason after they had pulled out all the stops to achieve a good result.

The profit forecast in the prospectus had been for £5.7 million before tax. There were a few worries from unforeseen problems with NMR magnets—one of the best profit-earners—and from changes in the types of MRI magnets required. But by March 1984 the forecast had been well overtaken; the profit was up by a huge 127 per cent on the 1983 figure at £6.14 million on a turnover up 68 per cent to £44 million. After the announcement in June the share price rallied to £2.90, but in July it dropped again and continued falling right down to £2.40. Profit performance, compared with expectations, was fine, but other factors had been at work. It seems that in July 1984 we suffered from a general loss of confidence in the short-term prospects for the 'Electrical Sector', where the Company was then listed. It was later to move to the 'Electronics Sector', but neither really suited its range of products.

Influences on the London stock market are legion: the state of world economies; political factors in the UK, such as uncertainty over the result of a forthcoming election; UK economic factors, such as interest-rate movements or predictions of higher inflation; and the influence of other stock markets, especially Wall Street. Then there are the factors influencing particular types of business or sectors; big rises in copper prices can affect most electrical companies, and their shares; exporters' share prices may suffer from an uncomfortably high pound. Smaller companies' shares sometimes go up when blue chips go down, because, in a falling market, stockbrokers selling top shares may seek opportunities for the proceeds among neglected lesser stocks; and this can be reversed in a rising market. A company's shares are affected by a cocktail of these influences as well as by their own results and news about their products and markets; and heavy sales of their shares, for whatever reason, is sure to depress them in the stock market.

The City always finds explanations for share price movements (see Fig. 13.1). A few press observations and comments from our advisers will give a flavour of the events that disturbed our share price in its first year of listing:

Oxford came to the market on a wave of enthusiasm for its high technology healthcare products and this confidence has since wavered in the face of healthcare spending cuts in the United States. In fact the market reaction seems to have been overdone.   (*Daily Telegraph*, 9 Feb. 1984)

The week has been an exceptional one for the market . . . There has been a marked movement into top quality stocks, which is reflected in the static prices of some of our less obvious tips . . . Oxford Instruments has managed to maintain its new peak of 333p. (*Financial Weekly*, 22 Mar. 1984)

'On Wall Street today, news of lower interest rates
sent the stock market up, but then the expectation
that these rates would be inflationary sent the
market down, until the realization that lower rates
might stimulate the sluggish economy pushed the
market up, before it ultimately went down on fears
that an overheated economy would lead to a
reimposition of higher interest rates.'

Fig. 13.1. A Robert Mankoff cartoon, © The New Yorker Collection 1981 Robert Mankoff from cartoonbank.com. All Rights Reserved.

The share price was down to 268 on May 25th. This slide of 32p in the week was more than twice the average for the sector. The reasons given by Flemings and our brokers are: (1) A large sale of shares by ICFC in April. (2) The current lack of buyers. (3) The impression of slow-down in the MRI market, felt in the US. (internal memorandum, 31 May 1984)

Profit more than doubled . . . exceptionally strong performance from our imaging magnet operations . . . Order books across the group show that growth looks set to continue . . . we expect the current year to show significant growth both in profit and in technical achievement. (press release for the Company's 1984 results, 20 June)

Oxford Instruments . . . looks unjustly neglected . . . at the year's low of 243p. (*Observer*, 15 July 1984)

Good value in Oxford Instruments. Only last month we highlighted them in our 'bombed out' list . . . There have been two large sellers in the market for some months . . . recently BOC has been selling the shares.   (*Holborn and City Guardian*, 24 Aug. 1984)

Our outside shareholders had been patient while the Company had grown through its adolescence and, understandably, reaped their reward when there was a market for the shares. The Company now had responsibilities to the thousands of new shareholders, who had paid a high price in the expectation of further rapid growth. We had warned in the prospectus that we were aiming for long-term achievement and not a flash-in-the-pan success. But, as far as the Company's performance was concerned in those early years, the public got everything promised and more, even if the share price was a less happy story. From 1983 to 1987 pre-tax profits went up by an average of 57 per cent a year. Few companies can keep up this sort of performance indefinitely, and after 1987 leaner years were to follow.

In November 1983, with the flotation out of the way, Barrie and the executive team started updating the 1982 plan further into the future. They needed to incorporate the current City performance expectations. The imaging-magnet business was the key growth area, but its market was still at a preliminary stage. Of the 107 magnet systems the Company had delivered so far, only around thirty had actually been installed in hospitals for trials. There were still frequent changes in customers' requirements, and these were not expected to settle down for some time.

That November the US company GE finally launched an MRI product, using a 1.5 T Oxford magnet. Through GE's influence, the market started moving towards these higher-field systems. The smaller companies, ahead of GE in producing complete superconducting-magnet scanners, had all used magnets with lower ratings of between 0.3 and 0.5 T. The higher fields had advantages and disadvantages for the Company. The magnets were harder to make, which would delay competition, but they would all have to be made in Eynsham, as the OAP had not yet been initiated in the high-field technology. These magnets also needed a different configuration of NbTi wire, and the New Jersey wire operation would have to work round the clock to supply OMT's needs. But the *most* worrying aspect of GE's announcement that November was its intention to manufacture its own magnets *by 1985*. Although expected, and foreshadowed in the prospectus, this was considerably earlier than Oxford Instruments had foreseen.

It was hard to assess what future market for MRI magnets would still be available to the Company. All our customers were secretive about their progress, their business plans, and their own expectations for the growth of the imaging market. Our own sales forecasts had to be based on the limited market research available; on hints dropped by our customers; and on discussions with doctors and radiographers about acceptance of the new equipment. Even if the technique proved of high value to the doctors, it was uncertain how fast hospitals would be able to buy MRI scanners so soon after the arrival of X-ray CT scanning.

The USA would be the largest market for the new imaging machines, but, until medical insurance companies there fixed the level of 'reimbursement' for an MRI scan, hospitals would not know what income they could expect to cover the high cost of a scanner. Current order books at OMT and OAP were no guide to our future market; with only one established supplier, our customers were all buying for stock. We had a bulk order from GE for 100 magnet systems, and we hoped to negotiate a continuing contract to supply *part* of their needs after they got their own magnet production running. GE in the USA, and Siemens in Europe, were expected to dominate these two MRI markets; but Picker, a US subsidiary of GEC of the UK, and the US companies Diasonics and Technicare, were then ahead in placing systems in hospitals for evaluation. Not much news came out of Japan; six or seven companies there had bought magnets from Oxford.

Oxford Instruments was in the business of making MRI magnet systems for the long term. The Company did not take advantage of its virtual monopoly to make a 'quick killing'. With a great deal of expensive development still needed, and our long-term experience of the fickle nature of superconductors, magnets were priced to give margins that were good, but not extortionate. The Company intended to stay ahead technically, and to create an unassailable reputation for the quality and reliability of its products—that was to take nearly eight years of ups and downs to achieve.

The new Company plan contained many ifs and buts; a reducing number of sales to Siemens was already included, but possible magnet production at GE by 1985 or 1986 was a new factor. When the latest information was evaluated, the Board could see a possible fall-off in MRI magnet sales as early as 1986 or 1987. According to the old rationale, by the time the bonanza was over, and Oxford's share of the MRI magnet market down from well over 90 per cent to the expected 30 per cent or so, the rest of the Group would have grown other substantial products to fill the gap. Projections of profits from promising products in development now looked inadequate to fill a rather earlier gap. If a good growth rate was to be maintained, and the City kept satisfied, further profit might have to come from somewhere else. The only source seemed to be one or more acquisitions of products or companies. Profits and cash were currently good for smaller acquisitions and, as long as the share price remained high, the Company would be able to finance a larger acquisition, without too much dilution of existing shareholdings, by issuing new shares.

The ability to acquire companies by an exchange of shares is one of the benefits of becoming a public company, although vulnerability to a takeover bid is the other side of the coin. The Board agreed that any target company should be in the high-technology area, should have some market or technology synergy with other parts of the Group, and should be within the capacity of the management available. As resources were then stretched, Peter Williams felt it might be better to acquire a larger company of up to our own size, with substantial profits, rather than several smaller companies. This was the 'big-hit' strategy, but, as Barrie concluded, 'we all agree that a badly executed move of this type is the quickest way to suicide'. A long list of necessary qualifications for such an acquisition made it clear that very few such opportunities were likely to come our way; but from that time onwards the Board kept a weather eye out for 'the big one'.

Large acquisitions were obviously very difficult, both in their negotiation and in the management needed afterwards. Some felt that, if the non-MRI companies in the Group could grow fast enough organically, or through smaller acquisitions, they might still be able to fill the turnover and profits gap threatening the Company from over the horizon. The original magnet company had already acquired a newly developed system for 'ion beam milling' that was looking promising. But corporate moves in the two years following the flotation were more in the opposite direction—in divestments.

For more than a decade when a new idea had been developed into a product with potential to stand alone, Group policy had been to form a company round its development team. We believed that 'small is beautiful'. In a new semi-autonomous satellite company there is scope for a young manager to mature and motivate his little group with single-minded concentration on one product area. This policy gives opportunities for promotion; many a manager would rather be a big fish in a small puddle, responsible for his or her own decisions, than a sardine lost in the ocean. In the 1970s Oxford Instruments had spun off Oxford Medical Systems, and later Oxford Research Systems and Oxford Magnet Technology. OMS in its turn had given birth to Oxford Dynamics.

These spin-offs were not always successful. Early in 1983 the Group had seen no shining future for *in vivo* spectroscopy, and had cut its losses, selling off ORS. But Oxford Analytical Instruments (OAI), formed in 1981 from the rump of Newport Instruments and products acquired from outside, was, by 1984, finally struggling out of losses. OAI now had a permanent MD in Tim Cook, who had moved within the Group, and it was to become a valuable subsidiary, which later became the Industrial Analysis Group.

OD ran a different course. While working in OMS, Julian Morris had developed 'Vicon', the computerized movement analysis system. This product and its team had been spun off along with the Unibed intensive-care monitor, at a time when OMS needed to work flat out on its own ambulatory monitoring products, uncluttered by these rather different medical systems. It was probably spun off too soon. No substantial market had been found for Unibed, and the product had been phased out. Vicon had been launched into a recession, a time when hospitals were buying only essential equipment, and orthopaedic doctors took some time to appreciate the value of assessing movement in this new way. For its turnover OD had rather a large staff and other overheads, but Vicon still needed investment in engineering and marketing. In the Group's honeymoon period following its public debut, the directors were concerned about reaching profit expectations, and OD was still making losses. When challenged to reach profitability sooner, Julian Morris suggested a management buyout. This was agreed, and in July 1984 the business was sold to its managers, and the name was changed to Oxford Metrics.

I am happy to report that Oxford Metrics, after a long struggle to survive, became successful and profitable. Julian Morris remained its leader. With ever increasing computer power, and new software, the fourth generation of Vicon for medical applications started to penetrate new markets; and new versions have been developed for film animation, for simplifying the making of cartoons.

The end of the story at Oxford Automation (OA) was not so happy. Formed from the Systems Division at Newport Instruments, this firm's technology had been in the lead, and had promised much at the turn of the decade when its System 86 was launched. The product had been well received, and the ratio of orders to quotations was unusually high. In spite of the large order book, by 1982 the delivery rate had fallen, and the Company was moving from profits into losses. The big problem here was the failure to recruit a good MD. The Company needed a better-thought-out product policy; and it needed to define and restrict its sector of operation because it was really too thinly organized to take advantage of all the opportunities offered. Lack of a good top manager was not the only deficiency: there was then a chronic shortage of software writers and systems engineers—at one stage the Company had to subcontract program design. There were huge potential contracts for millions of pounds, and OA still had a very good reputation and many successful installations behind it; but to compete for these large projects it needed far more top-rate engineers in place.

In October 1983 Paul Brankin moved from OMS to become MD of OA, and soon reported on the situation. The Company badly needed more investment, but the Group was preoccupied with the precipitate growth of the imaging magnet business. The Board started to consider whether OA would do better under some other organization that had the resources to capitalize on its excellent basic systems. In its order book OA had many difficult customized systems that needed a lot of software engineering; costs on one or two of these systems had started spiralling—almost out of control.

OA's technical lead was fast being lost; other companies were offering systems for similar purposes, often buying hardware units and standard software packages from OA. Paul pushed these less-problematical OEM sales, as well as a newly developed software package, 'Flexilogga', which could be used with the recently launched IBM PC. This combination allowed customers to organize their own systems to give a powerful data-logging capability. In spite of expectations, OA was still making losses, and in 1984 Paul had to make more people redundant before it could break even. It now became a steady low-level business, but, as its leading position had been lost, the Board decided that OA was of no further strategic importance for the Group. Paul returned to OMS under a new reorganization, and Mike Russell, an accountant, took his place. His task was to divest the Group of this company that might have been so successful. After talks with Automatic Systems Laboratories, a new Company was formed, Oxbridge Technology, part owned by ex-OA managers, and the new Company bought the business of OA. But it was never very successful.

Was the decline of OA inevitable? Between 1980 and 1984 the scene changed in the electronics industry, largely through computer developments. Competition grew; soon there were several good companies capable of developing systems comparable to System 86, and of improving on it, and reducing the costs by using the new PCs. This sort of business would always become more competitive than superconducting magnets, which need so much past experience for success. After overseeing this sad divestment, Mike Russell returned to the Group, and was for many years Finance Director of OM.

A lot of the funds generated in the flotation went to drive the MRI operations ahead. At the end of 1983 OMT formed a joint venture with a Japanese company, Furukawa, to manufacture imaging magnets near Tokyo. After the USA, Japan is the largest market for medical equipment, but established wisdom said Japanese scanner manufacturers would buy magnets only from a Japanese company. With an eye to the possible decline in our MRI magnet orders after 1986, the Company wanted an early share in this potentially large market. John Woodgate, Peter Williams, and Martin made multiple visits to Japan, and Paul Winson, then OMT's Marketing Director, spent two or three months there in serious negotiations.

Furukawa, a large electrical company, seemed a suitable partner. It was independent, did not compete with Oxford Instruments, and made superconducting wire. But Furukawa wanted the new joint venture to have a third partner, C. Itoh, a large trading organization in the same family of companies. These middlemen firms transact a lot of business in Japan, developing close relationships with both seller and buyer. Oxford Instruments could not see the need for a trading company for selling to the handful of medical equipment giants likely to be customers for magnet systems; it felt the company in the middle must take more than a normal sales and marketing overhead, which would lead to high prices. Furukawa reluctantly agreed, and Furukawa Oxford Technology KK (FOT) started trading in April 1984. Tony Ford, a senior engineer from OMT, moved to Japan to get production going under a Japanese President, and the first magnets were soon being built from kits sent out from Oxford.

A year or two later, there was concern in Oxford that sales from the joint venture were going very slowly. The Group may well have made an error in expecting a simple relationship between manufacturer and customer to work in Japan without the lubrication of intermediaries. Selling in Japan is quite different from selling in the West. It involves lining up with the customer in an overall corporate way. Through inexperience the Company had gone against Japanese culture, and probably suffered from the mistake. The whole magnetic scanner market took a long time to take off in Japan, partly owing to the low level of reimbursement from the insurance companies; the cheaper scanners with resistive magnets were, for years, more popular there than superconducting ones. Japanese competitors thus had a lot of time to catch up. The joint venture sold eight magnet systems in its first year of operation and only sixteen in 1985/86.

In August 1984 OMT moved from five different sites into its large new factory at Eynsham. Employee numbers were growing fast, and there were changes in management roles that year. At OAP in New Jersey, Jim Worth arrived to be the new President (MD); he brought to the Company the experience gained from a long career in Perkin Elmer. Soon after his arrival the joint venture changed its name to Oxford Superconducting Technology (OST). John Woodgate now spent all his time on important contacts and negotiations in the UK, USA, and Japan—he once said he had spent three months of his life in aeroplanes going to and from Japan.

In 1985 OMT won two Queen's Awards, one for Technology and one for Exports—the only company to win them both that year. The presentation by Sir Ashley Ponsonby, the Lord Lieutenant of Oxfordshire, the Queen's representative for the County, was again combined with the opening of a new factory. Lord Jellico, then Chairman of the Medical Research Council, performed this function by cutting a tape that caused the quenching of the Company's 500th imaging magnet, releasing clouds of vapour into the air in front of the fine new building.

Worries about the future market for imaging magnets persisted. There were repeated assessments of the possible 'cliff' ahead for magnet orders, and the best strategy to retain an adequate share of production. In May 1984 the Board appointed the international strategic consultants LEK to study the market, and to advise on the Company's best course of action. Outside consultants often get a bad press, with comments like, 'these people charge a fortune for telling you what you know already'. But they bring new ways of looking at a problem. They may have a clarity of analysis that those too close to the business lack, and they may be able to recognize the sacred cows and blind alleys more easily. And, without doubt, consultants are in a better position than busy company managers to obtain information on the firm's competitors and on its standing in the market.

LEK's first action was to review all existing market information and reports, business plans and forecasts, all customer contracts and agreements, OMT's technology and current developments, and its cost forecasts and analyses. This was followed by a study of customers and competitors and the end-user medical market; the work involved hundreds of interviews and telephone discussions. At the end of three months LEK produced more pessimistic forecasts than the OMT and OST managements, on both the total future MRI market and on the magnet market that would be available for the Oxford companies after 1986. Who was to be believed? LEK had made a very thorough analysis, which basically agreed with the Oxford managers' analysis, except that the numbers at the end were smaller. Nick Randall felt LEK had paid too little attention to the fast-moving technology and OMT's head start in some important new developments. Large superconducting magnets gave everyone problems, but OMT was better positioned than most for solving them.

LEK's advice to the Company was to try to cut the costs of production, to seek a strategic alliance with one of the principal players in the business, or to reduce capacity by closing one of the factories. If LEK's predictions were right, a closure would be necessary. If OMT was correct, one plant would not be able to cope. The strategic alliance was already on the agenda, and the main target was GE, with whom there had been inconclusive negotiations about future orders.

The OMT management received some cheer when estimates of the total market for MRI scanners started looking up. At the huge meeting of the Radiological Society of North America (RSNA) in November 1984, 400 out of 1,000 papers were on this new technique. It was proving of value in diagnosing more and more medical conditions. Crowds gathered round the handful of MRI scanners on display—almost all containing Oxford magnets. But the Company had to stay on its toes to keep up with the many changes in the market. An article in the journal *Diagnostic Imaging* produced just after the show talked about the rate of change in the equipment: 'MRI's clinical applications are quickly being defined. Occurring almost as quickly, however, are technical improvements in equipment that require investigators to revise their predictions almost as soon as they are made' (Ogle 1985: 87). The market, driven by the medical conditions for which it would be useful, was bound up with the technology.

With the continuing need for a strategic alliance there were several further discussions with GE. They had produced a clever prototype magnet to a basically different design from the Oxford ones, but they were taking longer than they expected to get them into regular production. OMT already had a limited technology joint venture with Philips, and, of course, a continuing dialogue with Siemens, which made magnets under licence from OMT. No acceptable major alliance emerged in the mid-1980s.

Meanwhile GE's production programme fell further behind and it needed more and more magnets from the Oxford companies. In all nearly 200 were made for GE before its own plant was running smoothly. OMT and OST were shipping as many magnets as they could make and producing well over half the profits of the Group. At OMT some sixty people were working on the development of new magnets and ideas for the future. In the absence of a strategic alliance, the Company was being forced into what became known as 'the technology route'.

The Group finished another reporting period, and in November 1984 the Board announced the profit for the first half of the year. It was not particularly good, as the period had included the costs, and loss of production, of OMT's move to the new Eynsham factory, as well as the start-up costs in Japan. The Company achieved profits of £2.3 million, up only a modest 28 per cent on the year before—and the City was used to at least 40 per cent:

The six month's profits . . . [were] a shade disappointing to the market, and perhaps to the management.   (*Guardian*, 14 Nov. 1984)

respectable year-on-year profits growth is in store . . . Group products are weird and wonderful . . . the financial message is easier to grasp . . . The enthusiasm which greeted Oxford when it came to the stock market last October . . . was certainly absent yesterday . . . but then high technology companies have come off the boil.   (*Daily Telegraph*, 14 Nov. 1984)

The trouble is too much has been expected too soon from this company.   (*Investors' Chronicle*, 16 Nov. 1984)

In the autumn of 1984 John Lawrence left to live in the north of England after twelve years as Finance Director. His last big task had been the organization of the first Report and Accounts as a public company. This included a colourful Review of Products to inform the new shareholders about Oxford Instruments and its technologies, and was widely acclaimed. David Craig left the Board at the end of the year, as BOC had by then sold all its shares. In February 1985 Martin Lamaison came from BOC to be the Group's new Finance Director, a post he still held at the end of the century. Martin had some knowledge of science, an economics degree, accounting qualifications, and wide experience in accounts and management in BOC, and before that in what was then the Wellcome Group. He brought new strengths to the Company's financial planning and to many other areas of its activities and he fitted in rapidly with the culture in Oxford Instruments, not always easy for people from a large-company background (see Fig. 13.2).

In February 1985 the pound reached its lowest point against the dollar, at $1.04, before starting to climb back up. For the remainder of the century it did not go so low again, nor did it reach the painful heights of 1979 and 1980, when it remained at well over $2.00 for more than two years. These major fluctuations in the exchange rate with the currency of our largest market could not be fully covered by forward transactions. When charting the profit performance of the Group over the years, a high or low pound, against the currencies of our major markets, has a marked effect. Where the distortion due to popular new products, or adverse events in a market, can be excluded, there is a lag of up to eighteen months between a strengthening or weakening pound and a trend of lower or higher profits in the annual accounts.

In the second half of the 1984/5 financial year, technical problems in several parts of the Group melted away in the last couple of months. With the low pound the medical and analytical instrument companies were performing well above budget; the research magnets business had come out with some new 'firsts' and was contributing well to profits through its NMR activities. At the new OMT factory at Eynsham, and in New Jersey, the monthly delivery count

Fig. 13.2. Barrie Marson, Martin Wood, Peter Williams, and Martin Lamaison in the imaging magnet factory in 1985

was rising. The final profit figure for the year was £9.2 million, an excellent 50 per cent up on 1984, and turnover had reached £59 million. This was well above the highest City expectations, and all the forward news was good, at least for the next couple of years. We expected the shares to rise from their depressed level—still below the flotation price of twenty months earlier. It was not to be. They did go up 5p immediately after the announcement, and press comments were all positive, but the price fell back again rapidly. How could this happen? The results had come at the worst possible time: the electrical and electronic sectors were again having problems. By July Oxford Instruments' shares had shrunk to below £2.50 in sympathy.

After the release of the results in June, there was more press comment:

Electronics stocks have not been the flavour of the month for some time, and companies supplying hardware to the health care industry are not very popular either . . . Oxford Instruments' activities are not well understood by investors at the best of times. Full year results to March 1985 demonstrated that it is dangerous to judge a share by the company it keeps in the Official List. Oxford is going great guns. (*Daily Telegraph*, 18 June 1985)

The City's corporate scanner sometimes goes on the blink. Oxford Instruments . . . is an unfortunate victim of this short sightedness.

(*The Times*, 18 Oct 1985)

But several comments in this vein did not stop the shares from falling further.

After more than two years as Executive Chairman, Barrie Marson now handed on the role of Chief Executive to Peter Williams, the Group MD. In his Chairman's Statement Barrie wrote, 'My confidence in the senior management team, together with its record to date, has allowed me to relinquish involvement in operational management. I shall continue as Chairman, but in a non-executive role from October this year.'

14 Seeds for Future Growth

Where was growth (beyond MRI) to come from? In-house developments? Or acquisitions?

If you can look into the seeds of time
And say which grain will grow and which will not,
Speak then to me.

(Shakespeare, *Macbeth* (1606))

On 18 June 1985 the *Financial Times* commented, 'The company's smaller operations are growing rapidly . . . This development could go a long way towards allaying City fears that Oxford is a one-product company.' The smaller divisions were all working hard to develop and launch new products, and three years later were collectively to overtake MRI in the generation of profits. But the prospect of a cliff in MRI turnover and profits still threatened. More growth was needed to satisfy the City—and the ambitions of the Board and management. It would have to come either from buying other companies with innovative products or from organic growth and improved marketing (see Fig. 14.1).

Organic growth comes from the updating of older products and their adaptation for new applications or markets, as well as from entirely new developments. In the early 1980s all operations were improving their established products; they were achieving higher specifications and they were upgrading instruments with microprocessors and computers, and with innovative software. All products were becoming more user-friendly; in magnet systems there could now be provision for the scientist's *experiment* to be controlled and documented automatically, as well as the system itself. Alongside this work on new generations of existing instruments, the scientists and engineers were working on exciting new proposals and projects. The Company has always generated new ideas, and there had been a list of potential projects for years. Another list now encompassed the products or small companies the Group might be interested in buying.

The original Oxford Instrument Company served the research market. A big part of this work was in providing specially designed equipment for scientists

Fig. 14.1. David Ellis and Barrie Marson studying a model of a new factory and Group headquarters in 1986

planning new experiments. There are always requests for something different: higher magnetic fields; more uniform fields, or fields with odd profiles; more research space inside a magnet; a wider temperature range; and so on. The designers would stretch current possibilities to the limit to satisfy customers, and these special systems sometimes led on to standard products. A passage in the Review of Activities with the 1985 Accounts describes the value of this process:

Many of our activities and products were conceived originally by research workers in university and other government funded laboratories. It is due to the original work of these researchers, both as customers and technical collaborators, that we have been able to develop market opportunities in many fields. For this reason our traditional market for scientific research equipment serviced by Oxford Instruments Limited, the founding company of the Group, remains important to us both in terms of its own profitability and as an incubator for future products.

The Group had become expert at this 'technology transfer'. The concept for the dilution refrigerator had come from AERE Harwell in the mid-1960s and in 1985 orders were booming. Refrigerators were now achieving temperatures down to 0.004 K (four-thousandths of a degree above absolute zero). A recent advance had been an online version for studying gamma ray decay of nuclei from an accelerator at the Government's research laboratory at Daresbury.

Another new first was a duplex superconducting magnet, made from NbTi and NbSn, with a persistent field strength of 16.5 T (see Fig. 14.2). These magnets with the highest possible field, often combined with a dilution refrigerator, were in demand for research on semiconductors. The *science* of these materials was then taking a great leap forward, and many physicists wanted to join this fruitful area of study. It was all about the movement of electrons in various configurations of semiconductor materials. New phenomena had been discovered with names like the 'Shubnikov-de Haas resistance oscillations' and the 'Quantum Hall effect'. To the layman these may sound like obscure scientific specialities going nowhere; but through these studies physicists were starting to make discoveries that would lead to extremely important applications in the field of semiconductor devices, and so in the whole growing area of information technology (IT). Our layman often accepts the latest IT device without realizing how much fundamental scientific research has gone into its conception and

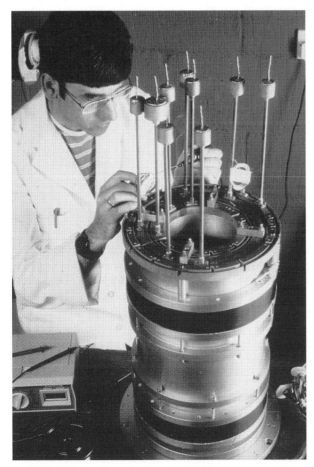

Fig. 14.2. A complicated magnet being assembled

development. In the late 1980s and 1990s there were huge advances in this area leading to the miniaturization, and vastly increased power, of microcircuits and microprocessors. The spectacular gain in function can be seen in mobile phones, in tiny powerful computers, and in many other innovative aids to living and working.

Professor Bill Mitchell was then head of the Clarendon Laboratory and wanted the Company to develop a 20 T superconducting magnet for extending this semiconductor work. In his application for funds he wrote of 'this most dramatic area of research as seen at the moment', and, later, 'the excitement felt by semiconductor physicists about this recent work, I would compare with the excitement produced by the discovery of the transistor'. The Company was finally to achieve 20 T magnets in 1990, made possible by improvements in NbSn conductors. There are always research groups waiting to order equipment with a significant advance in the specifications, and at each achievement step new orders came to Oxford Instruments.

Another enhancement for magnet systems was in the cryogenic area, in the use of closed cycle mechanical refrigerators or 'cryocoolers' for cooling the radiation shields round the liquid-helium containers. Several companies were making these coolers, which had, by this time, advanced a long way; they could now cool the radiation shield outside the liquid-helium vessel down to 20 K, eliminating the need for liquid nitrogen (which boils at 77 K). This colder shield dramatically reduced the boil-off rate of the liquid helium. Some systems that incorporated these cryocoolers could now remain operational for up to three years before they needed recharging with liquid helium; and they could be used in isolated sites away from laboratories, which encouraged their use in industrial environments.

One use for these cryocoolers was in a mineral separation system, in which a superconducting magnet could separate magnetic from non-magnetic particles. Electromagnets were already used for this in a few places, but the higher field and lower power needs of superconductors made these systems more viable economically. A conventional system may weigh 100 tons and take some 100 kW of electricity to operate, while a comparable superconducting magnet system weighs only a tenth of this and its refrigeration and extraction systems need only a tenth of the power. The magnets themselves, of course, when running in continuous mode, need no power at all, and their high fields enable them to separate out weakly magnetic particles. Oxford Instruments' first commercial system, completed in 1983, was commissioned for research on removing from coal the sulphur bound to iron in iron pyrites. Unfortunately the free sulphur, which is also present in coal, is non-magnetic, so only part of the sulphur dioxide can be eliminated from the emissions of coal-fired power stations in this way. No power company has yet thought this improvement worth the investment.

It takes a long time for a new technology to be accepted in an old industry. This is doubly so when it involves esoteric materials such as superconductors and liquid helium. These advances are usually driven ahead by a far-seeing 'champion', seeking technical advantages and willing to accept the risks inherent in change. The first systems to be adopted by the mineral processing industry were for improving the whiteness of china clay, used for coating paper and as a filler for toothpaste. This is achieved by the magnetic removal of the discolouring iron particles, and this process continued to be the main market for mineral-separation equipment involving superconductors. In the late 1980s Oxford Instruments came into contact with Carpco, a small company in Florida, run by Frank Knoll, a committed entrepreneur. This company supplied a variety of equipment for mineral separation, and had already become involved in trials with superconducting magnets from another source. Carpco soon started buying these large magnet systems from Oxford for trials at its customers' quarries, and the systems have since become more widely used.

During the 1990s Carpco installed one of these systems in Brazil, 1,000 miles up the Amazon, in an utterly remote industrial environment with almost no road access. Once or twice a year a thousand-litre liquid-helium container is flown up to refill the cryostat. As with other large superconducting magnet systems, such as those used in MRI, reliability is superb. This system presents a good argument to industries reluctant to adopt the no-longer-new technologies of cryogenics and superconductivity. We have had repeat orders for systems to go up the Amazon and by the late 1990s refrigeration was able to look after *all* their cooling needs. The Company has continued to develop mineral-separation equipment for other purposes, and this could become a substantial business. This is the first truly industrial application for superconducting magnets, but there are several others on the horizon.

A market started developing for large magnets for use in crystal growing, again for semiconductor work. All microcircuits or chips start as a very thin slice of a crystal, usually of silicon, called a wafer. The single crystals from which these are cut may be up to 20, or now even 30 centimetres in diameter; they are grown or 'pulled' from solution under tightly controlled conditions in special furnaces. The process achieves a better uniformity in the crystals if they are grown in a magnetic field, which must surround the whole furnace. Another market for large superconducting magnets was for gyrotrons. These are sources of extremely high-power microwave radiation, with applications in plasma-fusion research, short-wave radar, and communications.

Magnets were getting even bigger as well as more complicated. Oxford Instruments set up a Special Projects Group (SPG) to design and build these monsters—mostly needed in high-energy physics research. In 1983 there was a design contract from Cornell University for a large 1.5 T magnet. It was destined to surround a detector chamber where the field would influence the trajectories of short-lived fundamental particles with names like quarks and muons. The chamber would receive these products of collisions of high-energy particles coming from a new 'heavy-particle' accelerator. This would enable their paths to be mapped, and data on their mass, charge, and energy to be registered in the microseconds before they were lost. In 1985 the Company received a $1.6 million contract to make this 3.8-metre long, 2.9-metre bore solenoid, together with its cryogenic system (see Fig. 14.3). New techniques were needed for this giant detector coil, such as winding the specially developed superconductor on the *inside* of the former, which could then be used to contain the strong magnetic forces. The cold mass of the magnet weighed seven tons, and the total system twenty. As the 1986 Review of Activities states, 'This contract was won against strong international competition. We believe there will be a significant market for these large superconducting magnets in the future, and enquiries for other similar systems have already been received.' This magnet was successfully installed in 1987.

The reputation gained from the Cornell equipment led to other orders for these large magnets, the first from Darmstadt in Germany. Most of them were difficult to make and they varied a lot in size and shape. Magnets are needed in many areas of high-energy physics for focusing and bending beams of charged

Fig. 14.3. The large superconducting magnet for Cornell University; the size can be seen from the man, bottom right

particles and for analysing the products of collisions—which is what much of this research is about. One or two later magnets of this kind were very large indeed, causing queues behind them on the roads as they were driven slowly to their shipping points (see Fig. 14.4 and Plate 5). A few years later SPG won an order for a very peculiar shaped magnet for surrounding a target chamber. It was ordered for the new Continuous Electron Beam Accelerator Facility (CEBAF) in Virginia, USA. The magnet for this $6.5 million contract was a kind of torus, and consisted of six very large flat kidney-shaped coils, each in its own cryostat. It was too large to be built in the Research Instruments factory at Eynsham, but the team found a site nearby with a small existing building and space for the erection of a factory with adequately high eaves and doorway for this, and other extra tall magnets (see Fig. 14.5).

Fig. 14.4. The largest superconducting magnet to date—on the road

This magnet, known as the CLAS Torus (see Plate 11), was delayed by our old problem of defective superconductor. The second coil failed on test because of a faulty weld in the specially-made aluminium-stabilized conductor. Our Superconducting wire operation in New Jersey had been too busy on work for the huge Superconducting Supercollider project in Texas to make this very special conductor to the timescale needed, so it was obtained from another supplier. When SPG finally diagnosed the fault, the manufacturer replaced the material, but not before a long delay and heavy extra costs in labour and

Fig. 14.5. Preparing to wind the first coil of the unique six-coil toroidal magnet for the CEBAF Large Acceptance Spectrometer (CLAS)

materials had reduced the profit margin. This huge project was finished and the equipment installed in Virginia in the early 1990s, when the team was finally able to test the whole system for the first time; it proved very successful. Other orders were to come from the Virginian Laboratory. The SPG business was later to form part of a new Accelerator Technology Group.

The old original company's premises on Osney Mead, dating from 1972 and 1980, were again bursting at the seams. It was not until 1987 that the larger part of the operation, renamed Research Instruments (RI), moved to a 55,000-square-foot purpose-built factory on the edge of Eynsham, a mile from the large OMT facility. The successful and growing NMR Group, then run by Bill Proctor, remained in the buildings on Osney Mead. His clear objective had always been to stay ahead technically. At the lower fields there was competition from two or three of the spectrometer manufacturers, who were making some of their own magnets, and from at least one other small company. For most of the range Oxford's relatively large-volume production kept our systems competitive. The NMR Group was then developing a new 600 MHz magnet system; this demanding rating would need a magnetic field of over 14 T. Like the 500 MHz systems, the inner part of the magnet was to be made of niobium tin. The development progressed slowly, partly because of its technical complexity and partly because there was such pressure to get current models out of the door.

In 1984 Bill Proctor wrote about the programme in the internal newsletter, 'Oxford in View': 'Our best publicized secret is the 14 T (600 MHz) development programme. It only sounds a small increase [from 500 MHz] if you say it quickly, but the 600 MHz magnet is four times more difficult to build. So far the combined efforts of every department have led to 592 MHz, but the most difficult 8 MHz remain to go.' That last 8 MHz took longer than it should have done to achieve, and it was 1986 before the first system went to our old friends in Oxford University. The 600 MHz system was to become the international standard NMR system and remained the most commonly used high-field instrument through the remainder of the century (see Fig. 14.6). There are always buyers for the very latest model; every new step depends on improvements in the performance of superconductors.

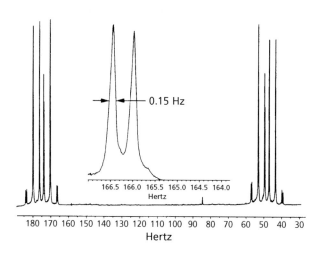

Fig. 14.6. Spectrum showing the resolution in a 600 MHz NMR spectrometer

Horizontal NMR magnets, with a 20- or 30-centimetre bore, were becoming a substantial part of this business in the mid-1980s (see Plate 5). They were used for imaging on small animals or limbs or for *in vivo* spectroscopy research. Often they were ordered for feasibility studies on some new technique before development teams, working on MRI systems, moved to full-sized magnets. The NMR Group's brochure at the time shows the growing range of possible imaging and spectroscopy procedures:

An Oxford Instruments' system allows the use of all the available imaging techniques, such as sensitive line and point, spin echo, multiplanar and three dimensional Fourier transform. Biomedical experiments can be performed not only on the more common species . . . but also on less usual nuclei. In this way it is possible to study organ metabolism, ischemia, and transplant viability.

Transplant viability was where the Company's interest in *in vivo* spectroscopy had begun back in the 1970s.

Through much of the eighties Oxford Instruments was also looking outside the Company for new products. Peter Williams had given this task to Peter Hanley after ORS, which he had been managing, was sold early in 1983. In the summer of that year Peter negotiated the acquisition of a newly developed 'ion beam milling' system from a small development company in Oxfordshire run by Dr Roy Clampitt. This system had been developed for etching or coating surfaces; its production required various techniques in common with cryogenic systems and it would give the Company a toehold in the research side of the huge semiconductor industry. For a modest £150,000 and a few years' royalty payments, the Company acquired the know-how, the potential customer list, one or two orders, and Tim Jolly, an experienced technologist. For the next internal newsletter Tim wrote a vivid description of what the system is all about (see Box 14.1).

Box 14.1. Ion Beam Milling
Tim Jolly

'For Managers and Engineers

If you take a jet of argon gas travelling at 100 kilometres per second and spray it at a surface, two things happen which should cause no surprise. First, the surface gets hot, but, secondly, and more interestingly, the surface is worn away . . .To make a beam of argon ions of high velocity is fairly easy. You ionize the atoms that make up argon gas by removing electrons and giving them an electrical charge. You then accelerate them electrically, using a potential of, say, 1,000 volts. If you accelerate them towards a charged plate with holes in, a sizeable number of them will come out the other side as the required beam. The difficulty is in producing an intense beam and in obtaining a reliable source. You also need to work in a very high vacuum. Thank goodness Oxford Instruments has been practising leak-tight welding for many years!

Ion milling has two major uses: machining or coating. The semiconductor industry in particular is keen to find better ways of machining patterns with dimensions of 0.25 microns or less. It's fairly easy to make masks this small by microphotography. So if you aim an ion beam at a surface covered with a mask, you can machine the required pattern . . . The material that you have milled off a workpiece leaves at a leisurely 10 kilometres per second, and thus sticks very hard indeed to the first thing that it hits. Thus, if you aim the beam at a target of aluminium, p.t.f.e., or whatever, you can coat nearby specimens with aluminium, p.t.f.e., or whatever.

For Salesmen

Ion milling is perfect for almost any insoluble problem that your customer has. It can clean surfaces, drill micro-miniature holes, and mill patterns into semiconductors, diamonds, and metal plates. It can roughen p.t.f.e. so that it can be copper-plated for high-temperature printed circuit boards; it can smooth copper mirrors for lasers. Ion Beam Sputter Deposition can make thin films of unequalled quality for use in optical coatings, medical implants, engineering tools, and, of course, in the semiconductor industry.'

Peter Hanley was to run the new Ion Beam Unit. His small team soon started work on a full range of systems and accessories; and a little later they set up an applications laboratory. Soon there were more orders for these systems, which could cost several hundred thousand pounds. In its first year the new venture exceeded expectations for both orders and shipments. Systems for optical coatings in the aerospace industry became important, and by 1987 the unit had established world leadership in this area (see Plate 14). The market for the semiconductor applications was slower to develop.

Another interesting outside contact was to lead to equipment on a larger scale. One day back in 1983 Tim Cook, then still Sales Manager for the old magnet company, received an enquiry from accelerator scientists working on the 120-metre-diameter synchrotron ring in Berlin, then nearing completion. The Berlin Electron Storage Ring for Synchrotron Radiation, known as BESSY, was a joint initiative between the City of Berlin and the Fraunhofer Institute, which is partly government- and partly industry-owned. BESSY scientists asked Oxford Instruments, recognized as the leading company in superconducting magnets, to consider the feasibility of making magnets for a miniature synchrotron they were designing. Peter Hanley, who had then recently started his search for new products, noticed the enquiry on Tim's desk. He brought it to the attention of Peter Williams, who had been involved with the team running the only British synchrotron ring, at Daresbury in Cheshire.

Synchrotrons are particle accelerators; the tubular ring round which the particles accelerate can be as big as a cricket ground. Both Peters were interested in the idea of using superconducting magnets, instead of the much lower field electromagnets, to shrink a synchrotron to the dimensions of a very small bandstand. Magnetic fields are needed to hold the beam of charged particles in the right orbit as they are accelerated round the ring to reach unimaginable speeds. In an electron synchrotron the highly charged electrons, as they career round the bends of the ring, radiate very powerful beams of light, rather like a car's headlights going round a corner. The beam consists of light over a wide spectrum of wavelengths, right down to X-rays. This 'synchrotron radiation' is the most powerful man-made source of X-rays, a hundred times more powerful than the X-rays produced by medical machines.

There was talk that these powerful X-rays would be required in the future for printing onto silicon wafers the minute designs needed in very large-scale integration (VLSI), for making tightly packed integrated circuits or 'chips'. Smaller accelerators would be necessary if this synchrotron radiation was ever to be used for X-ray lithography in a chip fabrication plant. Following the first meeting Peter Hanley wrote in his report,

'Pocket' synchrotron storage rings are by far the most exciting possibility I have seen. This could lead to an OEM business [making magnets for synchrotron manufacturers] comparable to the OEM magnet business in NMR. There is an immediate initial market for about ten units [for research]. Should X-ray lithography become the preferred process for VLSIs, a substantial market would exist for production machines.

This miniaturized synchrotron was an interesting idea. By the early 1980s some twenty large rings had been installed or were in process of construction, worldwide. Demand for the use of beams of this synchrotron radiation was growing fast, both for general research and for experiments in X-ray lithography on semiconductors. The Company was already becoming involved in magnets for equipment round conventional accelerators, like the Cornell detection chamber magnet. Could compact synchrotrons, and accelerators in general, provide the next substantial market, after MRI, for difficult superconducting magnets? The general view at that time was that optical lithography, even using the shortest wavelengths in the ultraviolet band, would not be capable of drawing patterns on silicon wafers with line widths of less than a micron (a millionth of a metre). Much shorter wavelength X-rays would be necessary in order to crowd more and more components and circuits onto each chip.

The Company was keen to be involved in the development of the magnets for these rings, but the task would need first-rate magnet designers as well as access to advice on the technology of accelerators. Peter Hanley had been Oxford Instruments' top magnet designer for many years, and his move to the Ion Beam Group left a gap. With other experienced magnet designers moving to the rapidly expanding OMT, those remaining in the original magnet company were overstretched. Martin rang his old friend Martin Wilson, who was head of the Applied Superconductivity Group at RAL, near Oxford. He wanted to know if RAL, which was then suffering from spending cuts, had any suitable scientists it could spare who might like to be seconded to Oxford Instruments to fill this gap. Martin Wilson, who already knew the Company well, came to look round the facilities, to discuss the projects in hand, and to assess the help needed. He was intrigued by the new synchrotron idea and promised to consider who might be available—he was just going on holiday. On his return he rang Martin and said, 'While sitting on the rocks in Spain with my feet in the sea, it occurred to me that, rather than finding someone to second to you, I'd like to do the job myself.' Peter Hanley's reaction on hearing this news was, 'It's like ringing up to book the local pop group for a disco and getting Beethoven on the line!'

This was a great *coup* for the Company. Martin Wilson is one of the top superconducting magnet designers in the world; he had built large magnets for high-energy physics at RAL and had written the definitive book on superconducting magnets. He has a deep understanding of the physics of the subject, coupled with a wide knowledge of materials science and technology. At RAL he had been feeling the burden of increasing administrative duties and spending more and more time assessing proposed experiments and monitoring them, rather than being directly involved in exciting developments. He was tempted to Oxford by the challenge of the synchrotron and of other interesting magnets under discussion with the Special Projects Group. In 1984 he became Development Manager for Oxford Instruments, and settled down to a mountain of work.

The first idea for a compact synchrotron, which could be sited in a laboratory, had come from scientists in Munich. They had proposed a small doughnut-shaped ring they called Kleine Erna. This would be made of a single large circular superconducting dipole magnet. Perhaps a rather tall hamburger would be a better description, with a large gap for the 'filling' between the two poles of the magnet (the two halves of the bun). The gap between the upper and lower halves was for the electron beam in its circular vacuum tube, along with other necessary equipment.

Kleine Erna was never built, but BESSY adopted the idea, calling their proposed compact synchrotron 'COSY'. Some of the shareholders in BESSY were large companies in the electrical or semiconductor fields; they formed a consortium to make and market compact synchrotrons for which they expected a considerable demand. The BESSY scientists and engineers were experts in accelerator design, but knew little about superconducting magnets. They asked Oxford Instruments, along with three other companies, to tender for the detailed design and construction of a magnet, associated cryostat, and power supply, for the provisional doughnut-shaped machine. They also asked, somewhat prematurely, for discount levels for an order for ten. Martin Wilson worked on the design and concluded that the cost would be prohibitive, but he had an idea for a better design (see Box 14.2).

Box 14.2, 'Fat COSY' and the Racetrack Design

In the early BESSY design, which the Company nicknamed 'Fat COSY', there was a 'warm' gap for the beam tube and other equipment between the top and bottom cryostat-enclosed magnet windings. Martin Wilson soon realized that the gap between the two poles of the circular dipole magnet would have to be very large to accommodate the layers of the cryostat walls, the RF cavity for supplying energy to the electron beam, and bridging structures needed for strength. Most of the expensively-generated high magnetic field would be wasted. The design would use nearly five tons of superconductor and magnetic stresses would be enormous.

Martin then developed an original design that looked like a small racetrack for the electrons. He suggested splitting the round dipole bending-magnet into two D-shaped segments, with the superconducting coils of each D fitting closely above and below the circulating electron beam, providing a more compact magnetic-field region. The beam would run in the cold space between the poles of the magnets, and inside the comprehensive cryostat for each magnet. The two D-shaped dipole magnets would be separated by straight sections of the electron beam tube, onto which all the ancillary equipment would be fitted. This design would use something like a twentieth of the superconductor of Fat COSY, many other parameters would be easier to manage, and there would be no reduction in the operational specifications. The magnets would be harder to make than for the doughnut, but the high-field region would be much more efficiently placed.

Martin Wilson presented his new ideas to BESSY, and early in 1984 it gave Oxford Instruments a design contract for the next stage. Meanwhile the Company was exploring the market possibilities. Among the research scientists there was a lot of enthusiasm, but production engineers in chip fabrication plants were not expected to embrace a novel X-ray technology until there was no longer an alternative. Peter Williams went to a conference on X-ray lithography (XRL) at the Brookhaven National Synchrotron Facility in the USA, where IBM was using two beam-lines for research into the possibilities. Peter discussed Martin Wilson's new ideas with Alan Wilson of IBM, in a meeting that was to bear fruit later. John Woodgate visited many companies involved in semiconductors, and in June 1984 reported, 'IBM, Texas Instruments, and some other semiconductor firms have said that if a compact synchrotron was available they would buy one immediately, and eventually equip all their relevant semiconductor facilities with them'. It was beginning to look a really exciting business.

The basic design study was finished in June 1984, with some assistance from scientists at RAL. Martin Wilson was confident that the whole machine could be built, but knew it would be a demanding task; the D-shaped magnets, at each end of the racetrack for the electrons, would be particularly difficult. Martin had some experience of accelerators as well as magnets, but knew he needed help from physicists who worked at the large Daresbury synchrotron facility in Cheshire, one of the laboratories run by the Science and Engineering Research Council (SERC). They soon agreed to act as consultants under arrangements with SERC.

The design of the synchrotron was bound up with the design of the magnets, and Oxford Instruments was now interested in making complete machines, not just the magnets. The team had been led to expect an order from BESSY, and continued to invest in more design work and in carrying out various tests and experiments on materials. A month or two later they were concerned to find that two other companies had been asked to quote on the basic Oxford Instruments' design while they had not been included in the new tender list. Just before Christmas they discovered that the order for BESSY's compact synchrotron magnets had been placed with Interatom, a Siemens subsidiary company 'in the national interest'. There was German government funding involved, and they wanted German companies to benefit.

The racetrack design needed a lot more development, and the Company was confident it had the best team for the work. They were keen to make a synchrotron, and all was not lost with the loss of one contract. A little before the BESSY rebuff, IBM had asked several companies, including Oxford Instruments, to quote for feasibility and design studies for a small synchrotron. Using the two beam-lines from the Brookhaven synchrotron, IBM was having some success in a development programme on the techniques needed for X-ray lithography and on the various equipment requirements. This included work on steppers for handling the silicon wafers, masks for the printing of the required patterns on them, and, for coating them, photo-resist materials that would harden when exposed to X-rays.

IBM was now planning to continue the XRL work at a new Advanced Lithography Facility (ALF) at East Fishkill in New York State. Future market predictions were still good, although it was becoming clear that optical lithography for chips could be stretched further than had originally been predicted. Everyone said this stretching could not go on for ever. A lot of the other semiconductor companies were experimenting with XRL, most, like IBM, hiring X-ray beam-lines on large conventional synchrotron rings. If the technique became established successfully, there could be a market for hundreds of these compact machines. There was growing competition. Some of the participants in BESSY had set up a new company, Cosy Microtec, to market their future synchrotrons. In Japan a committee had been formed to organize the development of all the systems needed for this promising new XRL technology. Their government was said to be subsidizing the developments heavily, and several Japanese companies were planning small synchrotron rings.

Through 1985, with frequent meetings, the teams at Daresbury and Oxford worked on the design of the racetrack synchrotron. There are many components to a ring. Some, like the conventional resistive beam-focusing magnets needed on the straight sections, were available commercially. Research was needed for other equipment, and to find materials with the demanding specifications that would be required. It was a very difficult project for a company of Oxford Instruments' size to undertake, but with a world-class accelerator team at Daresbury, and Oxford's expertise in superconducting magnets, the Company was confident that it could match any other team. So, in the mid-1980s, synchrotrons were seen as a major future product, although still some years off; they could not possibly contribute to profits until the 1990s. With the ion beam milling activity they might later form the core of a new division to provide high-technology equipment for the semiconductor industry.

Another possible future product was a superconducting cyclotron. This was another kind of accelerator, but considerably smaller than a synchrotron. Cyclotrons were the first kind of accelerator to be developed in the 1930s, when they produced particles with enough energy for 'splitting the atom'. A beam of charged particles is accelerated in a tight spiral path under the influence of a magnetic field and a very high and oscillating voltage. Cyclotrons are commonly used in the production of radioisotopes for medical research and clinical investigations. Many processes in the body can be studied by tracing metabolites (the chemicals involved in bodily processes) that have been 'labelled' with radioactive isotope 'markers' as they play their part in the biochemistry and physiology of the body.

In 1983 Amersham International, a producer of these radioisotopes, asked Oxford Instruments to consider the feasibility of making a cyclotron incorporating superconducting magnets. Large, heavy, and power-hungry resistive magnets were used in the current machines. Peter Hanley sketched a design, but a substantial development programme would clearly be needed to produce a working prototype. Discussions revealed two interesting applications for a small cyclotron; the first was the medical one, to produce short-life radioisotopes for Positron Emission Tomography (PET), another form of medical scanning (see Box 14.3).

Box 14.3. PET Scanning

Radioisotopes are atoms such as nitrogen, carbon, oxygen, and fluorine that have had extra positrons from a cyclotron forced into their nuclei making them radioactive. Those suitable for PET scanning must decay fast so as to minimize radiation damage to cells. The radioactive isotopes have to be substituted rapidly for common atoms in compounds that play a part in the normal metabolism of the body. The radio-nuclides, as the radioactive metabolites are called, are introduced into the body by injection or inhalation immediately after formation. As the isotopes decay, they release the extra positrons, which then collide with the first electrons they meet to produce pairs of gamma rays. These 'light up' the path of the radio-nuclides through the body, and the radiation is detected by banks of gamma ray counters, and converted into images by computers. PET scans can reveal areas of unusual activity in the brain or abnormal patterns of blood flow, and in many cases can display the very active metabolism of a fast-growing tumour. PET scans give very good contrast but poor resolution and definition.

Although PET scanning could diagnose abnormal physiology, the technique had grown slowly. For making the radioisotopes and radionuclides it was necessary to have a cyclotron very near the scanning centre. The old resistive machines were heavy and expensive, and had to be located in a radiation-shielded vault, usually in the basement of a hospital. A cyclotron based on a superconducting magnet would be smaller and lighter; it would still need radiation shielding, but might be sited nearer to an imaging suite. If available, it might give a good boost to this form of scanning.

The other possible use for a cyclotron was for producing neutrons for non-destructive testing. Neutrons have no charge, and behave in ways that seem strange to the uninitiated. Most metals are more or less transparent to them, while hydrogen reacts strongly, absorbing them. Radiography with neutrons can show up hydrogen-containing materials inside a metal container, corrosion cracks in aeroplane parts, lubricant penetration in a car engine, or blocked cooling channels. For security applications drugs or explosives might be traceable in luggage. Neutron radiography was not new, but the source of the neutrons was then a reactor, so its usefulness was limited. A small cyclotron, especially if it could be made mobile, might open up many new applications.

It was to be some time before thought was given to neutron radiography, but the medical possibilities came sooner. A scientist from the Japanese Company NKK, which was already involved in equipment for nuclear medicine, took notice of the idea at a lecture in Tokyo, given by Martin Wilson. NKK became interested in making PET scanning systems complete with a small cyclotron. Serious design and development started in 1985, but although the magnets—not unlike MRI magnets—gave little trouble, the team working on the cyclotron had a lot of problems with the accelerator physics. OSCAR was the acronym for the cyclotron. Oscar one, for PET, was completed very late, but was eventually commissioned successfully and gave a reasonable performance. Oscars two and three were assembled in the early 1990s and, when installed in Japanese hospitals, exceeded all specifications. But the development had been expensive, and the project became profitable only with later cyclotrons, which sold for about £700,000 each. NKK and Oxford Instruments were later to develop a complete 'Cyclotron PET Radioisotope Production System' named 'Isotrace'. With an integral radiation shield and automated chemistry modules, this could be sited nearer to patients, and could produce any of the commonly used positron emitting isotopes at the pressing of a few buttons.

When a synchrotron order finally arrived in 1987 preparations were made to build it in the former OMT factory on Osney Mead. Later still cyclotrons were made and tested in this factory. All this involvement with accelerators and large magnets for high-energy physics would eventually lead to the establishment of a new Accelerator Technology Group in the 1990s.

In 1988 the SPG was asked to design, and later to make, a 'wiggler' magnet for the large Daresbury synchrotron ring. Wigglers are insertion devices that are placed in the straight sections of large conventional synchrotron rings. Known as 'snakes' in Japan, they cause the electron beam, travelling round the ring, to 'wiggle' between alternate powerful magnetic poles provided by superconducting magnets. The quality and wavelengths of the X-rays produced at the sharper bends of the wiggler are different from those emitted at the normal bends in a synchrotron ring. The wiggler magnet made for Daresbury was delivered in 1991 and provided very short wavelength X-rays down to 0.2 Angstrom units (Å, a ten-thousandth of a micron), compared with the emission of 2 to 3 Å at the normal bends. The wavelength band needed for X-ray lithography peaks at 8 Å.

There were also seeds of new products in the non-superconductivity companies in the Group. In 1983 OMS moved to a fine new purpose-built factory in Abingdon. Ambulatory heart monitoring, by then more than a decade old, was becoming very competitive, especially in the USA. The next year the Company launched a new ECG system, the Medilog 4000 (see Fig. 14.7). Microprocessors in an 'intelligent real time' recorder analysed each heartbeat as it happened, before it could be distorted by recording onto magnetic tape; anything abnormal was digitally encoded onto the tape. Two channels were reserved for a complete twenty-four-hour analogue recording like the ones in the past. A powerful new microprocessor-based ECG analysis system could digest the analogue recordings and combine the result with the digital information. The end product was a complete printed report, including a printout of rhythm strips for any notable cardiac events. This advanced equipment became popular with the market, and was commercially successful.

Fig. 14.7. Medilog 4000 system, the 'intelligent' physiological recorder launched in 1984

In 1983 OM had introduced a sophisticated new nine-channel ambulatory brain monitor, the Medilog 9000, which won a design award. Beside normal EEG work, which was still principally in the diagnosis of epilepsy, there was a growing market for this equipment in sleep studies. With nine channels, this recorder could monitor other functions beside brain rhythms; snoring, ECG, and blood oxygen levels are significant factors in sleep disorders. This product was the precursor of several generations of instruments for studying sleep, then becoming increasingly important in research, and for clinical diagnosis.

The current interest in sleep was driven by new information on the harm caused by severe snoring, and the sleep apnoea (cessation of breathing) often associated with it. Some people who snore hard may actually stop breathing for up to a minute hundreds of time a night. They start breathing again, with a snort, as the carbon dioxide, building up in the blood, rouses them for a moment. Recent investigations on sleep apnoea in Sydney, Australia, had shown that this pattern could reduce the blood oxygen level substantially, perhaps dangerously. It could even result in a permanent loss of memory. In any case, sleep disturbances through snoring cause sluggish responses and tiredness by day, often with a poor performance at work. This can lead to short moments of sleep, and may be a substantial cause of road or factory accidents. The Australian team found that a simple facemask, supplying the patient with a small over-pressure of air, could correct the constriction of the air passage that was causing both the snoring and the apnoea. Sufferers were so relieved after a good night's sleep that most were willing to wear a mask at night; some also succeeded in changing their lifestyles, cutting down on alcohol and losing weight in order to recover full health—and to bring relief to the marital home.

OM also worked with doctors in Nottingham to develop a monitor for measuring the level of acid in the oesophagus. The objective was to distinguish between heart pain and acidity pain. The instrument succeeded technically, but was not a great success, as it was unpopular with patients. They had to swallow a tiny zeppelin-shaped device, containing acid sensors and a minute radio transmitter, which was anchored to a back tooth by a thread. The enthusiasm of the doctors had driven the development without enough discussion with potential patients. This pH monitor was sold to Oakfield Instruments, who eventually made it more user-friendly for a limited market.

OAI now had a well-defined policy to provide useful middle-range equipment, which would help in industrial production and quality assurance. These instruments could do a better job in less time than could be achieved using older methods, which had often involved wet chemistry. OAI had launched new versions of its two main product groups, the low-resolution NMR analysers and the X-ray fluorescence machines, and there were now many optional enhancements. One completely new type of product was developed with the British Non-Ferrous Metal Association. This was an online instrument, based on ultrasound, for monitoring the consistency of the 'grain size' of copper or bronze strip, as it was produced, in a continuous process. Grain size determines all sorts of qualities in the metal, and the slow old method of assessing it involved cutting out a small piece of metal, polishing it, and examining it under a microscope. The new instrument enabled the operator to check grain size and adjust the process without having to stop the mill and take a sample. The instruments worked well, but were quite expensive. The market was not large, and the industry was slow to move to the new method. It was not to be very successful as a product.

As in the medical field, the world of industrial analysis presents opportunities for scores of possible instruments for measuring particular parameters in industrial processes. OAI was involved in many industries including farming, food preparation, oil, paper, mining and mineral processing, cement, and plastics. It was later to focus on a smaller spectrum of industries. With both the oesophageal monitor and the grain size instrument, enthusiasts in the field convinced our scientists that a large waiting market needed these products. It is usually hard to pick winners, even with good market research, but the Company had already achieved some notable successes. We had recognized early the winning points of Medilog; in the 1970s we had understood the analytical need that would sell the NMR spectroscopy magnets, and later we could hardly miss the huge 'market pull' the medical imaging magnets would meet if magnetic scanning proved its diagnostic worth. Since those days in the 1980s, when a few mistakes were made, the Company has developed a better methodology for deciding which of many possible products to choose. But it is important not to turn away too many good ideas.

Beside these in-house developments with growth potential, the Group also had its eyes open for more new opportunities that might arise in the world outside. In the mid-1980s the takeover boom was warming up. There were companies whose products would fit in well with our range, which the Group would have liked to acquire, and there were companies up for sale, but the two hardly ever seemed to coincide.

15 Boom Years

In a time of many mergers the Company sought profit stabilization through some corporate relationship

If to do were as easy as to know what were good to do, Chapels had been Churches, and poor men's cottages princes palaces.

(Shakespeare, *The Merchant of Venice* (1600))

In the thriving business environment of the mid-1980s there came a spring tide of mergers and acquisitions. The financial press bubbled with stories of takeover battles and bids and rumours. Recognizing a possible slowdown in its imaging magnet operations after 1987, Oxford Instruments sought to acquire one or more good profitable companies or products to fill any gap in profits. But good profitable companies or products were much sought after, especially substantial quoted companies of roughly our own size. This was the 'big-hit' target that, it was thought, could fill the profits gap in one move. Premiums of up to 50 per cent above the current share price might be needed to secure a sale by the shareholders of such a company.

The Group had ambitions to take over VG, where Peter Williams had been Deputy Group MD a few years earlier. The Board felt the 'fit' would be excellent; the companies had similar structures and complementary science-based products. The combination would make industrial sense and would result in a significant world-class business. But most of the equity of VG had, long since, been sold by its founders and early shareholders and it had come under the control of the Eagle Star Insurance Company through a subsidiary. In the winter of 1983 a takeover battle was raging for possession of Eagle Star. In an attempt to put a higher valuation on its subsidiary, Eagle Star prepared VG for flotation, and it was launched, rather hastily, on the London market that December. The tobacco company BAT won the battle for Eagle Star, and became the ultimate owner of 69 per cent of VG. Oxford Instruments was unable to persuade BAT to sell this company, although premiums above the market price were offered on more than one occasion. Finally, in 1989, the BAT stake in VG was sold for over £200 million to Fisons, which was then building up a Scientific Instruments Division. The Company could not possibly have matched this price.

Several times between 1983 and 1987 Oxford Instruments considered a merger with the instruments group United Engineering Industries plc (UEI). One of the UEI subsidiaries, Link Systems, had held talks with the Group in 1982 in the hope of acquiring OAI, because the two companies had complementary products. Oxford Instruments had decided not to sell OAI, but Peter Michael (now Sir Peter), then MD and the driving force behind UEI, had become interested in the whole Oxford group.

Peter Michael had been managing a successful private group of electronics companies including Quantel, of which he had been a founder, which made products for manipulating television images by computer. This group had been 'reversed' into UEI, then a small public company, and he wanted to make the public group into a substantial 'club of high technology enterprises'. The more old-fashioned engineering companies were gradually being sold, and he was on the lookout for more leading technology companies to join the club. In 1983 Peter Michael approached Oxford Instruments suggesting a merger of the two groups. There was little in common between them except the complementary products in Link and OAI, but UEI was an impressive group. Oxford Instruments was then in the throes of its preparations for a public flotation, and momentum was building up towards the launch. It was a bad moment for an approach and little time was spent considering the offer.

Not long after our launch the possible early decline in MRI magnet sales started the worries about the security of future profits. In the autumn of 1984 the management consultants LEK confirmed the possibility of MRI magnet orders falling over a cliff in two or three years' time. An exciting new imaging magnet was already in development and, if technically successful, could prove irresistible to our large-company customers. If it failed, or was long delayed, the 'merchant' market for our MRI magnets would certainly shrink. The Company was then engaged in ongoing negotiations with three possible partners, which the Board hoped would lead to a strategic alliance and a steady market for magnets; but the outcome was uncertain. LEK started to look out for a suitable acquisition to soften the effect of any fall in MRI magnet sales. VG was one of the companies on the list of only four it identified in the UK. There were very few independent British instrument companies of a suitable size and fit for the Group. Many promising companies had already been taken over. After spending some energy in discussions over VG, LEK looked at the next most interesting company on the list, which was UEI.

Towards the end of 1984, at LEK's prompting, the Company made a new approach, and found that Peter Michael was eager to reactivate the talks. Marriages between people vary, and so do marriages between companies. Richard Kennett of Laytons, the Company's solicitor wrote,

there is a considerable variety of different forms which an acquisition or merger can take
... I take acquisition to mean a transaction in which one company is totally dominant so
that after completion its management has full control over the enlarged entity . . .
Although many transactions are dressed up as mergers it is usually the case that one company is dominant, and many so-called mergers are really disguised acquisitions.

In order to make a *real* acquisition of a company like UEI, the Group would
probably need to pay a high premium over the share price. A merger between
equals would be different, but as we can see from Richard's remarks, much
more difficult to achieve and manage successfully.

There were several problems in the way of a merger with UEI. One was the
marked difference in company cultures. In UEI, as in Oxford Instruments, the
various subsidiaries had a good deal of autonomy, but there, group executive
power was concentrated more firmly in the hands of the MD. Oxford
Instruments' Board was more open and consensual. Then, our advisers in the
City were against a merger with UEI; they felt it would be unpopular with our
shareholders. Oxford Instruments' markets were perceived as high-technology
medical and research, with the potential for producing occasional blockbusters
like MRI magnets; UEI's markets were seen as entertainment and leisure, with
a lower entry barrier. They felt 'neat Oxford', as they called it, should not be
diluted.

For any good public company there must be a dilemma in negotiating a
merger with another well-managed, healthy, independent company of roughly
the same size. The company with the higher market capitalization at any time
will try to take its window of opportunity and force a conclusion, while the
other will seek to delay negotiations until it can again see an advantage. Oxford
Instruments' results for the half-year to September 1984 had been unexciting,
and, as various difficulties emerged, there was not much incentive to go on negotiating. Sporadic talks still went on over the next two years, but the future shape
of a merged UEI and Oxford remained opaque, and the future need for *any*
acquisition or merger was still not clear. Although a strategic alliance continued
to elude the imaging operation, on the positive side the new and very complicated type of magnet, which would save a lot of problems in hospitals, had been
shown to work. Best-case and worst-case projections ranged from a growth rate
in MRI magnets of 10 per cent a year, if the new magnets could be launched
soon, to a serious decline within two years, sharp enough to affect the whole
Group's results.

The good results for the year to March 1985, with the 50 per cent rise in profits,
hailed in a golden period for the Company, a mini boom. The nagging worry
over MRI did not go away, but a solution might still come in time to forestall
the cliff. The results in the half-year to September 1985 were again good, and the
Company's shares rose above 400p on the announcement. Large numbers of
imaging magnets were being delivered, and the second half of the year was even
better, with results way above City expectations. For this year, to March 1986,

Fig. 15.1. The Executive Committee in 1986: left to right, John Woodgate, Tim Cook, Jim Worth, Martin Lamaison, Peter Williams, Paul Winson, Jack Frost, Nick Randall, and John Pilcher

the pre-tax profit, at £17.2 million, was a brilliant 88 per cent up on 1985. It was an exceptional year, with every part of the Group performing well, and all achieved while ploughing back some £7 to 8 million into R & D (see Fig. 15.1).

The year 1986 was also good for awards and publicity, with important visits and other events. OAI won a Queen's Award for exports that year—the other three companies in the Group had all won these awards previously. Then, in the autumn, John Woodgate, Bill Proctor, Ian McDougal, and Barrie Marson received the MacRobert Award, the premier engineering prize in the UK, for their work of national importance on magnets. In his Chairman's statement with the accounts, Barrie wrote,

We have also been privileged by the attention received from the Royal Family during the year. This has included a reference to us in the Queen's Christmas Day speech, a visit by the Duke of Kent, the special interest shown in our exhibits by the Queen Mother during her visit to the Industrial Soirée of the Royal Society and finally, in June 1986, the knighthood of our founder, Martin Wood.

At the Royal Society Soirée everyone was dressed up in evening clothes as they had been back in 1962, when we had exhibited our very first magnet on a similar occasion. This time the Company displayed the Cryorama and, most interesting to guests, a large functional imaging magnet. This had been charged to give a real magnetic field inside, albeit a weak one. Martin demonstrated the field using an iron chain that visitors were invited to hold at one end to feel the tug as the magnetic forces pulled the chain towards the centre of the magnet. The Queen Mother was fascinated, saying the pull on the chain felt rather like taking the corgies for a walk (see Fig. 15.2). She also floored all the professors and magnet specialists around her by asking just what a magnetic field was. No one had a simple answer.

Fig. 15.2. HRH Queen Elizabeth the Queen Mother feeling the tug of a magnet on an iron chain at a Royal Society Industrial Soirée

In 1987 Martin was elected a Fellow of the Royal Society.

In the first quarter of 1986 the Company's shares had already risen to over 500p on the back of a sharp rise in the stock market in general. The top figure they reached that summer was 576p; they were only once to come close to this level over the rest of the century. The world of industry and the City were confident places in those heady days of 1986 and 1987. The business scene was soon to suffer great changes, scarred first by the 'meltdown' of world stock markets on Black Monday, 19 October 1987, and, after it had recovered from that, buffeted in the early 1990s by the worst recession since the great depression of the 1930s. In the late 1990s the world again experienced raging bull markets, but, as the new century began, there were fears that the tide would soon turn.

In those optimistic years back in the 1980s, mergers and acquisitions were very much the fashion, but many company boards had not foreseen the difficulties of managing new subsidiaries through a deep recession rather than a boom. Although still disapproving of a marriage to UEI, the Company's City advisers urged the Board to use its high share price to secure other strategic acquisitions. Looking at the reverse side of the coin, they advised the Company to set up a defence committee with defined procedures and pre-rehearsed roles, to be activated in the event of an unwanted takeover bid *for* the Group. This was unlikely to occur so long as the share price remained high.

The Company was to complete a couple of tidying up acquisitions before seeking other possibilities. One of these concerned the New Jersey operation. Any joint venture tends to become frustrating for the parties after a few years, and Jim Worth and his executives wanted full control over management decisions. In April 1986 the Group bought the BOC share in OST. Although the long-term fate of our imaging magnet market was still uncertain, OST was then making good profits from both MRI magnets and the superconducting wire business, which was expanding fast (see Figs. 15.3 and 15.4 and Plate 8). The operation was supplying trial materials for pilot magnets for the immense 'Superconducting Supercollider', the ambitious American project to build a fifty-four-mile circumference particle accelerator, or 'atom smasher', underground in Texas. This would have provided vast orders for all the superconductor manufacturers in the world, among whom OST was one of the leaders. Several more large orders were to come to OST over the next few years before the escalating cost forecasts for the project proved too high for the continued support of a US government trying to balance the budget.

The other tidying up acquisition in 1986 was much smaller: OAI bought its American agents, Analytical Marketing Inc. The purchase included sole rights to a different X-ray fluorescence instrument called Chem-X to add to the Lab-X range. This larger £30,000 instrument was more sensitive and could analyse up to eight elements simultaneously. It needed upgrading and, in 1987, was replaced by a new product with novel features, the QX, for multiple analysis in quality control. QX was a very useful and successful instrument, and was only replaced in the late 1990s.

The Company was not having much luck finding a suitable large acquisition, and Paul Winson, then in charge of business development, made a series of presentations on medium-sized and smaller companies, where a purchase might be easier to negotiate. This would be less risky and easier to manage after acquisition. One small purchase was frustrated at the last minute when several key people refused to move from the north of England because of the much higher price of housing in the Oxford area. In one or two other cases interesting companies went to purchasers willing to pay what seemed to be extraordinarily high prices. But there was one small acquisition that was to prove a success.

Hexland, a small company based near Wantage, south of Oxford, already had connections with the Company. It was making a range of accessories for electron microscopy, for both the scanning variety (SEM) and the transmission variety (TEM). The products were complementary to the Oxford range of cooled microscope stages, sample manipulators, and other accessories made by the standard products operation at Osney Mead. Some Hexland products were quite sophisticated, including a liquid nitrogen-temperature specimen

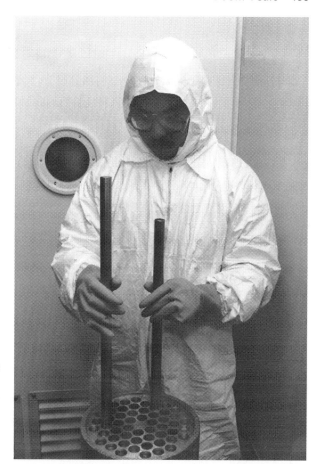

Fig. 15.3. Packing rods of superconductor in a copper billet at OST—the beginning of the process of making multifilamentary niobium titanium

Fig. 15.4. Cross sections of different multifilamentary conductors—the end of the process. Each of these wires has now been reduced to 1mm diameter from billets as above

preparation system for SEM, which had many applications in biological and medical research and in medical pathology and the food industry. Hexland had a turnover of about £600,000, a recently bumpy profits record, and twenty-three employees. It needed more working capital, and its directors decided they would rather see it come under the financial umbrella of Oxford Instruments than go through the hoop of seeking more permanent capital as an independent company. The Group acquired all the capital over two years for a relatively modest sum, and the operation remained at Wantage until the new Eynsham factory was finished. By the end of the 1990s the section providing attachments for electron microscopes had a profitable turnover of over £2 million a year (see Fig. 15.5).

Fig. 15.5. An ultra-high precision computer-controlled silicon wafer probing stage for testing sub-micron integrated circuits in an electron microscope

In 1986 the Group succeeded in taking over one sizeable company, Plasma Technology (PT). David Carr and John Ball, its founders, had looked into the possibility of a flotation on the secondary market, then the Unlisted Securities Market (USM). Having learned about the onerous and expensive process involved, the directors hesitated, and then came to talk to Oxford Instruments.

PT had been formed in 1981 to develop and manufacture plasma etching and deposition equipment (see Box 15.1). This is not so far removed from the technology of ion beam milling; both are needed in the semiconductor industry. Most plasma products used in chip fabrication are large and costly and serve a

Box 15.1. Plasmas

Plasmas are gases in which the atoms or molecules are separated into positively charged ions and free, negatively charged electrons. Often associated in the public mind with very high temperatures and nuclear fusion experiments, plasmas are actually extremely common, and are present in most houses—in fluorescent light tubes. Pulses of electricity 'ionise' the gas in the tubes, making them glow. In the semiconductor industry a powerful plasma beam can be used, instead of wet chemistry, to etch away parts of patterns on a wafer of silicon following lithography. Plasmas can also be used to deposit very thin layers of materials on semiconductors. These are applied in order to build up arrays of minute transistors and other components on silicon chips.

competitive market. The PT directors had seen a niche market in smaller and cheaper systems for research. By 1986 they had developed a modular range of versatile systems, marketed under the name 'Plasmalab'. These products involved vacuum technology, gas-handling systems, microwave generators, high voltages, and all the usual electrical and electronic elements and control and monitoring facilities needed for sophisticated equipment. Among other innovations, PT had developed advanced reactive ion etching (RIE) systems that produced very small and clean features on silicon wafers.

PT had grown rapidly, concentrating on the development of new processes and products. The chemistry of plasmas was not well understood, and the young Company's links with its customers, the semiconductor research laboratories, gave its scientists an insight into which techniques would be of value. The Company participated in collaborative R & D projects under UK- and European-funded programmes. From one such joint study had come electron cyclotron resonance (ECR) etch and deposition systems, then still in development in PT. An advance on RIE, this involves magnetic fields of up to 1 T, and can speed up the etching process many times without loss of sharpness and definition, even for features below 1 micron.

In 1986 the Company's turnover was about £5 million, and its pre-tax profit nearly £1 million. PT had recently moved to a spacious, if old, freehold factory at Yatton near Bristol, and employed a staff of about seventy-five, thirty of them engaged in R & D. With its close ties to semiconductor companies and its fast growth, PT presented an attractive target. Its products were complementary to those of our smaller operation producing focused ion beam equipment. It was just the company to put some substance into the objective of building a division to serve the semiconductor industry. Agreement was reached in August 1986. PT was acquired for £9 million, and left in peace to complete its financial year.

Orders for MRI magnets were already falling, but this was not the cliff—
Medilog orders were also falling. These reverses were due to the 'turmoil in the
healthcare world, particularly in the United States', as David Ellis described it.
Healthcare spending had been getting out of control in several developed countries, and new hurdles were being erected to restrain and delay purchases of
expensive equipment. These were not the only reverses to hit the Group. The
price of oil had fallen, and the petrochemical industry, a major market for OAI,
was curtailing capital expenditure. So orders were falling in three separate divisions.

After the exceptional results for 1985/6 the Company found itself on something of a treadmill, with over-high City expectations for the 1987 performance.
As the Chinese proverb puts it, 'If you're riding a tiger it's hard to get down!'
(see Fig. 15.6). The half-year results announced in November 1986 showed a
profits growth of 28 per cent, which should have been acceptable, but came at
the bottom end of expectations. With development problems as well as the market situation, the Group foresaw a worse second half, especially in MRI. At the
announcement of the results Barrie Marson warned that the US market for MRI
scanners would 'move into a period of consolidation next year'. The shares
promptly fell by about 80 points, and went on falling until they were back below
400p by December.

Fig. 15.6. A Chinese proverb

This fall in the share price damaged the Company's negotiating position
should it find a potential acquisition. At the same time the fall in MRI magnet
orders showed a more urgent need for an alliance, or for some corporate action
in advance of any steeper fall in orders—the in-house competition from at least
one magnet customer was expected any day. The Company was still talking to
two or three of these customers, but they seemed prepared to go on talking for
ever, without reaching any conclusion. There were, for them, too many uncertainties for the future. They were no doubt also waiting to see how the innovative new magnets, launched rather prematurely in November 1986, would
perform. These complicated magnets were not really fully developed, and were
suffering from technical problems; they were very difficult to make and, with

pressure for trial orders for the new system, the engineers did not have enough time to debug them thoroughly before they went out.

The Company looked again at any possible profitable acquisitions to help plug the looming profits shortfall. Nothing looked remotely likely. There was still sporadic contact with UEI, although the difficulties of forming a relationship had not changed. Indeed, Oxford's purchase of PT, and UEI's acquisition of a company in advanced audio equipment, had widened the market and technology gap between the groups. UEI was still keen, and Peter Michael had been seeking a meeting. The Company decided it was worth discussing again how the two groups of semi-autonomous subsidiaries might fit together in a merger; so talks began.

By January 1987 the two companies seemed to be closing together rather precipitately, with teams of advisers on each side, and a series of detailed planning meetings. The talking had turned into serious negotiations. Martin, like most of the Oxford directors, was ambivalent about the prospect of a merger. He saw the need, but instinctively shied away from the diminution of independence. There was one meeting, in particular, in which he was on his own with UEI advisers, snow having prevented a second director from attending. It was an uncomfortable meeting in which he felt unduly pressurized, as the other side's advisers tried to make him give way on various points. He did not want the Company to be railroaded into any union other than a friendly agreed merger.

If the share price of a company engaged in negotiations such as these moves by more than 10 per cent, the stock exchange decrees that it must put out a press release saying what is going on. UEI shares had gone up, and they felt it necessary to put out a press release about talks with a similar-sized company 'with a view to a merger'. The City and financial press got the impression that a merger was virtually a *fait accompli*. There being so few similar companies, they immediately homed in on Oxford Instruments, which responded in a more cautious tone saying the Board was 'reviewing the merits of a merger proposal'. By and large the financial journalists who responded to the releases were sceptical about the logic of a deal, and saw it as a defensive move by Oxford Instruments and an opportunistic bid by UEI.

With expectations of an announcement in three days' time, the Oxford directors involved spent most of their waking hours closeted with their advisers in Flemings offices in London. There were major points of difference between the companies on matters like board composition and the division of executive power; and the less fundamental difficulties of merging two substantial groups of companies were legion. The Board had to weigh the pros and cons, and decide the minimum acceptable position for Oxford Instruments.

Not all the directors were in England. David Ellis was in Seoul, with Jim Worth, trying to negotiate a complicated and important deal with a Korean company wanting to buy a number of MRI magnet systems. The deal being negotiated over several days turned on the security the Seoul company could offer for extended credit. David's voice came crackling over the line from a night-enshrouded Seoul, giving his critical assessment of the merger proposals. Then he switched subject; suddenly the whole solemn assemblage in London was thrown into hilarity by David asking advice on the acceptability, as security for several magnets, of a Korean laundry in California. The juxtaposition of concerns was so bizarre.

On Friday, 23 January, the Oxford directors present were determined to reach a recommendation for the full Board. As well as the main dilemma and still unresolved details, there were indirect problems. If agreement were reached, the directors would have to make a commitment to support the deal; what would happen if a better competing bid were to arrive soon after? If the Company were to turn down UEI, would they make a hostile takeover bid to our shareholders? Why was the Board of Oxford Instruments even contemplating a deal in which positive decisions seemed so difficult? Basically it *was* a defensive move; UEI was a very good company that had been driven forward by the man who would, if agreement was reached, be working also for Oxford's interests. But the press commentators, our advisers, and two large institutional investors the Board had consulted were all against a deal in which they saw little industrial logic; they still wanted 'neat Oxford', even if a reverse was in view.

Sometime in the middle of that Friday a consensus against a deal began to crystallize, probably inevitably. When it came, the decision was unanimous. Strone McPherson and Richard Kennett immediately started to plan how and when to make the announcement. It had to be seen in its true light as Oxford Instruments turning down UEI, or the City would imagine all sorts of skeletons in the cupboard. Any idea that UEI had turned Oxford down could send our shares into free fall, giving UEI the chance to make a hostile bid at a low price. At 3.00 p.m. a press release was transmitted to the various screens in the City saying,

The Board of Oxford Instruments announces that, following discussions with shareholders, advisers and customers, it has concluded that the proposed merger with UEI plc would not be in the best interests of the Company, its employees or its shareholders.

Although the Board acknowledges that there could have been some industrial logic in the proposed merger, the merits of this were insufficient compared with the long-term benefits of Oxford remaining independent.

The Board is confident that the scope for further significant development in the medium term of its core technology-based businesses, including the compact synchrotron, is considerable.

So ended this saga, to some relief in Oxford. The attempts to make a large acquisition had failed, and no new possibilities were to come into view for some time.

16 Challenges and Changes in 1987

The year brought the cliff, a profits warning, and major changes in the world outside

I cannot praise a fugitive and cloister'd vertue, unexercized and unbreath'd
. . . that which purifies us is triall, and triall is by what is contrary.

(Milton, *Areopagitica* (1644))

The ever whirling wheele
of change, the which all mortal things do sway

(Edmund Spenser, *The Faerie Queene* (1596))

In 1987 there were happenings of historic importance in the arena of science and of economics. In the smaller world of Oxford Instruments the year brought several notable events and significant developments. Any idea of a merger with UEI finally ended that January; then there was the growing urgency of the MRI Division's problems leading to the cliff; there were important orders for a synchrotron and a cyclotron; there was a small and successful acquisition in August; and there were changes in the Board, and changes in divisions. In the world outside we were to witness superconductivity at liquid nitrogen temperatures—far higher than previously known—and, on Black Monday in October, to experience the first stock-market crash to affect the Group.

After the end of the UEI discussions the Company had to face up to the reality of the imminent cliff in imaging magnet sales. Early in 1987 the US company GE, which had been our largest customer for MRI magnets, finally got its own magnet production on stream. GE had also recently taken over the MRI operations of two other groups that had been good customers for Oxford magnets, and orders were dropping fast. Siemens was still taking half its requirements from the Group, and making the rest under licence from it, but it had announced its intention to make all its own magnets after 1990. Both Philips and Picker, a subsidiary of the British company GEC, were still ordering some magnets. But the whole MRI market was in retreat following health-care spending cuts, while more competition from independent producers was depressing prices.

The big preoccupation throughout 1987 was the Imaging Division's problems. Was the Group to see this exciting and profitable business decline as rapidly as it had grown? Looking back, was the major drop in sales in 1987 inevitable? It is hard to say in retrospect. Apart from the general malaise in the market, it was largely due to the loss of the GE business in the USA, and GE managers had early decided to make their own magnets. Even with a strategic alliance with one of the main players, the Company's percentage of the total market would have fallen. But those were not the only causes of the problem.

Between 1983 and 1987 the Division's turnover rose from £5 million to £60 million. This headlong expansion was hard to manage. Nick Randall and Jim Worth achieved wonders in getting an escalating number of large magnets out of the door. But the Company had little previous experience of the production of large systems in quantity. Over this period the products themselves had to accommodate frequent changes as the demand in this exploratory early market changed. With strong pressures for delivery, product engineering was not always very thorough, and corners were sometimes cut in manufacture and testing.

Perhaps the only way the Division *could* have retained GE as a substantial customer would have been by completing the most important new development *before* GE got far with its own magnet project. But in the hectic activity of those years, the R & D priorities had probably not been sufficiently well analysed. In OMT there were many programmes for generic improvements as well as for new products; the number of scientists working on the various projects grew to sixty, and new ideas kept coming and spawning new proposals. Two early product ideas were good, and one of them was particularly innovative, and would later prove the salvation of the Company; but for some time the necessary focus eluded the R & D groups. In the autumn of 1986, with fears of falling sales for current systems, these two new products were hastily launched. They were not really ready for the market; they had not been fully debugged; technical hitches appeared, and they had to be taken back into development.

In the winter of 1986 Nick Randall's talents in production were needed for a new potential acquisition. Nick was chosen to be its future manager and transferred to the Group offices to study the task ahead, and prepare to move to the USA. At the last moment this potential acquisition was lost to a higher bidder, and Nick then left, soon to become MD of a new company. (He became head of Airtech, a company in the business of improving radio coverage by 'smart' amplification, which floated on the London market in the 1990s, and was taken over by Remec Inc. in 1999.) In his place came a promising candidate from another company with good experience in managing development. He was expected to rationalize the R & D activity at OMT, and speed the completion of the important new products.

Soon after the new MD started, he and John Woodgate carried out a major review of the situation, and in March 1987 they produced a hard-hitting report and recommendations. The position was very serious; with the new products launched prematurely, some customers no longer wanted the old ones, and all the companies involved were ordering the new magnets for assessment. The problem was that they could not yet be produced with confidence or in any quantity. In order to recover the Company's position, it was vital to complete the full range of the new systems that had been announced, and get them into efficient production fast. A quarter of the magnets budgeted for 1987/8 were expected to be for the new models; with the many gremlins still to be eradicated as well as the normal 'learning curve' of a new product in production, this would be difficult. As John wrote in his March report to the Board, 'It is highly likely that each new system will take, on average, at least twice as long to process through OMT as the current models. The mechanical complexity of the product far exceeds that which we currently produce in volume . . . there will be a major margin erosion.' Orders for the older simpler systems were dropping fast. As Hamlet's stepfather remarked, 'When sorrows come, they come not single spies, but in battalions'.

So in March 1987 the Imaging Division's outlook was uncertain, at its most optimistic. At the previous half-year's announcement Barrie had warned about consolidation in the industry, and a pause in the market. In spite of these fears, at the financial year end the effects of both market and product problems on its performance were still only marginal; shipments of MRI magnets were down 10 per cent and orders somewhat more, but profits were not very much lower than in the previous year. The results for 1986/7 announced in June, showed Group profits actually up by £2 million to £19.5 million. This was due to the performance of the other operations, which had collectively overtaken the MRI companies in profits. This was the 'fifteenth consecutive year of rising profits' (see Fig. 16.1).

In the press release for the results Barrie Marson wrote, 'The growth of the total magnetic resonance imaging market was clearly slowing down during the latter part of the year, and the market is now waiting for the next phase either of lower cost systems or systems with significant technical advantages.' The new prematurely launched products had these 'technical advantages', which the Company hoped would bring the market back. Barrie then pointed out that the current slowdown should be kept in perspective and listed the significant positive factors affecting the Company's future in the MRI business:

first, that the expectation of total market size remains similar to that for X-ray CT scanners, i.e. around 10,000 to 12,000 units world-wide—this means that less than 10% of the market has been exploited so far; second, that MRI has been confirmed as the recommended clinical approach for the diagnosis of an increasing number of medical conditions; and third, that the Group remains in a technically leading position with substantially greater development and production capacity with regard to magnets than any other scanner or magnet manufacturer.

But later in the press release came the warning about the level of profits in MRI in 1987/8; the profits of the other companies in the Group, although

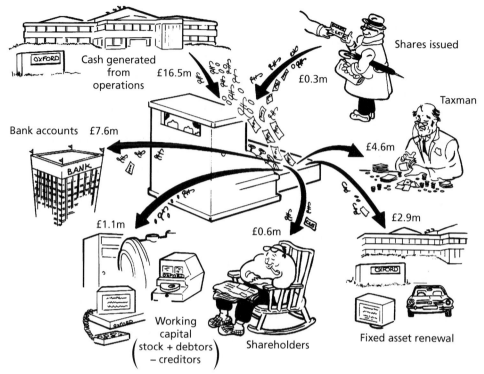

Fig. 16.1. Accounts for employees from 'Oxford in View', September 1987

growing fast, would not be enough to compensate for the MRI shortfall. With a backward glance towards our warning in the prospectus he added, 'The continuing importance of our R & D programme is likely in the current year to take priority over the demands for short-term profit potential'. The shares promptly fell some 35 points to about 390p. The *Daily Mail* (18 June 1987) commented, 'The City would have preferred plenty of profit growth at the expense of the research necessary to generate long term profits. But Oxford refused to comply.'

The new MD of OMT found the technology and the stressed environment very different from that in his old company, and, in the summer, he accepted an offer to go back. The Board saw only one thing to do; they asked John Woodgate to step in as MD, as well as coordinating the whole Imaging Magnet Division. The summer of 1987 was an anxious time. Should the Division be sold to a joint venture proposed by two of our large customers, GEC and its subsidiary Picker with Philips? At the very least, should magnet production at OST be closed down? This was the obvious textbook solution to over capacity, but, apart from an emotional reluctance to take that route, there were concrete reasons against such a move. There were quite a few orders coming in for the new systems, and if the bugs could be zapped they could prove real winners, and the

Division might then need all the capacity it had. And any closure of magnet production at OST might destabilize the wire-making operation on which the
Group now depended for much of its work.

John Woodgate looked at the obvious practical steps in the factory. His priority was to get the new products right, and production better organized. That
summer was the real turning point in the fortunes of the Division, although
1987/8 would be a year of pain. It would see many fewer magnets shipped, a
descent into losses, and distressing but inevitable redundancies before the
Division started to recover. With the multiple problems, outside help was
needed, and 'total quality' consultants were brought in. The future of OMT was
so closely studied and its manufacturing systems so carefully planned that it was
to climb from this low point to become the most efficient unit in the whole
Group.

The continuing technical problems in OMT prompted the Board to make a
closer examination of development throughout the Group. Between £7 and £8
million were being spent each year on R & D; was the Company getting value
for money? Many technical developments, from military hardware to computer
systems for government departments, overrun their budgets and their time
schedules; the closed world of industrial R & D has the same problems, whether
the projects be modest or immense. It is of the nature of a new development that
the exact outcome cannot be known in advance. But at least there can be a plan;
progress can be monitored; shrinking money and time budgets can be assessed;
and early warning signs of probable failure can be heeded.

Early in 1987 the Board asked Martin and Rex Richards to review the technologies and the systems for new developments in all the operations. After preliminary discussions with the MDs and R & D managers they carried out a
survey, with a score board, awarding grades for a handful of significant criteria.
There was much that was good, but looking critically they found some compla-
cency; some ignorance of what was going on in their technology in the outside
world; even a failure to see what was going on in the same technology elsewhere
in the Group. And there was a little 'not invented here' arrogance. A major cause
of inadequate technical development in some operations was the pressure for
monthly output; the most creative people were often switched to help in 'trouble shooting' in current production. It had been a long-standing and successful
part of Oxford Instruments' tradition that the same people were responsible for
development and production. This made sure that development programmes
were focused onto clearly defined customer needs and in-house production
know-how. The problem was that, under delivery pressures, development often
took second place to the monthly output budgets.

Martin and Rex were asked to form a new Technology Advisory Committee
(TAC). Martin was in the Chair, and, in 1988, Alyson Reed joined the Company
to help in the task of promoting improvements in development methods
throughout the Group. TAC soon recommended that each division should, like

OMT, use a consultant to help in a drive for quality. This was not to be for the products alone, but for quality in organization, in the management of design and production, in outside relationships with suppliers and customers, in after-sales service—in fact in total quality, a term that was already being used in many good companies.

In January 1986 the Challenger spacecraft had exploded just after take-off, cutting short the lives of its crew and schoolteacher Christa McAuliffe, who should have been the first 'citizen in space'. The cause of this catastrophic failure was a fault in an O-ring that linked two connecting pipes. The report on the accident the following June discovered that the failure was due to a faulty design. It came to be seen as a failure of total quality, including a failure in team-work.

Alyson organized a forum on technology and quality, which came out with three phases of objectivess for each division: (1) to make the goods the customer wants on time, to quality and to cost; (2) to develop the next generation of products; (3) to look to the future (i.e. what next when the current markets are satisfied?). These objectives may look glaringly obvious now, but the need to spell them out showed they were not always in focus. Alyson's questionnaires and seminars on the innovation process, and on 'new technology', helped the divisions to think more analytically about their technology needs, and where their developments were leading (see Figs. 16.2 and 16.3). TAC held periodic in-depth reviews with the executives at each operation; the presentations were followed by discussion with the non-executive Group directors on TAC, and sometimes with outside scientists. A better development methodology was laid down. But, with continuing pressure for profits, some divisions still complained that it was hard to put enough time and effort into the effective management of development.

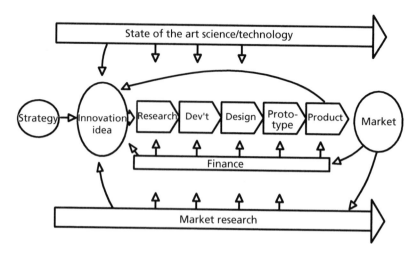

Fig. 16.2. The innovation process

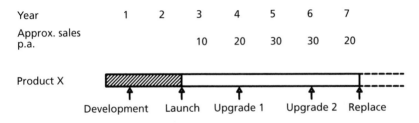

Fig. 16.3. The life cycle of an average product at Oxford Instruments

TAC continued to be a success, and several eminent scientists have served on it, including Professor Alec Broers (now Sir Alec), who was appointed Vice Chancellor of Cambridge University in 1996. Mike Brady, a Professor of Engineering at Oxford, became its Chairman in 1996, and a non-executive director of the Company. But technology never stands still. A new and better method of choosing and developing potential products was launched early in 1999.

In August 1987 the Group bought a small company, Sonicaid, to provide a new business for the Medical Division. Based in Chichester, Sonicaid made equipment for foetal monitoring—that is, for checking the condition of unborn babies. The technology it used was ultrasound, but it did not produce imaging systems. Sonicaid had indeed spent a lot of time and money trying to develop such a system, but it had not been very successful, and had been dropped as the market became more competitive. Although profitable, Sonicaid had run into cash-flow difficulties, and needed a new parent. Oxford Instruments acquired the company for £3.5 million, and provided a further £1 million as working capital. With adequate finance in place this new medical unit did very well; in its first seven months in the Group it contributed to profits. Oxford Sonicaid, as it became, continued to operate from Chichester for some years, and made a welcome and profitable addition to the patient-monitoring products.

In the late summer of 1987 Barrie Marson resigned as Chairman, and from the Board. He had asked to be relieved from office a year earlier, but had waited until the successful conclusion of the long search for a successor. He had been in office for sixteen years, first as Group MD and, since 1983, as Chairman. In that time the Company had surged forward from an unpromising position to become a substantial public company (see Fig. 16.4). As Martin wrote in the September 1987 edition of the internal newsletter, 'Oxford in View', 'We have been through some rough times, but no one could have asked for better leadership—employees, customers and shareholders alike.' The Company owes much to him.

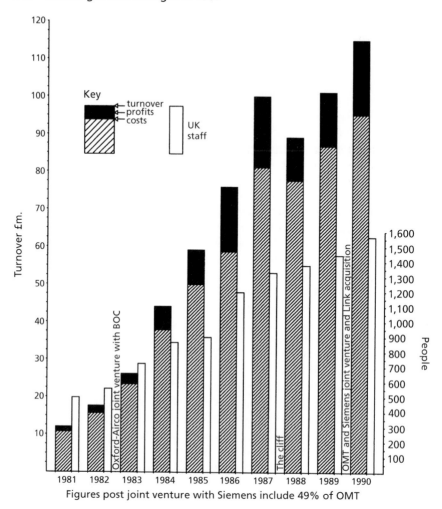

Fig. 16.4. Oxford Instruments' progress in the 1980s

The new Chairman was to be Sir Austin Pearce (see Fig. 16.5), the retiring Chairman of British Aerospace. He knew of the impending problems, the profits warning in June, and the share-price fall, and was prepared for a stormy sea as he took the helm. Martin's welcome to him, in the same newsletter, reads:

Of course, we welcome you for your skill in guiding large technical programmes and developing international strategies—at ESSO in the seventies and more recently at British Aerospace. We welcome you for what it takes to be a director of several major British public companies; for your experience with City institutions; as a chemical engineer and as Chairman of the Trustees of the Science Museum. More than all that we welcome you as an individual and we look forward to working with you.

Oxford Instruments' shares had fallen sharply after the profit warning in June 1987, and over the next few months fluctuated between 300p and 400p. But world stock markets were booming. Several years later *The Economist* was to comment on the 1982–7 bull market, noting that it was driven further and faster than any before it because of market deregulation, institutional shareholders with swelling coffers from insurance and pension funds, and wider share-ownership following a series of privatations of state enterprises. In the summer the *Financial Times* Index of the top 100 shares in London (*FT* 100) peaked at over 2,400—up 50 per cent since January. From a few commentators came warnings of overvaluation and comparisons with the situation before the 1929 crash. But there seemed no real reason for markets to fall; economies were still booming.

Fig. 16.5. Sir Austin Pearce

Then, during the second week of October, Wall Street became nervous. On the 14th, a Wednesday, the Dow Jones Index lost 95 points, then a record. On the 15th it was down again, although not so far. But on Friday, the 16th, it suffered another record fall—of 108 points. The falls came fast enough to trigger 'program-selling', the automatic selling of shares on orders from computer programs when prices drop by a certain percentage. The growing number of 'sell' orders clogged up the system and dealings were suspended for a while. By the time the selling became alarming, European stock markets had closed for the weekend. All, that is, except London, which had not been able to open that day.

In the early hours of that Friday, 16 October, a very real and catastrophic meteorological storm, said to be the worst for 300 years, ripped through southern England. Seventeen people died and the hurricane-strength winds uprooted or snapped off millions of trees, pulled down power cables and telephone lines, and left devastation in its wake through several counties. By dawn almost all the railway lines and roads into London were blocked by trees and the City of London was without power or communications and remained virtually deserted.

All round the world that weekend people in the financial sector were fearful of what would happen when the markets opened on the Monday. It came to be known as 'Black Monday'. The London market, which opened five hours before New York, was still considerably disrupted by the storm; the *FT* 100 index fell about 11 per cent that day. But Wall Street suffered a major crash, losing 23 per cent of its value, down 508 points, of which 300 were lost in the last hour of trading. After this the main index in London went on to fall another 12 per cent on the Tuesday, and more in the following weeks. By early December it was down to 1,600—over 700 points below its peak. It had lost 30 per cent of its value.

In this crash the shares of many smaller companies lost half or more of their value—investors unloaded their lesser stocks before the blue chips. Just a week before the crash Peter Williams had written, 'our most turbulent period in the stock market is now hopefully behind us'. As the Company's shares dropped he claimed the booby prize in forecasting—although the weatherman who failed to predict the hurricane probably deserved that. Our shares went on falling to reach a low point of 150p—just a third of their highest value that year. By Christmas they had struggled back up to 200p, but the following year, although the blue chips were recovering, there was a virtual absence of trading in the shares of smaller companies. Oxford Instruments' annual November Strategy Meeting in 1987 was surprisingly positive in spite of all the problems. The stock-market crash had resolved a few uncertainties; it was a time to 'batten down the hatches' and conserve cash resources. Wall Street had suffered worse than London, and the $US was very low; there were fears of a US-based recession developing. The USA was our largest market, and Peter predicted a £5 million drop in the Group's profits if these fears were to come true.

With a very low share price and a cash reserve of about £20 million, the Board feared an unwanted takeover bid. James Lupton of Barings gave a useful and convincing presentation on how to defend the Group against an attack. The Company was already making use of the skills of another Barings partner, Bernard Taylor, for advice on possible acquisitions, so the Board engaged them as corporate advisers. Everyone knew the year to March 1988 was going to show a big drop in profits. Towards the end of the financial year, remedial actions at OMT and OST were having a good effect, but the Imaging Division still made a loss of £2 million that year after some large provisions. The other manufacturing companies in the Group made up for part of this huge adverse movement by contributing profits of £12 million—a fourfold increase since the flotation. After consolidation, the final Group tally was a profit of £11.2 million—seriously down from the £19.7 million of the previous year. This result was quite well received by the City; investors probably feared worse news and were surprised by the growth of the non-MRI companies.

In March 1987, after a long wait (see Chapter 14), the Company finally received the order for a synchrotron from IBM. Through 1985 the team in Oxford, and the Daresbury scientists, had worked on the design, and studied the feasibility of making a compact synchrotron. The study, with cost estimates, had been presented to IBM early in 1986. There was then still a lot of design, computation, testing of materials, and development of winding techniques ahead, but Martin Wilson was confident that the synchrotron on the drawing board could be made to work. IBM had warned that it might take a further year before it could place an order, and this posed a problem for the Company. Daresbury scientists would be able to continue their assistance in this work for the next six months or so, but after that they would be fully occupied in upgrading the large synchrotron ring at their own laboratory. If the development programme in Oxford were to be put off to a date in the future when an order might be expected, our team would be on its own, and they still needed the accelerator specialists.

Oxford Instruments has always been willing to take a calculated risk in an exciting new business. The Board gave the go-ahead for continuing the work with Daresbury, and for the hardware development programme that would lead to a working D-Magnet—the most difficult part of the whole project. The Company sought funding for this work from the UK Government's 'Support for Innovation' scheme. Governments were subsidizing the development of X-ray lithography in Germany and in Japan. The Department of Trade and Industry (DTI) could assist to the tune of 25 per cent of allowable expenditure, and the Company was grateful to receive a grant of a little over £1 million towards the heavy costs of the programme.

The X-ray lithography team at IBM were impressed, both by the design study and technical competence of the teams, and by the Company's commitment in going ahead with the development. They were worried by the relatively small size of Oxford Instruments, and whether it would have enough available resources for this major project. Much later Alan Wilson of IBM wrote one of the articles for his company's research publication dedicated to the whole subject of X-ray lithography. This article was on the development of the syn-chrotron, and it shows how far the Company had won IBM's confidence:

Developing a ring for IBM was, to say the least, an ambitious project, such as is usually done only by government-sponsored laboratories with vast direct and indirect resources. As we began to consider a number of vendors for this project, we were seriously coun-seled by synchrotron leaders at various laboratories that it would not be possible for us (IBM and vendor) to design, build, commission, and operate a ring! I am glad we did not take this advice too seriously.   (A. D. Wilson 1993: 313)

In winning the IBM order in 1987 the Company was in competition with the Swedish company Scanditronix, which had experience of larger synchrotrons, and with Cosy Microtec, which was having trouble with its magnet develop-ment, patterned on Oxford Instruments' original design for BESSY. Other com-panies had earlier been eliminated from consideration—the Japanese ones working on small synchrotrons were mostly IBM competitors. With market sur-veys and other sources of information showing a big market ahead, the Company had been proposing to produce a prototype synchrotron even with-out the IBM order, but in that case other sources of finance, or a joint-venture partnership, would have been necessary. IBM had wanted a price below $US10 million—the Cosy price—but was willing to compromise at £8.75 million. It also wanted assurance that the Company would make future synchrotrons for it in preference to any orders that might come from Japan and it was willing to pay $US0.5 million in advance for each option slot—it expected to need nine compact synchrotrons. For the first, the prototype, it wanted delivery by March 1990, three years off. This would be possible to meet provided there were no unexpected problems in this innovative new development.

It was also in 1987 that the Company received the first order from NKK for a superconducting cyclotron (see Chapter 14). It was to be built in the new Oxford Instruments' factory at Eynsham, which was completed in 1987. The Company's range and reputation in these low-temperature magnets were grow-ing; but an important breakthrough towards higher-temperature superconduc-tivity was leading to waves of new research in universities and companies throughout the world.

The new discovery had come in April 1986. Professor Alex Müeller and Dr Georg Bednorz, working in the IBM research laboratories in Zurich, reported that they had found superconductivity in a ceramic-like substance, lanthanum barium copper oxide, which remained superconducting at 30 K. This was about ten degrees higher than for any previously recognized superconductor. The sci-entific world was sceptical; claims of at least 25 K had been made before and had not stood up to further tests. This time the experiments were repeated in Japan and the USA and the result confirmed. It was a major step, which caused great excitement. Many scientists joined the 'gold rush' search for yet higher-temperature superconductors, substituting different related elements, and preparing them by different recipes. An important target was to get to 77 K, the boiling point of liquid nitrogen, which had become a ubiquitous industrial commodity. If good superconductors could be found, which remained superconducting above this temperature, all sorts of new commercial developments might be possible.

In the UK in 1987 the SERC and the DTI jointly set up a National Committee on Superconductivity, of which Martin became the first Chairman. Its brief was to coordinate research in the academic and industrial areas. Many other new bodies and journals were launched around the world to discuss and publish the huge amount of research in progress on these strange new superconductors. The number of scientific papers on the subject rose 900 per cent. The work needed high magnetic fields, so there was also a surge of orders for magnets, cryogenic equipment, and temperature measuring and controlling instruments.

These new materials are known as perovskites; they are mineral oxides, usually containing rare earth elements. They often look metallic, but they behave like ceramics. During 1987, the 77 K target was achieved by the development of yttrium barium copper oxide. YBCO, as it came to be called, had a critical temperature of 90 K. But ceramics do not lend themselves to being drawn out into wire conductors, and, although the critical magnetic fields were high as well as the critical temperatures, the currents these materials could carry were very low. Hundreds of research groups continued to hunt for better materials, and in 1988 a report came from Tsukuba, the Japanese 'Science City', of a new material with an even higher critical temperature. This was bismuth strontium calcium copper oxide, known as BSCCO. These two materials were the basis of most research work during the 1990s.

Clearly there was no imminent threat to the technology Oxford Instruments had been developing over twenty-five years. Both in the UK and the USA our companies participated in collaborative research projects on these materials, and kept in touch with the literature. Dr Seung Hong, head of R & D at OST, was allotted finance for work on the least unpromising materials for magnets. It was to be a long time before any group anywhere achieved actual small experimental magnets. Conductors for making coils proved extremely difficult, and were just beginning to look a little more promising by the end of the century. This is not really surprising, and parallels the development of niobium tin. Berndt Matthias had identified this brittle material as a superconductor in the mid-1950s, but it was not until 1960, when Eugene Kunzler studied its properties in a high magnetic field, that it was shown to have outstanding qualities. It took four or five more years before the first fairly stable magnets could be made from it, and another ten before reliable multifilamentary niobium tin had become available and the joints problem had been solved. Periodically improvements in ways of making conductors of this material still lead to new records in the magnetic field achievable.

The technology of superconductors is difficult. The first practical uses for the new 'high'-temperature superconductors were expected to be in electronics, for microcircuits and microwave antennae. Thin films for these applications proved easier to make than conductors that could be used for winding magnets. In 1988 studies in Japan and the USA, on the commercial impact of the new materials, predicted medicine, transport, and electronics as the areas likely to see the first applications. The Japanese were to forge ahead with several developments. They were very optimistic over the products they thought had a high probability of being made with 'high' temperature superconductors within ten years. These included imaging magnets and particle accelerators. Lower down on their probability list came motors for ships and supercomputers. More than ten years have passed since the list was drawn up and at the beginning of the twenty-first century most people would predict at least another ten years before some of these items will be made from the ceramic superconductors.

There are understandable difficulties in trying to apply a new technology to an established industry with well-developed engineering methods. Superconductivity has always found its best markets in situations where the job can *only* be done, or can be done much more practically or cheaply, by superconductors. Our first market had been in high fields for research, and most of our early customers worked in laboratories that could not afford the major electrical and water-cooling installations needed for running high-power copper solenoid magnets. The next big application had been in magnets for high-resolution NMR spectroscopy, for analysing large molecules and developing new drugs. Here the higher fields, combined with high homogeneity and stability, could be achieved *only* using superconductors.

The application that has taken this technology closest to the public, and has become by far the largest use for superconductors, is magnetic imaging. This extension of NMR to medicine is possible with electromagnets, and even with permanent magnets, but for the best images the high fields and stability of superconducting magnets are needed. In the growing market in mineral separation, superconducting magnets are so much lighter and more reliable, and use so much less power than conventional magnets, that they are becoming accepted in this very conservative industry. Where will the next significant application come? There are several important industrial possibilities, and, following imaging magnets and magnetic separation, perhaps the new applications will be accepted more rapidly.

17 The Renaissance of Oxford Magnet Technology

Following the cliff OMT won a soft landing and rose to be named among the most efficient manufacturing companies in the UK

> Out of this nettle, danger, we pluck this flower, safety.
> (Shakespeare, *Henry IV Part I*, 1598))

The cliff fall came in 1987. The drop in shipments that year was steeper and deeper than ever expected. At the 1980s peak of production in 1985/6, the three imaging magnet operations, in Eynsham, New Jersey, and Tokyo, between them shipped 220 magnet systems. In 1986/7 only 196 systems were delivered and during 1987/8 the numbers fell fast, bottoming at only 118 systems, just ninety-nine of them from the two main plants. OST, which had been making most of the magnets for our fast-reducing US market, felt the pain first, and had to lay off thirty-five people at the end of 1986. But in March 1987 fifty had to go from OMT—some finding places elsewhere in the Group.

The 1987 reversal, as I have explained, was due to OMT's problems in its difficult technical developments as well as to its declining market. In its field, Oxford Instruments has a worldwide reputation for tackling projects for its customers that are ahead of the current cutting edge of technology, and not giving up until it has delivered. In our own development projects it is often difficult to foresee which early lines of study will succeed, both technically and in the marketplace. Even when a future winning product has been identified, the Company has not always committed sufficient resources for a rapid completion of the development, the vital production engineering stage, and the planned transfer to manufacture. It was this 'technology route' to salvation that OMT had not driven fast enough to have any chance of avoiding the fall.

Keeping ahead in its technologies has always been vital to Oxford Instruments. Since 1980 the development teams working on MRI magnet systems had been experimenting with many new techniques for improving manufacturing methods and magnet performance. In the rush for production, not much work had actually gone into optimizing the design of the standard and reliable Unistat systems and cutting their cost of manufacture. From about 1983, development work on new magnet systems had been leading in three directions: magnets for mobile scanners; economical compact magnet systems; and active-shield systems. It was the third of these projects that was later to keep OMT in the forefront of the MRI industry, and lead to its successful and profitable business in the 1990s.

First, I must say a word on these other two developments. The small company IGC in the USA, our main 'merchant' competitor, beat us to the first magnets for mobile scanners. These were introduced in response to the US market, where the commercial provision of outside services to hospitals is much commoner than in Europe. MRI scanners were so expensive that many smaller hospitals could not contemplate buying one and instead hired time on a mobile system. Large vans, containing a scanning suite and patient reception and preparation areas, would be driven from the previous clinic and parked in the hospital grounds. The electricity cables would be linked up to the van; the magnet, still cold, would be charged up; the imaging system would be prepared; and the waiting patients would be scanned one by one. These systems had to be rugged and vibration proof, and to begin with used only lower-field magnets. Early in 1985, after accelerated ten-year simulation tests on the prototypes, OMT delivered its first mobile system. These systems presented special design demands, and the first two versions were apt to develop faults, although later models were as reliable as static systems.

The Compact product was to be a small, low-cost magnet and cryostat, and the version launched in 1986 was initially developed as the Mark 2 Mobile. It was over-hastily developed and turned out to be complicated to produce—and it was certainly not low cost to OMT. OST was going to take over the building of the Compact systems for US customers, and Jim Worth, who would be accountable for its efficient production, wanted more US input in the engineering of the product. Glen Epstein came over from OST and became Production Engineering Manager for the Compact. He worked with Graham Gilgrass and his design team to rethink the whole basis of MRI magnet and cryostat design and construction, looking much harder at manufacturing processes and costs.

Really streamlined production systems, for this kind of product, are capital intensive and may prove inflexible. In such a rapidly changing field a half-way house had to be found. From the rethinking process at OMT a new generation of much better MRI magnets and systems emerged, which were probably ahead of any others. The GE magnets, starting to come off their production line in 1987 after many delays, were clever, but, from published data, they looked over-engineered, and their production must have involved a major investment in expensive tooling. They could not easily be changed to a different design as demand changed, and a while later GE reverted to an arrangement closer to that at Oxford. The team examining OMT's design philosophy installed more computer aided design (CAD) systems, and better software programs for the analysis of magnetic field and stress. These systems demonstrated the inefficiencies and unnecessary costs of the older products. Following this focused work, the new Compact was relaunched late in 1987. When compared with the Unistat, it still had the same sized central bore for the patient handling system, the same range of fields, with even better uniformity; but it was otherwise much smaller, lighter, stronger, cryogenically more efficient, more flexible, and simpler and more convenient to use and to service. In fact, it had been transformed into a successful product.

The active-shield system was designed to contain the serious stray magnetic field round a magnet, which could extend some distance beyond the imaging suite. American regulations put the limit for the magnetic field exposure for the public at 5 gauss, so hospitals had to have ways of shielding the affected areas. Some hospitals sited their MRI scanners in new small units in the grounds, away from the main hospital buildings, but this was expensive, and inconvenient for both staff and patients. Others installed mesh screening in the walls, floor, and ceiling in old buildings. Some companies, including OMT, offered 'self-shielded' magnets, absorbing the stray magnetic field with specially designed outer shells made of many tons of iron, which made other difficulties for hospitals. The cost of overcoming this extra irritation could be hundreds of thousands of dollars.

The idea for an elegant solution to this stray-field problem came up in OMT back in 1983. The basic concept was simple. If a second set of windings was to be placed round the main magnet, with its field direction reversed, it should be able to cancel out much of the stray field while having only a minor effect on the central field in the magnet. It looked theoretically viable, but no one was certain it could be made to work.

Patents were taken out, but it later transpired that one related idea had been patented just before the Oxford Instruments' deposition. This was to cause subsequent arguments, with claims and counterclaims, and with considerable sums going into the pockets of American lawyers. As it was the only really novel idea since the early MRI magnets, it was an important matter for the Group. OMT scientists were later to argue that their patent specification could actually be made to work—a criterion for acceptance of a patent—while the other one, as described, would not be able to do the job. OMT had a good case, but the Company has never liked litigation; it is always expensive and it distracts those involved from their normal work. Patents are only as strong as the holder's willingness to defend them. They *do* ensure that a company will be able to go on using its own inventions without paying royalties to someone else, but they also tell competitors what one is doing. In this case a compromise was reached for free cross-licensing. Eventually, some fourteen or fifteen years later the other patent was disallowed in the USA, and the Oxford Instruments' patent was reinstated.

In June 1985 Peter Williams reported to the Board that the active-shield experiment had worked at 0.5 T, bringing the 5 gauss contour line much closer to the magnet (see Fig. 17.1). The magnetically affected area had been reduced to the size of an average imaging room. The news was passed, in confidence, to our larger customers, GE, Picker, and Siemens, and encouraged the latter to renew its agreement with OMT, which was due to run out in 1986.

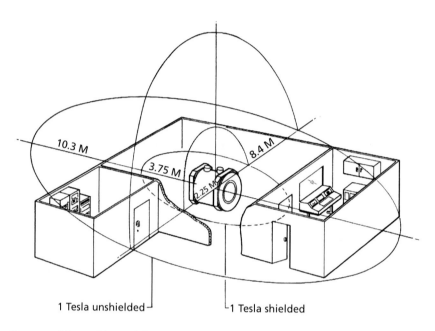

Fig. 17.1. The positions of the 5 gauss contour lines around unshielded and actively shielded MRI magnets

The active-shield magnet was always seen as a winner, and its development was given priority, but probably not enough planning, or resources of people and development funds. In August 1986 a 1.0 T prototype was tested successfully. The standard Unistat cryostats were not capable of accommodating the larger new magnets with their additional windings, and the new cryostat was not yet fully developed. But the deteriorating market precipitated their premature launch at the large RSNA conference and exhibition that November. A full range was announced: at 0.5 T, 1.0 T, and 1.5 T. No 1.5 T active-shield magnets had yet worked, and it was at the higher field strengths that the shielding mattered most.

These magnets were very complicated and there were difficulties at all field strengths. They were really made of two nesting magnets, and took a lot more superconductor than the older types, using up to 70 kilometres of wire each. They would have to cost more than the standard systems, but the hospitals greeted them with great enthusiasm—they were already suffering the high extra costs of shielding their sites by older methods. Soon the scanner manufacturers caught the excitement, although for them new magnets might involve the redesign of the complete scanners with further development costs. By the end of 1986 they had all ordered at least one system for evaluation. As the technical problems at OMT multiplied and delivery dates lengthened, they got more and more impatient and doubtful about the whole concept.

Pauline Hobday, who had been Marketing Manager at OMT since 1985, now became Product Manager for the active-shield systems. Through 1987 the development and production-engineering work continued. The magnets really were very difficult, in development and in manufacture. Large forces pulled the two nesting windings in opposite directions. Ian McDougall and David Hawksworth, who were in charge of different aspects of development at OMT, published a paper on these systems in 1986. They showed that, for a 1.0 T magnet, a 1-millimetre axial misalignment between the sections would cause an out-of-balance force of 1.5 tons. Then there was an extra 30 per cent stored energy in these magnets that needed to be dissipated safely in the event of a quench.

By September 1987 several of the lower-field systems were being shipped each month, and started to be used in both static and mobile scanners. But there were still to be new problems ahead, even in the middle of the range at 1.0 T, where prototypes had performed well. This was a serious blow and held up deliveries for months. There are so many variables in a complicated system like this that the cause of a problem may take some time to identify. In this case it was found to be due to the relative resistance ratios of similar types of wire from two suppliers, which led to a need for different impregnation regimes. The 1.5 T prototype magnet again failed under test—the stresses were too great at that field, and it needed a complete redesign. It was to be a further sixteen months before a satisfactory 1.5 T active-shield system was completed.

By March 1988 most aspects of the business at OMT were looking up, although the delays continued, and there was still too much reworking of magnets in the factory, and remedial action needed after delivery. Picker International had soon seen the virtues of the active-shield system, which was starting to give the new Picker product a competitive advantage. Both Philips and Siemens were ordering these new systems at the, now successful, lowest field strength. Siemens was also developing at least one new magnet system of its own, and it was reassessing its options for the future, after 1990, when the second contract with Oxford Instruments was due to end. The managers concerned wanted to expand their own magnet production facility at Erlangen in Bavaria, and make *all* their own magnets.

Back in 1987 the objective of the new drive for efficiency, assisted by the outside consultants, had been 'to establish the feasibility of increasing the throughput of the Eynsham facility to 200 units by April 1988'. By that date the gains already achieved had increased the capacity of the OMT factory considerably, and John Woodgate and his management team now considered that all likely production needs could be accommodated at Eynsham. If orders were to rise further, there would be plenty of room for expansion on the eleven-acre site. The downsizing of OST started to look inevitable, and the Group Board decided to stop magnet production there in October 1988. OST would then concentrate on the service side of magnets for the US market, and on the growing wire production business.

At Carteret Jim Worth was concerned about customer reactions, and whether some of his best people would choose to depart, possibly to competitors. But he and his executives managed the difficult transition so skilfully that the Company continued viable through the changes. Before any rumours could escape Jim talked to his two top executives, Phil Sanger and Ron Hynes, and got their backing; all the senior managers supported the decision and agreed to stay, at least for a while. Unhappily, the workforce had to be reduced by seventy-two; a few took early retirement and the rest redundancy. There was little reaction among US customers as the magnet 'staging-post', installation, and servicing were still to be the responsibility of OST.

The Siemens development team working on MRI had produced its own high-technology magnet system, the 'Helicon', which it launched in 1987. This offered field strengths of up to 2 T. A 'Linde' helium liquefier formed part of the system, eliminating the need for regular recharging with liquid helium. These magnets were shielded by a large iron shell, which weighed some 30 tons. After the launch of this product arguments arose over the numbers in our contract and the proportions of the various models to be supplied from Oxford. OMT was not very pleased to be allocated all the straightforward, low-field, and low-value systems, while Helicon became Siemens' main high-field, high-value product. But this may have been a blessing in disguise for OMT. I must have said many times how difficult it is to get complicated high-field superconducting magnets to perform to specification, at least until a lot of experience has been built up on a particular model. Our own higher-field active-shield magnets were a case in point. Siemens was secretive about its progress with Helicon, but we heard rumours of problems, especially in Japan, where it was particularly successful in selling its MRI systems.

In the autumn of 1988 John Woodgate made a presentation to Siemens seeking a new contract after 1990, when the current one was due to end. His objective was to persuade the Siemens managers that an agreement with OMT would serve their company better than expanding their own production of magnets. Scanners containing active-shield magnets were growing in popularity with hospitals; the 1.0 T problem had recently been solved, and the development of the new 1.5 T version was at last looking hopeful. John offered Siemens the new OR40 model at 0.5 T, which was also being offered to Picker. It was shorter and cheaper than the earlier version, and had a one-year 'hold time' for liquid helium. The specification and performance of this magnet set the Siemens team thinking hard. Apart from the suspected problems with Helicon and its liquefier, the philosophy of shielding with tons of iron was probably now out of date. The active-shield system was the product of the future (see Fig. 17.2). Siemens now suggested a joint development for a new active-shield design, to be made for it exclusively.

Fig. 17.2. Final inspection of an Active Shield magnet

Peter Williams and John Woodgate visited Germany again before the end of the year to discuss the various options. At this meeting Siemens unexpectedly proposed a joint MRI manufacturing facility in California. (The DM was then very high against the $US, and its MRI systems were in danger of becoming uncompetitive in the USA.) Peter and John rapidly countered this suggestion by proposing a joint venture for MRI magnet systems based at the OMT factory at Eynsham. The Siemens managers were surprised at our manufacturing costs, which were much lower than those in Bavaria, and they agreed to go through all the options with the management at Erlangen. For OMT there were no real alternatives in view; could the Company at last be finding the sound strategic alliance it had sought so long, to enable it to keep this business alive in a world of Goliaths?

Firm plans for a joint venture began to take shape. There were, of course, many necessary elements to be discussed, and opposing views to be reconciled before agreement could be reached. Peter Williams and John Woodgate conducted the negotiations for Oxford Instruments. The discussions covered the name of the new venture; how it was to be funded; how magnets were to be priced; the new Board, and what to do in a deadlocked situation; how to value the assets at each of the current operations that would be bought by the new joint venture; necessary changes in the patent agreement; the Pension Fund and other employee matters; and many other administrative details.

Siemens would be giving up magnet production at Erlangen, a move that would be strenuously opposed by the local management. Even if jobs could be found for all the employees affected in the vast Siemens Group, it would mean major disruptions to people's lives. Very few would move to Eynsham. Oxford Instruments agreed early on that Siemens needed to have 51 per cent of the equity and with it control of the new operation, but the name 'Oxford Magnet Technology' was to remain. As Sir Austin Pierce noted at the Board meeting in March 1989, the joint venture would give the Company and the Group a soft landing from its troubles, but it would *also* benefit Siemens. It would be able to develop complete new MRI systems in close collaboration with the acknowledged leaders in superconducting magnets, which had already delivered over 1,000 whole-body magnet systems. OMT would remain intact, with no further redundancies. There would be a big reduction in the large investment Oxford Instruments had built up in imaging magnets, and the Company would still be able to participate in a profits stream from MRI, the valuable diagnostic system that our original magnet developments had made possible.

The agreement with Siemens, satisfactory to both parties, was signed on 2 May 1989, subject only to shareholder and regulatory confirmation. OMT was to continue to supply its other existing customers as well as Siemens, which must have pre-empted any complaints from the Office of Fair Trading, the Monopolies and Mergers Commission, and their counterparts in Europe and the USA. A press release went round the City; approval and congratulations were almost universal. The new joint venture started trading on 1 September 1989. Dr Eric Reinhardt, General Manager of Siemens' Magnetic Resonance Division, was the first Chairman. John Woodgate, the MD, faced the difficult and complex job of integration. Only one person from Siemens became an executive of the new Company, Wolfgang Katzensteiner, who took the position of Finance Director. David Hawksworth—who later became the MD—Pauline Hobday, and Tony Groves made up the rest of the senior management team.

Unlike the experience in many joint ventures, relations between the partners have remained good and cooperative in this mutually advantageous alliance, not an easy matter between a giant like Siemens and a little group like Oxford Instruments. John Woodgate once told a company gathering that the joint venture was like going to bed with an elephant. The second Chairman, Dr Lothar Koob, replied that OMT was a mouse, quick to move and respond, running circles round the elephant. Siemens, he continued, appreciated its tiny partner; it had completed some 200 mergers or joint ventures, of which only a handful had been really successful. OMT was certainly one of them, and the only joint venture to make a profit in its first year (see Box 17.1).

Once the problems of amalgamation were over, OMT forged ahead. The basic development of the new generation of systems was complete; the 1.5 T active-shield magnet, the most difficult the Company had ever made, had finally proved itself. Development was then concentrated on an Active Shield Compact product, combining the advantages of the two systems. The great majority of all the MRI systems sold to hospitals in the 1990s used this intrinsic shielding method rather than iron shielding.

At the time of the joint-venture agreement OMT was already introducing some forward-looking ideas on manufacturing efficiency and quality. Siemens was then suffering from a high DM and strong competition from the USA and Japan, and had started to look more closely at costs in all its businesses. The cost of magnet production at Erlangen had been very high, and the OMT drive for better quality and lower costs fitted well with Siemens' objectives. The efficiency drive had already achieved improvements at OMT since the dismal days of 1987, and the Company had returned to profits by September 1988, but much still remained to be done. The yield was still too low, with too many magnets failing at the first test, and there still had to be too much reworking.

John Woodgate's top objective was 'to provide our customers with a competitive advantage through skilful design and total quality'. New proprietary Mark 2 versions of the active-shield magnets were soon being developed for Siemens and for Picker (the proposed joint venture with Philips had not gone ahead). The one for Siemens was the OR41, a 1 T product that was the making of OMT, and gave Siemens the competitive advantage John was aiming for. In all, some 1,250 OR41s were sold to Siemens before it eventually gave place to a newer system. Older products, including the over-complicated Helicon, soon started to be phased out. With the additional work from the closed-down Erlangen factory, more space was needed. In 1990 Siemens, having purchased the whole Eynsham site as part of the agreement, added another 70,000 square feet behind the existing 120,000-square-foot factory.

The story of the development of OMT over the following five years reads like a textbook case of how to move from being a clever, but rather disorganized, small company to achieve the status of a world-class manufacturing operation. The techniques used are now widely known, but their success was phenomenal. The message was 'continuous improvement'; the main objectives were to achieve 'lean' production methods, a reduction in 'time to market' for new products, and the quality accreditation ISO 9001. The management techniques used were not new even then, but at OMT they were applied with such speed, enthusiasm, and ubiquity that the transformation was breathtaking.

OMT attracts results

Oxford Magnet Technology's difficult ride through the 1980s could have ended in the company withdrawing from the world market that it had helped to establish. In 1982, OMT led the world in the development of large-bore magnet systems used for magnetic resonance imaging (MRI), a revolutionary new medical technique for whole-body scanning. In the mid-1980s, the company was the darling of the City, hailed as a shining example of how British innovation could lead the world. By 1987, however the euphoria had gone as OMT fell into loss in an unequal struggle to compete with multinational heavyweights such as GE and Toshiba.

At this stage, however, OMT departs from the experience of countless small entrepreneurial businesses in Britain which have given up the fight and allowed others to enjoy the fruits of their innovation. Today, with a healthy order book and growing sales, OMT has come back with a vengeance, the world's largest manufacturer of large-bore magnets, exporting 97 per cent of its output to North America and Europe.

Key to the recovery was a joint venture formed in 1989 between Siemens and Oxford Instruments Group, OMT's parent. According to OMT managing director John Woodgate: 'We had already had a long and fruitful association with Siemens, to whom we had previously licensed our technology. We had relationships with Siemens people at many different levels, and they knew our strengths and weaknesses'.

In discussions, the rewards of a joint venture seemed obvious to both parties. Siemens would benefit from OMT's long-established lead in MRI; OMT would have the essential backing of a world player which could afford to take a long-term view and assist in vitally needed investments in manufacturing capacity and R & D.

In practice, the benefits of the joint venture have been more than realised. Since the agreement was signed, nearly £6 million has been invested by the company, £0.75 million of this for the purchase of a new helium liquifier. A top priority has been to improve and upgrade manufacturing capacity. The site at Eynsham, near Oxford,

has been expanded by 70,000 square feet to accommodate the growth in business resulting from Siemens' decision to close its German operation and concentrate all manufacturing at OMT.

At the same time, the two partners have pooled their research and development resources, the better to maintain the company's position as technology leader in large-bore magnets. As Siemens' worldwide centre for research and development into MRI magnets, OMT has further strengthened what was always a very strong R&D capability.

As a result of the joint efforts, OMT is confident it can compete in an intensely competitive sector, where survival will depend on technological innovation and high-quality manufacturing, coupled with long-term commitment. Notes Woodgate: 'The last decade has seen a tremendous fallout of small entrepreneurial companies. Through the joint venture with Siemens, we are well placed to prosper in an industry which now consists of a handful of world players with the resources to take a long-term view'.

● **Marion Devine**

Apart from pruning overheads, the cost-cutting efforts included buying more components from outside, but for particular orders and not for stock. The Company adopted a 'one-vendor' policy for subcontractors instead of competitive tendering. In exchange for a commitment to long-term business, the suppliers had to deliver on time, to quality, and to price. They were considered part of the team and got to know the Company well; they were brought into strategic discussions, and given support over their own planning. Beside information on the next few weeks' deliveries and any near-term changes, the medium and long-term requirements would be reviewed. Between 1990 and 1995 the total number of suppliers went down by half, but, because of the Company's insistence on quality, the number of suppliers whose incoming deliveries needed *inspection* went down from 210 to about ten. This sort of relationship is now known as 'partnership sourcing'.

The next technique adopted was 'just-in-time' supplies to the work centre. Originally, most incoming material delivered to 'goods in' had to be identified, registered, put on a conveyor to the right stock location, and booked to inventory. When it was needed, it was withdrawn from the stores to a 'kit box' for a project, booked to a 'work-in-progress' order, and delivered to the work centre (see Fig. 17.3). In the new system most materials were ordered for a particular job, ready to size, or the right number for that job. If the order were for superconducting wire from OST, it would be in kits of the right types and lengths for each particular magnet; this saved time and a lot of the 5 per cent wire wastage at OMT. Some suppliers were asked to pack materials in their vans in the right order to match the layout of the factory, so they could be delivered directly to the location where they would be used. For all this extra work and care the

You mean we used to
do that!

Fig. 17.3. An OMT cartoon

suppliers were paid a premium, and most were very happy with the new system; not having to re-tender at regular intervals was a big bonus. Meanwhile, the stores stock was in the process of going down from £7.2 million to the £2 million it reached by 1994. With more buying-in of subassemblies, the number of parts also went down—by about 40 per cent. In stage two of stores reduction, between 1994 and 1995, the stores area was physically reduced from 11,000 square feet to 6,000.

The new total-quality techniques permeated through every section of the Company, with training and involvement for all the staff; the transformation was quite extraordinary. All this sounds rather Japanese, and there have been supercilious remarks on this sort of continuous improvement, but in OMT it certainly works, and that is what matters. At the date the joint venture was consummated, the turnover was down to about £35 million, and the profit was marginal. A year later, September 1990, the new financial year-end for OMT, the turnover was up to £51 million, and pre-tax profit to about £5 million. By 1995 annual sales were up to about £75 million, and the number of magnets shipped had risen to 350 a year, about one a day. A few statistics show the startling improvement. Each magnet now took an average of twenty-six days to build and ship instead of the pre-renaissance total of 120 days. In 1989/90 190 magnet systems were produced by 300 employees; in 1994/5, 360 systems were produced by a staff of 380—a productivity gain of 50 per cent. This was a period of strong pressures on prices, but, through these advances, costs were cut so far that the index of OMT's magnet prices went down by 53 per cent, while profits rose by 120 per cent—and this was for improved products. Some of the cost reduction came from the smaller inventory—work in progress, stores stock, and finished goods—which together went down from £16 million to £7.5 million. This reduced the number of staff needed, and the time and paperwork involved in the office, and it released a lot of space and working capital for other purposes.

The other big gain came in the time it took to develop a new product—the 'time to market'. A working party had the responsibility of recommending the product structure and how to reduce the 'lead time'. 'Concurrent engineering' was introduced, involving people from all areas of the business. Project teams started to use production resources and methods for their prototypes, which helped to remove any difficulties in a new product from a production point of view. With enthusiasm and cooperation, the time to market for one important product was reduced from the twenty-four months it would once have been to just eight months. The product range was rationalized. By 1995 only three types of standard product were in regular production including the new C magnet.

The C magnet was developed for Siemens for a new scanner called the Magnetom Open, which was launched early in 1994. This patient-friendly open system is based on a powerful electromagnet, a departure from the cylindrical superconducting magnets of recent years. The system is especially useful for children, people who suffer from claustrophobia, and very sick patients needing life-support paraphernalia. Patients can be moved about between the two large flat poles of the magnet, which are connected by a sturdy iron yoke (see Fig. 17.4). These new resistive magnets are far more powerful than the old cylindrical resistive systems made in the 1970s. The Magnetom Open scanner was well received by the market; it enabled hospitals to buy a good scanner for about half the price of a low-field superconducting MRI system. The C magnet soon accounted for some 20 per cent of the business of OMT by volume (10 per cent by value), but it never spoiled the market for the higher-specification superconducting magnet systems, which the Company was continuing to make in quantity at the end of the century. Among these magnets there were some special systems, and some very powerful magnets for spectroscopy, which could achieve fields of up to 6 T (see Plate 5).

Fig. 17.4. A 'C' magnet being assembled at OMT

In 1995 the National Manufacturing Council of the Confederation of British Industry (CBI) arranged a visit to OMT as part of its Competitiveness Forum programme. Its 'Best Practice Focus' leaflet on the Company stated, 'Oxford Magnet Technology is a world-leading example of applying manufacturing technologies to a very sophisticated product'. OMT was formally recognized by the CBI as one of the best twenty-five manufacturing companies in the UK. It had indeed come a very long way from its hectic beginnings.

There is a postscript to this story. Nothing is for ever. In 1997 and 1998, ten years after the cliff, OMT was again plunged into crisis, but this time round it recovered at the eleventh hour. The potential disaster was caused by a combination of factors: a surge of orders for a brilliant new product was followed by unexpected technical problems that emerged late in the day. OMT got over this threatening situation through exceptional management and the use of carefully structured analysis of the problems and of their remedies, with company-wide agreement and participation in the necessary changes.

Since 1994 OMT had been working on a family of new products. These were to be shorter and lighter superconducting magnet systems in 1 T and 1.5 T versions, and with no diminution of the high technical specifications of the earlier products. A few familiar technical problems delayed the launch date; but the really serious headache was caused by the system's unforeseen sensitivity to outside vibrations—a problem not encountered in previous models, and quite unexpected. In fact it was not discovered until the prototype magnets had been delivered to the two main customers and installed in their new MRI systems. This novel fault had to be assessed and measured before it could be cured by mechanical solutions, and all this took some time. To add to OMT's woes it then encountered failures in the epoxy impregnation process. By agreement with OMT, Siemens and Picker International had launched their new products, already later than intended, in the summer of 1997. Deliveries were supposed to start in January 1998, but the Company had to correct the early magnets for the vibration problem at a late stage, as they were going through production. This was a slow process, and when the epoxy problem struck OMT found itself way behind its schedules.

Three managers wrote up the whole episode six months later, in an application for the Siemens Engineering Excellence Award, which they won. According to their report, 'Oxford Magnet Technology was in crisis' (Reynolds et al. 1998: 1). The new products had been launched to a very receptive market, and the budget for deliveries for the year to September 1998 had gone up from 412 to 483 magnets. They were so popular that this total was later to increase to 515 magnets. With the slippage in the early months of the year, OMT would need to make seventeen magnets a week to meet this order book—a rate of 884 magnets a year—more than double anything achieved to date.

The MD, David Hawksworth, asked Bob Graham, the Production Director, to work out and organize a way for the Company to deliver these 483 magnets within the OMT financial year, which ends in September. The endeavour became known as 'The 483 Order Fulfilment Project'. Production Engineers Kevin Reynolds, Steve Quick, Les Day (the three authors of the report), and Brian Harris joined Bob in the Project Team. They brought into a steering committee other managers from R & D, Purchasing, Quality, Personnel Facilities, and Customer Liaison. Core teams and sub teams worked to analyse all activities and identify responsibilities and possible timescales. The technique known as 'goal-directed project management' (GDPM) was to be used to monitor progress. All employees were brought into the picture through a letter from the MD making the urgency of the situation clear and their enthusiasm was kept up through participation in the analysis of their particular processes. They all discussed how they could double their team's rate of production, and the management kept the action going by providing additional machinery, more staff, and new shift patterns.

The solution came from a series of initiatives. It would have been no good increasing the capacity of some of the ten work centres if other vital processes still held up completion of magnets in bottlenecks. Priority was given to processes where the analysis showed a low maximum capacity. In fact *all* the manufacturing work centres were able to increase their 'overall equipment effectiveness' (OEE), which can be measured as a percentage. For world-class OEE a company should aim for about 85 per cent. By June 1998 the OEE for the whole factory had increased from 41 per cent at the time of the severe problems six months earlier to 69 per cent. By the end of another year they expected to achieve 80 per cent. After a first quarter when OMT shipped only seventy-one superconducting magnets, it succeeded in shipping 515 in the full year. OMT took on 150 new employees; some of them were only on one-year contracts, as this maximum rate of production might not be needed in the future. By June 1998 the Company was using 1,500 kilograms of superconducting wire a day, costing about $US75,000. The number of magnets coming out of the factory each week had increased from five in December, to sixteen, and the hours needed to process a magnet had gone down from 1,100 to 627.

This is a tremendous success story, and rescued not only OMT, but also its main customers, who would have been severely embarrassed if their newly launched products had failed to be delivered. These magnets continue to be very successful because of their low weight, their small stray field, which makes for easy siting, and the patient-friendly contoured profile, which reduces the feeling of claustrophobia.

When the whole Group can launch equally popular products and match this increase in productivity, the bottom-line profits will be transformed. 'Bugs' and manufacturing problems often arise at the introduction of a new product, the time when it may be most in demand, and most profitable. Nowadays there are 'tool kits' of leading practice techniques available for analysing and tackling these problems, and for the difficult task of managing change in general. Cultures, work practices, and attitudes are always difficult to change, but every organization must move ahead if it is not to fall back. There will always be new problems; no solution fits every case, or lasts for ever. This is of the nature of innovation.

18 Link Scientific and a New Japanese Initiative

A major acquisition on the threshold of the 1990s and a successful new sales company in Japan

Union is strength.
Opportunity seldom knocks twice.

(Old proverbs)

The Company was still on the lookout for a good acquisition to widen its technology base and give access to new markets. From the joint venture with Siemens, announced in May 1989, the Oxford Instruments Group was to receive a net inflow of £18.5 million in cash. The 1989 results, to be announced that June, would show the Company already had 'net liquid funds' of £23 million. By September, when the new joint venture at OMT was to start trading, the total cash funds would amount to over £40 million. The Board was already anxious about hostile bids. Leading-company shares had recovered quite fast after the 1987 stock-market crash, and takeovers were back in fashion. A 'cash hillock' of £40 million, even if nowhere near a 'cash mountain', could be tempting to a predator. The last thing anyone wanted was for the Company to be acquired in a hostile bid for the sake of its cash, and then, possibly, be broken up and sold off piecemeal.

The obvious solution was the rapid use of this money, generated by hard work and risk-taking in MRI, to buy a substantial, suitable, and willing company. This was easier said than done. Through the 1980s Oxford Instruments had been trying to make a major acquisition and had been unsuccessful. The two companies of any size that had been acquired, bought in 1986 and 1987, were responding differently. PT was contributing to profits, but at a disappointingly low level, and had started to suffer from technical and management problems; the much less expensive Sonicaid was performing unexpectedly well and, in 1989, had contributed nearly £1 million to the profits of the Medical

Division. With strict criteria for acquisitions, the Board could see no obvious and willing high-technology candidates in the UK. Making a hostile bid would be alien to the Company's culture. An overseas acquisition in a different technology would be difficult to manage with the existing resources, although quite a few had been considered through the 1980s.

The thought of the cash hillock spurred the Board to new efforts. Peter Williams said, 'We should look for a new area of technology; the Company needs to take a risk to achieve a significant repositioning.' It was felt to be their duty to the shareholders to use the money from the reduction of investment in MRI to establish a new 'leg' to the business. Just at the right moment, a very suitable company unexpectedly became available. In June, Carlton Communications made a bid to buy UEI, the other player in the abortive merger negotiations between 1983 and 1987. Most UEI subsidiaries were by then involved in equipment for advanced television or audio recording; but the Link Scientific Group, with products akin to those of OAI, would not be of interest to Carlton. If successful in its bid, Carlton would probably sell this Company, and one or two other UEI subsidiaries, to raise as much cash as possible. Link was a profitable and substantial science-based group. It would be expensive, but it was the opportunity Oxford Instruments had been waiting for.

The Link Scientific Group, managed by Rob Sareen, had recently issued a brochure describing its range of activities based on 'a common technology in the area of leading edge products for radiation detection and analysis'. Like some Oxford Instruments' businesses, Link was research based, and aimed to be the technical leader in its field. The products were designed for quantitative and qualitative analysis of the elements present in a sample. As the brochure said, 'Scientists, technicians, product development engineers, quality controllers, medical researchers, archaeologists, geologists and forensic analysts all want to know what their particular samples are made of. They increasingly need more accurate analysis to identify the presence of very small quantities of individual physical elements.'

Link Analytical was the principal company in the Group. At its rather dilapidated rented factory in High Wycombe, under an hour's drive from Oxford, an impressive R & D department worked on new ideas for advanced components and systems, in order to extend the analytical range and accuracy of products for this wide market. The sophisticated multichannel instruments made there were based on the principle of X-ray fluorescence (XRF) analysis. This is the principle used in the much simpler Lab-X products then being made by OAI for quality control in industry. The XRF technology was the main source of synergy between the groups, and the products were complementary, but there were other matters in common. Link products needed liquid-nitrogen cryostats for cooling X-ray detectors, which might be improved from the fund of experience at Oxford. Link's deep knowledge of X-ray technology could be of benefit to the team developing the synchrotron. And Link had markets in common with the Hexland Group at Eynsham, which made accessories for electron microscopes.

The most important and promising products made by Link Analytical, accounting for 70–80 per cent of turnover, were XRF systems for microanalysis, for use with electron microscopes (see Box 18.1 and Fig. 18.1).

Box 18.1. Microanalysis as Explained in the 1989 Link Brochure

'Microanalysis is a technique where selected pinpoint locations on the surface of a sample in an electron microscope can be very accurately analysed. The electron beam in the microscope is similar to that in a domestic television tube. However, when it strikes the sample, not only can the image be detected for display on the microscope's screen, but it produces characteristic X-rays from the atoms in the sample. The X-rays from each element have their own specific wavelength. Link Analytical's detector in the vacuum chamber of the microscope senses these and generates a signal which, after amplification, is analysed by a special computer and the results displayed on the screen.' (See Fig. 18.1.)

The other products were stand-alone general XRF systems used for rapid simultaneous analyses of many elements. These sensitive instruments cost between £20,000 and £60,000, compared with the simpler Lab-X products made by OAI, which were usually dedicated to one element for a particular task, and cost from £10,000 to £15,000.

Link's unique and patented detector technology was sensitive enough for its newest systems to register the signals from lighter elements that emit low energy X-rays. This enabled the instruments to detect important contaminants such as sodium or aluminium in cement production; river water could be tested for minute traces of pollutants, and the source could often be identified; and jet engine wear could be measured by the amount of bearing material in the lubrication oil. These microanalysis systems were being used increasingly in forensic laboratories. Materials from different factories almost always contain their own unique, and very specific, percentages of each main element, with telltale traces of others. Analysis of gunshot or bomb fragments, flakes of paint, bits of headlamp glass, or fibres from a garment could often lead to the detection and conviction of criminals.

The market for microanalysis systems was then a buoyant £100 million a year, of which Link's share was about 30 per cent and growing. Recently the Group had restructured its marketing and sales organization in the USA, and the moves were bearing fruit in more orders. Oxford Instruments and their advisers set about the examination of the Company, its products, financial records, and prospects. The study was quite difficult, as Link had lately undergone extensive corporate changes as well as launching new products. As recently as 1988 it had

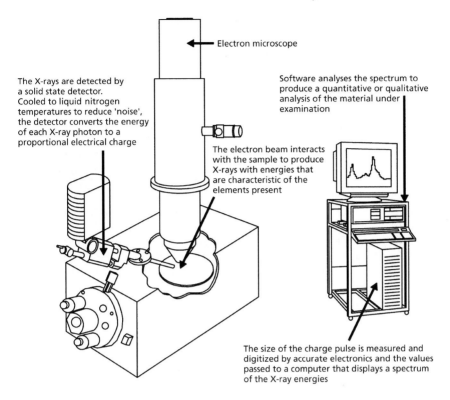

Electron microscope

The X-rays are detected by a solid state detector. Cooled to liquid nitrogen temperatures to reduce 'noise', the detector converts the energy of each X-ray photon to a proportional electrical charge

Software analyses the spectrum to produce a quantitative or qualitative analysis of the material under examination

The electron beam interacts with the sample to produce X-rays with energies that are characteristic of the elements present

The size of the charge pulse is measured and digitized by accurate electronics and the values passed to a computer that displays a spectrum of the X-ray energies

Fig. 18.1. Diagram of a microanalysis system

taken over—for $20 million—a nuclear instrumentation company in Oak Ridge, Tennessee. This was the home of a large US National Radiation Research Laboratory, where the founders of the nuclear instruments company had originally worked. This company had itself resulted from the 1987 amalgamation of two firms making related radiation products. The Nucleus and Tennelec had been jointly owned for the past ten years, but had been run as separate units until this integration, which was not physically complete in 1989. In 1987 Link had acquired X-Tech, a very small company in California that made X-ray tubes and power supplies, in particular special tubes for the stand-alone XRF systems made in High Wycombe.

Tennelec/Nucleus made many kinds of detectors, components and measuring instruments for the recognition and analysis of the various sorts of radiation. These were used for research and teaching, in hospitals, and for pollution monitoring. The US Government was then starting to clean up areas of radioactive pollution left over from research and testing during the Second World War and the cold war, and there was a good market for these instruments. One speciality, shared by only two other companies in the world, was expertise in growing the high purity germanium crystals needed for making sensitive germanium detectors, which Link needed. This Group's sales had added £8 million to Link's turnover, and £1.5 million to its profits in the year to January 1989. The American companies looked a valuable part of the Link Group.

In July 1989 the Board agreed to make an offer for Link, subject to more detailed 'due diligence' research. They gave quite a high price limit to their negotiators, as the opportunity seemed unique, and there were other bidders. Oxford Instruments was the top choice of the Link management. After some bargaining with Carlton, a price of £47.5 million was finally agreed, with an extra £10 million to be paid later, in new Oxford Instruments' shares, but only in proportion as the profits of the Link activities should exceed £4.75 million over the next twelve months.

In the press release announcing the acquisition, Sir Austin Pearce expressed the strategic significance of the move: 'The acquisition of Link represents an exciting opportunity for the Oxford Instruments Group and, following the acquisition, Oxford's analytical instruments business will be the Group's second largest activity. This will be our fifth major business and represents a significant step towards the achievement of the Group's strategic objective of becoming a major international group in advanced instrumentation.'

The Link acquisition from Carlton was announced on 11 September and received favourable press comments. On 1 October, after shareholder ratification, Link started to trade as part of the Oxford Instruments Group. This was just one month after completion of the joint venture with Siemens. Although Link's profits had gone up to £4 million in 1989, a substantial part of this was due to its recent acquisitions. The profits of Link Analytical itself had actually gone down a little, but there was a reasonable explanation for this, and the local management was confident in its forecast of nearly £5 million for the full year to 1990. As the press release at the time of the announcement explained, 'Recent results of Link (excluding Nucleus) have been affected by the costs of establishing a direct sales force in the USA and by the transition between its established products and its new range. Since 31st January 1989, a number of new products have been successfully introduced and the current order levels have increased significantly.'

Link's profits held up for the next six months, while most of its shipments were in the older established products; but there were problems ahead. The recently launched Link Inc. in California had high overheads, and the US market was becoming more competitive. The American sales team lowered their prices to retain an adequate market share and the operation soon became unprofitable. To add to this reverse, the new products now being ordered had the sort of teething problems all too familiar to Oxford Instruments; they had lower profit margins than the older experience-refined products, and would not recover respectable margins for at least two years. Although Link's optimistic forecast had been considerably discounted, the full extent of these problems could not be foreseen. There was no extra payment to Carlton at the end of the first year.

In 1990 a new Analytical Instruments Division was designated, a new 'leg' to the Group, to include all the Link companies and OAI. The stand-alone XRF products were moved from High Wycombe to join the simpler Lab-X and QX ranges at OAI in Abingdon, and that company was renamed the Oxford Instruments Industrial Analysis Group (IAG). Paul Winson had been managing OAI since Tim Cook had left to join another company, and he now moved to the head offices in Eynsham to set up a new Group-wide marketing service. Ray Bailey from Link came to Abingdon to manage the new IAG. The High Wycombe operation was renamed the Oxford Instruments Microanalysis Group (MAG), and was to concentrate on its world-leading products, sold either on an OEM basis to electron microscope manufacturers or as 'retrofits'. Ron Jones became its MD, while Rob Sareen headed the whole new Division.

The integration of new member companies into a Group takes some time. There are inevitable differences of culture and reporting systems, and the staff at High Wycombe took longer than expected to absorb the Oxford Instruments' way of operating. Although Link had been part of a public group, its financial control and information systems were not so well developed as those at Oxford. Its US companies, in particular, had suffered a lot of changes in quick succession, including the recent death of a top manager, and they were very unsettled. But there had to be yet another reorganization and rationalization of sales companies and agents, which led to a reduced order intake for a while. The Oak Ridge Company, when fully amalgamated onto one site, became the Oxford Instruments Nuclear Measurements Group (NMG). As the Link companies became more integrated, but before the benefits of this had led through into higher profits, the world started sliding towards the worst recession since the 1920s and 1930s.

Before long the climate for acquisitions cooled, and many company boards, who had bought expensively in the 1980s, looked with chagrin on the poor performances of their new subsidiaries. For Oxford Instruments, the Link story was to have a happy outcome, but not until after years of trials and changes and low, if not negative, profits. But as the 1990s began these troubles were still in the future. OMT, now the joint venture with Siemens, had started to prosper again, and it looked as if the Group was sailing into calmer waters.

The Company was now looking to improve its market coverage, and Japan was a target. At the end of the 1980s the Japanese economy was some four times larger than that of the UK, and second only to that of the USA. With its forward-looking manufacturing industry, perfecting and using the latest technology, Japan was a world leader in many industries. These included microchip memory (DRAM) production and electron microscopy, both of great relevance to Oxford Instruments. In Japan, R & D and investment for future market share

are commonly considered more important than profit. In 1989 Japan spent a staggering $US90 billion on R & D. Several Japanese companies regularly register more patents in the USA than any American company, and, in Japan itself, 1990 saw 357,000 patent applications filed—over twice the number in the USA.

This thrusting technological development activity should have been a key market for a science-based group like Oxford Instruments, but by 1990 we were doing only some £7 million of business in Japan. Our largest market there, which we had been cultivating since the 1960s, was in specialist one-off systems for research, bought mainly by the large well-funded Government laboratories, but also by university and industrial research groups. The analytical instrument companies and PT also had some business there, but Group-wide sales to Japan accounted for only about 7 per cent of our output. Paul Brankin went to Japan to find out why, and how the Company could best rectify the deficiency.

Back in 1969, Oxford Instruments, then just the original magnet-company, had appointed Columbia Import and Export, headed by Mr Tobita, as our first agent there. It had served the Company well for many years, but recently our share of the market for magnets and cryogenic equipment had been slipping. As the Group had grown and had spun off other units in different markets, new agents or distributors had been appointed, and other agents had been inherited with recent acquisitions. By 1990 there were ten separate outlets in Japan, some achieving very little. Companies in the Group also had direct OEM sales agreements—with JEOL for NMR magnets and with Hitachi for microanalysis attachments for electron microscopes. These accounted for almost half the £7 million of sales there.

The Group's other foray into Japan had been in the Furukawa Oxford Technology (FOT) joint venture making imaging magnets for the Japanese market. Joint ventures rarely last very long. FOT was successful in manufacture but less so in cost minimization or in marketing its MRI magnets. In 1990 Oxford Instruments reduced its shareholding from 49 per cent to just under 20 per cent, and was eventually to pull out altogether. Overall the Company made a profit on the operation because of the original technology transfer fee and the royalties on all magnets sold, but it was not a very substantial payback.

As soon as Paul Brankin arrived in Japan he engaged the help of Tony Ford, the only Oxford Instruments employee there. Tony had been working in the Furukawa premises fifty miles south-east of Tokyo, and had recently moved to a tiny office in the City, with one secretary, to fill a liaison role for the Group's businesses. With advice from the British Embassy, and from a Japanese consultant, Paul set about assessing the current complicated situation and deciding on the best way forward for the various operations.

The Company's market research had predicted a good market for compact synchrotrons in the medium term, and early orders were expected to come from

Japanese semiconductor companies. A stronger representation was needed, and none of the existing outlets seemed suitable. Paul visited all the current agents and distributors, many customers, contacts in the semiconductor industry, and the Japanese operations of a few European and American companies. Changing an established pattern of representation is not easy and can be costly, but the arrangements in place were not working well. There was confusion of the Group's image and its impact was reduced by this fragmentation.

The Company had one trump card in Japan. The name 'Oxford', associated with a prestigious university, is well recognized there, where prestige matters. Apart from its academic fame, everyone in Japan seemed to know that their Crown Prince, Prince Naruhito, and his younger brother Prince Ainomiya, had both studied at Oxford University, and Oxford is the Company's home city on the River Thames. The Crown Prince had made a scholarly study of the eighteenth-century commercial traffic on this famous river, and we are now proud possessors of a signed copy of the published work, which he gave us.

It was important to use the Oxford name more positively to show the strength of the Group, and to win recognition for its wide range of products. After an exhaustive study Paul counselled the Board to launch a Japanese subsidiary company, a Kabushiki Kaisha, or KK, equivalent to a limited liability company. The executive committee was enthusiastic, and the Group directors were prepared to take a long-term view of what would be a major investment—offices and staff do not come cheap in Japan. It was a high-risk/high-reward strategy. Paul was appointed the Director of Operations, and would manage the new business, with the help of Tony Ford and temporarily seconded engineers from the UK, until he succeeded in appointing a high-flying Japanese President.

Oxford Instruments KK (OIKK) was registered in June 1991. In July the tiny staff arranged a high-profile launch, supported by Peter Williams, who speaks some Japanese, Martin, and others from Oxford. They were joined by senior people from the British Embassy and by existing agents who were being brought into future plans. Representatives of many companies, universities, and government departments had been invited. For such an event a prestigious gathering and an expensive party are *de rigueur*, it reassures Japanese customers and friends and warms the atmosphere.

The new sales company was to start slowly, taking on only the marketing of the synchrotron and the business of PT, whose previous distribution arrangements were about to come to an end. The other divisions would be phased in over several years as the Company could negotiate disengagements from the various existing agreements. New offices were found in Hatchobori, near Tokyo Station, and the Company started trading on 1 October 1991, with an expatriate engineer experienced in PT products. Paul travelled to and fro between the UK and Japan, coordinating the new enterprise. It was an unusual commitment for a relatively small Western company.

Paul learned a lot about doing business in Japan, and was later seconded to the DTI for two years, to help other companies with their Japanese plans. In

1993 he wrote an article for the publication *Engineering Management Journal* about establishing a sales operation there:

For an engineer Japan is a wonderful country in which to work—a country where science is widely understood and where technology is recognised as a positive benefit. For a UK business man Tokyo is a refreshing city in which to live, a city where there is real enthusiasm for building new enterprises and a confidence in the future, even in times of recession.   (Brankin 1993: 185)

He had also learned a lot about selling in Japan:

The first step is to throw away the book of conventional sales wisdom and think about Japanese customers. Although attitudes are changing, the first things on their minds are unlikely to be price and detailed specification. More likely they will be concerned with the supplier's 'name' and 'image', particularly its reputation for quality and customer service and whether it has products to offer which are not available from a Japanese company.   (Brankin 1993: 187)

After a long search for a Japanese President, Paul found Jiro Kitaura, who had worked for fifteen years in marketing and strategic planning at Mitsubishi Materials, and later for an American advanced materials company in Japan. In January 1992 he flew to meet the Board, but was held up somewhere on the way by bad weather at Heathrow. Although he missed the Board meeting, he managed to tour five factories, meet twenty managers, and dine with Peter and Martin, all within thirty-six hours; and to learn enough about the Group to accept with enthusiasm the role of President. A major worry had been whether a really good manager would be willing to work for a small Western company in that land of lifetime employment where 'big is beautiful'. An experienced and respected President would be the key to recruiting good sales engineers.

With the arrival of Jiro Kitaura as President in April 1992 the Company was fully established, and soon grew to a headcount of nine. In spite of the deep recession, which was by then affecting Japan, sales of PT equipment began to grow. In the UK the Group held seminars on Japan Awareness, and video training courses on the culture and the language. Soon there were many volunteers offering to go for a period to help establish sales or servicing operations for their products. An expatriate engineer was provided to help Columbia, with whom there was an amicable agreement to take over the business in two stages, by region, when some of its staff would transfer with the products. The first region was the north and east of Japan, and in May 1993 the Company acquired several sales engineers from Columbia. These included Nishio San (Mr Nishio), who had been selling Oxford Instruments' equipment for twenty-four years. With the arrival of temporary staff from England, and the ex-Columbia engineers, OIKK needed more space, and moved to larger offices in Kudanshita, another central suburb of Tokyo. A small new office was opened in Tsukuba, the 'Science City' north of Tokyo, and soon there was a Technology Centre, for demonstrations and servicing operations, at Wakaba in Tokyo's western suburbs.

In the 1992/3 financial year the Japanese Company, selling only PT equipment, made sales of £400,000. By the time of the half-year accounts, September 1993, it had become 'a material contributor to the overall improvement in new business, partly as a result of the Japanese Government's supplementary budgets'. These budgets were an attempt to improve the skewed trade between Japan and the West by importing capital equipment, some of it for use in university or government research laboratories. The book *Nippon: The Land and its People*, written in Japanese and English side by side, was originally produced by Nippon Steel in order to explain foreign views on Japan to its employees, because its company was becoming more internationalized. It was found to be even more useful in interpreting the Japanese culture to the growing number of foreigners working or doing business there. The 1993 edition covered the many changes since the mid-1980s, and, among other new matters, it explained why some of the supplementary budget money was spent on research equipment:

[Japan's] scientific and technological contribution to the international community has been criticised as not being proportional to the country's economic, scientific and technological capabilities. Specifically, Japan is accused of focusing too much effort on research linked to the development of new products and not enough on basic research, which is the kind that produces results beneficial to all mankind.   (Nippon Steel 1993: 87)

Many of Oxford Instruments' products are sold for this basic research.

For the full year to March 1994 OIKK sold £5.7 million of equipment, and, with the sales from the remaining distributors and the direct OEM sales, Japan overtook a sluggish Germany as our second largest export market after the USA. In the following year OIKK's sales rose to £16.5 million as the Company took over more and more of the Group's products. The staff had by then grown to thirty-five, with only three expatriate members. OIKK grew from there to account for 20 per cent of the Group's business.

As the 1990s wore on, the problems in the Japanese economy brought a new exchange-rate situation. The yen weakened and the pound strengthened. OIKK found its markets contracting. Sales in the products of several of the Group's businesses shrank, although the supplementary budgets still helped to keep up the orders for the research equipment made by RI. In 1997 both RI and NMR Instruments moved to new factories, which caused some disruption to production. Added to this there were technical difficulties that held up the completion of many projects. In Japan on-time delivery is very important, and the Company started to lose some of its good reputation. Changes were introduced late in 1998, and a more extensive reorganization started during 1999. By early 2000 things were starting to improve, but it was still expected to take a year or two for the Company's superconductivity-based businesses to recover their good standing.

Oxford Instruments has a very long history of working with physicists in Japan, especially those doing research in solid-state physics. In 1998 Jiro

Kitaura of OIKK, with the help of Professor Miura of Tokyo University, a very old friend of the Company, promoted a Millennium Science Forum. The British Embassy in Tokyo supported this as a method of increasing scientific exchanges between Japan and Britain.

In March 1999, at its inaugural meeting, HRH the Princess Royal presented an Honorary Millennium Science Forum Award to Professor Yasuoka for his exceptional work over a long career. One of the functions of the Forum is to select an outstanding young Japanese research physicist in this field as winner of the newly established annual Sir Martin Wood Prize. This includes a cash prize and a visit to the UK to give a series of lectures at British universities. On 17 November 1999, at the second Forum, in the presence of the Japanese Minister of Education, Science, and Technology, the British Ambassador, other VIPs, and senior OIKK staff, Martin awarded the first prize to Mr Yasunobu Nakamura of the Fundamental Research Laboratory of NEC in Tsukuba. The Forum is also supported by the British Council, the *Yomiuri Shimbun* (daily paper), and by many eminent Japanese scientists. We hope the Forum, and prize for younger physicists, will cement Oxford Instruments' relationships in Japan.

19 Helios—a Product Ahead of its Time?

How the Oxford team developed a synchrotron for IBM, why the market lagged behind, and further work for high-energy physics

Give me the splendid silent sun with all his beams full dazzling.

(Walt Whitman (1865))

In research the horizon recedes as we advance.

(Mark Pattison, *Isaac Casaubon* (1875))

Helios stands for High Energy Lithography Illuminated by Oxford Synchrotron. After the order for Helios 1 was announced in June 1987 (see Chapter 16), David Fishlock wrote an article in the *Financial Times* (10 June 1987) headlined 'IBM Plugs into British Brilliance', a play on words for this very bright light source. He went on: 'The new kind of superconducting lamp that IBM has ordered from Oxford Instruments for trials in a US microchip factory . . . will generate a light a million times more intense than the light sources used to make microchips today.' Some lamp! Peter Williams used a different image when he described a synchrotron as 'an orchestra of different magnets and energy systems performing in harmony under an electronic conductor to get the effect the physicist wants'.

The more I read about the development of XRL at IBM between 1980 and 1993, the more surprised I am that the order for the key X-ray source was given to a smallish company in England. Although central to this large and expensive project, several other important systems had to be developed in parallel before the synchrotron could be used for chip production. IBM was developing some of the systems itself, and others were contracted to companies experienced in making equipment for optical lithography (see Fig. 19.1).

One IBM team was developing masks for XRL, a particularly difficult task. In optical lithography the masks used are like photographic negatives; each mask contains the pattern to be printed on one of the many layers in a chip.

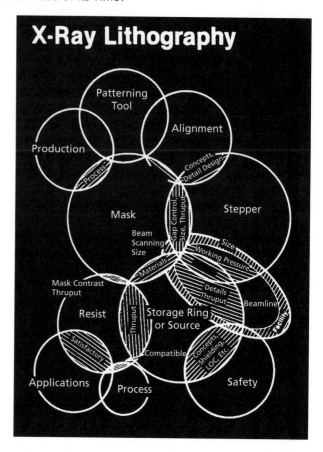

Fig. 19.1. The many systems needing development for X-ray lithography

Source: *IBM Journal of Research and Development*, 37/3 (1993).

They are made quite large, and can be focused down to the necessary small size on the chip by glass or quartz lenses. X-rays cannot be focused by a lens, so the mask pattern has to be exactly the same size as the pattern to be drawn on the chip, which is extremely small. The mask must almost touch the wafer, and the X-rays then shine straight through the clear parts of the pattern on the mask to change the chemical structure of the 'resist' material that coats the wafer on which the arrays of chips are constructed. The unchanged areas of the resist are washed away to reveal the patterns on the bare chips below. Other processes follow, such as chemical or gas treatment, plasma coating or etching, interspersed with lithography stages. In the fabrication of complex microcircuits on the wafer, up to 100 steps may be necessary in order to build up all the layers and connections required.

For this complicated process unimaginably accurate handling and alignment systems were needed. Known as 'steppers', these control the relative positions of mask and wafer. The pattern for each layer has to be inscribed at exactly the right place, or the tiny components on the chip will not be aligned or linked up perfectly. Then special 'beam lines' had to be developed for the task of accommodating and controlling the X-ray beam between the synchrotron and the wafer exposure point. New resist materials were needed for X-rays, and there were many other lesser, but necessary tasks that had to be completed before a chip production line could be contemplated.

The synchrotron, the essential heart of the system, would cost less than 5 per cent of the whole XRL programme at IBM's new complex, ALF, in New York State. Oxford Instruments won the contract for this difficult and innovative prototype because of its expert project leader, Martin Wilson, its unique experience in superconducting magnets, and the advice available from the world-class synchrotron team at Daresbury. No other group, either large or small, had yet got far with a superconducting synchrotron, and IBM, one of the largest companies in the world, had the wisdom to recognize a winning combination in a small independent company.

For such a complicated system, I am told, the development of the Oxford synchrotron went relatively smoothly. Watching from the sidelines, there seemed to be plenty of problems, but that is to be expected in a development of this complexity (see Fig. 19.2). Martin Wilson insisted on getting each stage right rather

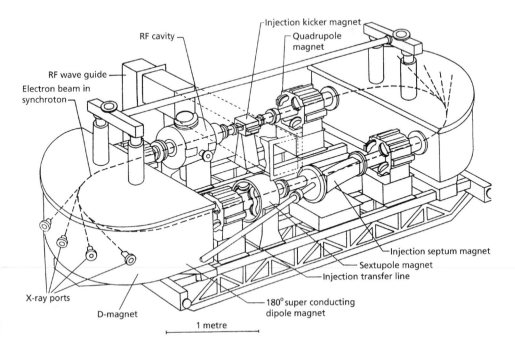

Fig. 19.2. Diagram of Helios I 1987

than rushing to finish by the original delivery date, which he had always felt left little room for unexpected obstacles.

By September 1987 'Stage A', the detailed design, was complete and accepted by IBM. It then started to press the Company for more investment in order to complete the project rapidly, but the Board could not justify more engineers and duplicated facilities until further orders came. In any case, numbers are not everything; a really good and experienced small team may do better than a larger group, where the managers may spend too much time getting agreement on procedures, and maintaining the quality of the work. Oxford Instruments had accepted a low price for the prototype because of its strong expectation of future orders from IBM—it originally told us up to eight more would be needed for its production facilities.

The Synchrotron Group moved to the factory on Osney Mead that had originally been built for OMT. For testing Helios and the linear accelerator, which was to inject the electrons into the ring, the team would need a radiation-proof vault. A quotation for special radiation-absorbing bricks seemed high, and they found that a wall of ordinary heavy building blocks, twice as thick, would serve the same purpose and would cost less than half as much. So, making a typical small-company saving, they bought 2,000 tons of heavy concrete blocks and built an enclosure in the factory. This would shield the rest of the building and the personnel from synchrotron radiation when testing began.

Martin Wilson's next benchmark would be the successful completion and testing of one of the D-shaped dipole bending magnets. These key magnets were the most difficult part of the whole development, and their success would give IBM confidence in the project, perhaps enough confidence to order a second system. These strange and complex-shaped coils had to be wound slowly with very great care. Ingenious solutions were needed for the challenge of holding them in place against the strong electromagnetic forces the magnets would generate. Inevitable difficulties had to be overcome, and the timetable started to slip. While the magnet team was working to get the intricate coils constructed to very fine tolerances, other engineers were making or purchasing components and parts, and assembling and testing the many subsystems needed for the rest of the complicated accelerator. There was much to be done to develop the electronic test, monitoring, and control systems, and to adapt or write the software for 'conducting' this large orchestra of subunits, which would have to work in concert.

In June 1988 Alistair Smith joined Oxford Instruments from VG Scientific, where he had been MD. He was to head a newly designated Semiconductor Processing Division, which was to include PT as well as the Synchrotron Group.

In June 1989, soon after the announcement of the successful joint venture with Siemens, the team cleared the first vital hurdle. The prototype dipole D-magnet exceeded its design field at its first cold test. It also produced an excellent magnetic field profile for its task of keeping the electron beam in place in the middle of its lane and forcing it round the bends at the ends of the racetrack where the X-ray ports were located. This profile is critical for the performance of the whole ring, and the team had devised a 'mouse' to measure it. This was before the days of Windows software and computer mice. This mouse was more like a tiny curved train made to move by remote control round the small gap between the two sets of magnet coils where the electron beam would eventually run. The mouse bore minute instruments for plotting the magnetic field throughout the gap; it was a delicate device, and had been giving trouble. As Peter Williams reported to the Board, 'The "mouse", which gave problems in the last test series—and is in danger of being rechristened the "rat"—has been reassembled and retested and will be ready for the October tests.'

After one or two modifications the team completed the second D-magnet and both were welded into their cryostats. The rest of the synchrotron ring was being assembled in its heavily screened vault. The linear accelerator, a standard piece of equipment for injecting the electrons, had cost over £1 million and now underwent its tests. The focusing and other special resistive magnets had been bought in and were in place on the developing structure. The radio frequency (RF) cavity, for transmitting energy to the beam, was in place. The various power supplies were finished; the online helium liquefier was functional; and the monitoring and control system would soon be ready for its intricate task of coordinating the whole synchrotron. Testing would take some weeks, and the engineers expected the system to need small modifications between tests. By June 1990 Helios was complete, assembled, and pumped out to achieve the necessary high vacuum; the magnets were cooled down, and soon all was ready to go. It was a complicated system and there could still be pitfalls ahead in spite of the confidence of the staff. Tension ran high.

'We have a beam!' is rather like 'We have lift-off!' Many anxious people were waiting to hear the words. To everyone's relief the cheer went up early in June, when a beam of electrons was first stored in the synchrotron, travelling round and round the ring. It was not a spectacular beam, only 100 milli-amperes (mA) at the injection energy of 200 million electron volts (MeV), and it did not last long; but for the first test it was more than enough for rejoicing. By early July the test team had achieved several good 'runs', with the system operating up to the full energy of 700 MeV. As the energy goes up, the beam current and its life span tend to drop. However good the magnetic focusing of the beam, a few stray high-energy electrons will react with the tube walls or hit those few atoms that remain in any vacuum space. This is inevitable, however well the system has been pumped out, even when 'cryopumped' by the liquid-helium temperature of the electron tube through the magnets, a great advantage in maintaining a good vacuum. But it still slowly deteriorates, leading to some decay in the current. By mid-July the synchrotron had succeeded in storing a beam of 50 mA at the full energy of 700 MeV for an hour. This was really good progress; it was time to announce the success to the world. (See Box 19.1.)

IBM accepted the synchrotron as commissioned in October 1990 (see Fig. 19.3). This was seven months after the agreed delivery date, but IBM was very relieved

Box 19.1. Storing a Beam, and the Helios Performance

The linear accelerator injects electrons into the ring through a special 'septum' bending magnet, in bursts of 10 mA, at an energy of 200 MeV. Repeated bursts are added until the desired beam current is reached. The speed of the electrons is then raised by injecting energy, via the microwave cavity, to that required to produce X-rays of the correct wavelength for lithography.

The specification for Helios called for a guaranteed beam of 200 mA at the full energy of 700 MeV, with a beam lifetime of more than five hours, and X-ray total emitted power of 8.2 kW. After the Oxford Instruments' team had been operating Helios at IBM for a while, and had 'tuned' it for maximum performance, a beam of 297 mA at full power lasted for an average of nineteen hours, with emitted X-ray power of 11.2 kW. The electrons were travelling at 99.5 per cent of the speed of light. 'Up time' for the equipment was then approaching 100 per cent, a remarkable achievement of reliability.

by its successful performance, especially when it was compared with reports on the progress of the compact synchrotrons being developed elsewhere. Only three were, by then, getting anywhere; one rather large and heavy Japanese system had indeed produced a beam, but after many months of testing it was still of limited energy and short duration.

Fig. 19.3. Helios I during final assembly at Oxford Instruments

Many IBM engineers came to see Helios before it was shipped from Oxford; its success was crucial to the next phase of the $500 million development project at the new facility. The achievement was also important in the internal politics of IBM, which was by then suffering from a worldwide recession in the semi-conductor industry. An internal conflict was developing between advocates for continuing investment in long-term strategic developments and those concerned with short-term financial imperatives. There would be many redundancies and cuts in programmes before IBM regained an even keel. There were doubts about the survival of the whole XRL project, which would never have been sanctioned in the harsher new climate. Had Oxford Instruments failed to deliver, there might have been very wide repercussions at IBM.

A team of Company engineers went ahead to East Fishkill, and assembled and tested the subsidiary systems. Early in 1991, after final adjustments, the ring itself was reassembled onto its rigid steel base, bolted into an open metal crate, and loaded onto a flat-back truck for its transatlantic journey and its delivery to upstate New York. Helios arrived on its truck on 29 March, and the same day was slid on air pads into its place in its new vault. The installation team from Oxford got to work at once, and the first beam was stored in the ring within six weeks. Peter Williams, who could not resist being there, reported to the Board, 'The synchrotron is working! . . . on Saturday, 5 mA was stored at injection energy at the first attempt . . . IBM are delighted with progress.' The atmosphere was as if the Houston Mission Control team had seen a new space laboratory, with much depending on it, safely into orbit after an unexpectedly trouble-free launch. Large expensive systems of this type usually arrive in pieces and take months to get going. One IBM executive said he had never known such a com-plicated system come onstream so fast with so few problems.

The final performance was a lot better than the original specifications; this was much appreciated by those who worked with the system. In 1993 Chas Archie wrote in the *IBM Journal of Research and Development* that the gain over the beam lifetime specifications 'has allowed a single fill [start-up] in the morning to meet the facility's needs for eight hours of operation with an aver-age current exceeding the specification of 145 mA'. He went on to say, 'The actual performance should help convince observers that this is a reliable techno-logy for a semiconductor manufacturing setting' (Archie 1993: 383).

But there was no new order for Helios. What had happened to all the old IBM enthusiasm? In truth, optical lithography had proved capable of much more than had been expected of it. IBM's programme to develop XRL had been dri-ven ahead in the early 1980s by its perception that it would be essential for mak-ing chips after the next two, or at the most three, generations. Generations are measured in 'bits', the capacity to store small fractions of information. Alan Wilson, who led this development, wrote in the 1993 IBM research publication on XRL, 'Optical lithography, in 1980, was very poorly understood by the experts. On the basis of historical trends and current difficulties with existing tooling and technology, the limits of lithography were thought to be about 1–1.25 μm for optics. X-ray lithography, consequently, was targeted for entry around 1 μm, the perceived limit of optical lithography' (A. D. Wilson 1993: 300). (μm stands for micron, which is one millionth of a metre. A human hair is about 75 μm thick. One-μm line-widths correspond to a 1 Megabit memory chip, holding one million bits of information.)

In 1980, when IBM began its XRL programme, the current production was of 16 kilobit chips, which could hold 16,000 bits of information. A doubling of the number of minute components in each direction on the chip gives a capacity four times as large, which is usually counted as a generation. So 16 kilobit chips were followed by 64 kilobits, which were just being introduced in 1980, and later by 256 kilobits. It was the next generation of 1 Megabit chips that the IBM team assumed would need X-rays for making the very fine lines. A new generation seems to appear with surprising regularity every three years, which Dr Moore, the founder of Intel, adopted as 'Moore's rule'. The early perception at IBM that XRL would be needed within three generations, was shared by companies in Japan, where the government was subsidizing similar developments.

The market survey Oxford Instruments commissioned in 1986 added to our expectations with a forecast of a market, over ten years, of at least 200 machines for XRL and another thirty for research and other industrial purposes like micro-machining. In 1987, when IBM awarded the contract, it was still concerned about Oxford Instruments' capacity to make the next eight machines it expected to need. But even by then optical lithography had been 'stretched' to make those 1 Megabit memory chips, with minimum features of 1 μm, which could hold sixty-two typed pages of information; 4 Megabit chips were just being introduced with 0.7 μm 'ground rules'—their patterns *still* drawn by optical lithography. Japan, which makes a great many of the world's memory chips, clearly wanted to stay ahead. Many there believed XRL would be needed before long, and seventeen new synchrotron rings, conventional and superconducting, were being constructed or planned. Peter Williams described the atmosphere in Japan as 'SOR Fever' (Synchrotron Orbital Radiation); on visits there he and Martin were asked into many offices and pressed to talk about the progress of Helios.

In November 1989 the American journal *Electronic Business* produced a wide review of the whole XRL scene. The journalists gave their impression of what was happening in Japan as the 'five biggest chip-makers . . . racing to develop an advanced technology that, within two DRAM [memory chip] generations, could give them the power to manufacture commodity integrated circuits far cheaper than anything available in the United States' (*Electronic Business* 1989: 26). Japan's view of future progress was presented in a timetable, with startling entries such as '1992—first Japanese chip maker takes delivery of compact synchrotron for X-ray lithography production', '1996—Japanese chip makers begin commercial production of DRAMs using X-ray lithography', and '1997—X-ray becomes mainstream lithographic technology in Japan' (*Electronic Business* 1989: 26–7).

When Oxford Instruments announced the early success of Helios in July 1990, Bruce McInroy of Hoare Govett wrote in a stockbrokers' commentary on the Company:

We believe that the key factor accelerating the adoption of X-ray lithography (and thus the synchrotron) will be economic rather than technical. X-ray lithography does not just produce denser chips but also improves yields. X-rays go straight through most of the common dust particles, particularly human skin, and thus the pattern transferred to the chip is less likely to be disrupted by contamination. X-ray lithography also has a larger depth of field and allows a larger area of the wafer to be exposed at a time, simplifying production. IBM has done a series of unpublished experiments and we understand that these have demonstrated excellent results, with yields improving by 10% plus (compared with the 1% improvements normally thought of as exciting by the chip industry).

From his researches Mr McInroy expected about a third of the twenty major chip manufacturers to order synchrotrons for R & D quite soon, and, if XRL became the norm for production, 'demand could well be over 100 units by the end of the decade'. With the higher yield expected from XRL as chips became more and more crowded, he concluded that the price of Helios would not be a factor in the decision: 'It is important to recognise that the cost of a chip fab [fabrication line] is $600M and rising, with three-quarters of that cost going on equipment . . .We anticipate that there will be at least one Japanese competitor by the mid-1990s, but we expect Oxford to dominate the Western market as well as selling useful quantities in Japan.'

In spite of all these predictions of a large future market, in 1990 short-term prospects were gloomy. IBM was already suffering badly from the worsening semiconductor recession and the top management had decreed a freeze on capital spending. No further orders were now expected from IBM until at least 1992. As an insurance policy IBM did relax the rules enough to order a spare dipole magnet, costing £1.5 million. There were also many meetings over a possible order from a projected lithography centre in Germany, but the whole project there was cancelled and the funds reallocated to an entirely different scientific development in the newly reunified East Germany. There was still interest from Japan, where no commercial rings had yet been ordered from any company, and there were proposals for Oxford Instruments' participation in joint ventures or consortia there.

In order to keep the successful synchrotron team together, the Board agreed that they should start work on Helios 2, confident that a buyer would be found before it was completed. Considering the economic advantages in making chips with X-rays, the Company saw this lack of orders merely as a delay in the market. The Board still worried more about future capacity to build several at once if IBM were to come back sooner than anticipated. So, in the meantime, while Helios 2 was slowly taking shape on the back burner, the Company worked on orders for ancillary equipment for the large accelerators then being built in Grenoble in France, in Chicago, and in other scientific centres.

At the end of 1991 IBM ordered a beam line for Helios 1. Beam lines take the synchrotron radiation from one of the ports on the curved end of Helios to the point where the X-rays are applied; for XRL this is a wafer exposure chamber. Costing nearly £0.5 million, the XRL line for IBM was a complicated high vacuum tube containing several valves and shutters for the beam, mirrors for collimating it and other equipment. Lines for research need some different components. The Company was to win good orders for beam lines, and their various elements, for attachment to the new conventional synchrotrons. This type of equipment would once have been made in a laboratory's own workshops, but over the previous decade there had been a trend towards curtailing the work of these costly in-house facilities and going out to industry for apparatus needing advanced engineering design and construction.

Later, work started on a complete and relatively small electron accumulator ring for a joint European laboratory near Frascati in Italy, using only conventional magnets. From CERN, the major European Nuclear Research Centre in Geneva, came substantial orders for prototype equipment with strange names, like cold iron quadrupoles. There a large team of scientists and engineers has, for many years, been planning the Large Hadron Collider (LHC), a very large accelerator ring that will eventually be used to probe the very smallest particles of matter. The LHC will use the same tunnel as the earlier ring at CERN, running under the frontier between France and Switzerland. It is considerably smaller than the Superconducting Supercollider (SSC), with its fifty-four mile underground ring, which had been planned and started in Texas, and was cancelled by the US Government because of escalating costs. The magnets the Company was designing in 1999 for detectors at the LHC facility were so large they would have to be built on site. When this powerful accelerator is finally commissioned in the first decade of the twenty-first century it is expected to yield up many of the remaining secrets of quarks, mesons, and other fundamental particles.

In 1991 cyclotron production was transferred to the synchrotron building on Osney Mead, where the radiation-shielded test facility was standing empty, and where a lot of experience was now available on accelerators and systems integration. By the end of 1995, some half dozen cyclotrons had been delivered for PET scanning, but the market was slow to grow as the equipment was inevitably expensive. Recent developments in functional magnetic resonance imaging, which can demonstrate physiology as well as morphology, have overtaken PET scanning for many purposes. At the end of the 1990s there was renewed interest in this 'modality' for studying active areas in the brain. But superconducting magnet cyclotrons are inevitably more expensive than resistive ones, and their advantages are not always sufficient to supplant the older type.

Neutron radiography was the other potential use for the cyclotron, identified back in 1983 (see Chapter 14). It is contrary to our layman's perception of materials that neutrons go straight through metal, but are absorbed by hydrogen-containing matter, such as explosives or cheese or nylon tights, which can thus be shown up by an imaging system. Because of this property, neutrons can reveal hidden substances such as rust, or oil, or water in the wrong place, or drugs *inside* a metal container. A plane crash in Amsterdam in October 1992 was thought to have been caused by a failure of the locating pins on the 747's engine mounts. This kind of problem, if caused by corrosion, could probably have been revealed by neutron radiography. The Company wanted to develop a mobile system suitable for this type of imaging, and sought partners for a complex multifaceted project. In 1991 Government support was obtained for a DTI 'Link' Scheme, in which Rolls-Royce and Birmingham University participated in a joint project with Oxford Instruments. By the end of 1992 the cyclotron had been refined to produce an adequate beam of protons that would generate neutrons when they hit a suitable target. The Company started using this system for neutron experiments on aeroplane parts provided by Rolls-Royce.

The Group's 1993 Review stated, 'The potential of the cyclotron as a compact neutron source was demonstrated in a series of trials at Oxford. Remarkable quality radiographic images were obtained of turbine blades and of corroded sections of airframes. Potential applications lie in the area of non-destructive testing, waste analysis and security.' The Link partners developed this system into a comprehensively equipped demonstration facility on Osney Mead for exploring the many possible applications. Using large Rolls-Royce imaging systems, the team obtained astonishing pictures of light hydrogen-containing materials inside heavy containers (see Figs. 19.4 and 19.5). Among other examples, they scanned a row of bullets and the image revealed all except one properly filled with explosive, while the last was empty. In X-ray scans they all looked the same—the contents did not show up at all.

This system was named 'Neu-SIGHT'. It is an 'intense switchable neutron source' ten times more powerful than any conventional neutron source except a reactor. Combined with handling and detection equipment it can provide neutron radiographic images and chemical identification. More than forty companies visited the facility in its first year for feasibility demonstrations. As well as the air-frame work, the system was used to image the lubricant flow round a car engine *while it was running*; the distribution of fibres in composite materials; blockages in the cooling channels of turbine blades; and the integrity of complex castings. Several contracts were obtained from manufacturers and air-safety groups for providing these images from specimens they supplied. This is the only type of cyclotron that *could* form the basis of a mobile neutron radiography system.

In 1994 British Nuclear Fuels ordered a cyclotron as a pulsed neutron source. This was to be used for the detection of low levels of fissible materials in reprocessed nuclear fuel containers. Delivered in 1995, this has been a success. But a regular market has not yet developed for any of these applications. The cyclotron, as a compact neutron source, is a tool; although the markets

Fig. 19.4. Studying a neutron radiograph of a bearing rig

identified at the start have not yet grown, other uses may emerge for this unique facility.

In 1993 the Synchrotron Group, the Cyclotron Group, and the SPG joined to form the Accelerator Technology Group (ATG). In 1994 Martin Wilson left for a two-year secondment to CERN in Geneva, and Alan Street, who had been managing the third of these groups, became MD of the combined operation. The synchrotron was still the jewel in the crown, and Helios 2 was very slowly taking shape. But 1991 had passed and 1992 and 1993, and 1994 would give way to 1995 and 1996 and 1997 before another order came.

The recession had taken its toll of many new developments, and optical lithography was still fighting back. Semiconductor company laboratories were working on clever optical tricks to push this major move to X-rays further into

Fig. 19.5. A neutron radiograph of a metal aerosol can

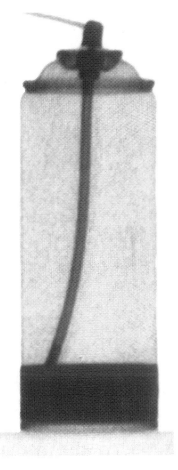

the future. One line was to use the lowest possible wavelength in the ultraviolet band; another was to develop compensating 'phase shift' masks to prevent blurring at the edges of patterns on silicon wafers. The goalposts were moving; the predicted physical limits for optical lithography went on shrinking. Back in 1988, 0.5 μm had been thought the limit. Late in 1989, when IBM opened ALF, it had gone down again.

John Markoff wrote an article in the *New York Times* (19 July 1991), as Helios was beginning to show its paces at ALF. He talked about IBM's hopes that its new XRL system would be 'vaulting it past Japanese competitors in the race to build a memory chip the size of a fingernail that can store the equivalent of four thick novels'. In the second half of the 1990s, IBM executives were arguing, the X-ray process would become essential to American competitiveness in fields ranging from consumer electronics to military weapons: 'now they acknowledge that their target is moving. Today IBM researchers are betting that X-rays will be essential to make chips that can store 256 million bits of information . . . [which] will require circuitry a quarter of a micron wide, and X-rays have already demonstrated that they can etch lines a tenth of a micron wide.'

A little later Peter Heinz of Karl Suss America, a company working on 'steppers' for XRL, wrote an article in the journal *Microelectronics Manufacturing Technology*. He commented, 'Although X-ray advocates were few in number, their semiconductor device results were quite impressive . . . These results have convinced most device-producer strategists that X-ray lithography in manufacturing was certainly inevitable, but not necessarily imminent.' Later in the article he wrote, 'Most will admit . . . that the laws of physics will inevitably prevail and much shorter exposure wavelengths will be necessary. But when?' (Heinz 1991: 6).

DRAM chips are now a commodity, and are produced by the million, many of them in Japan. But there are more sophisticated types of microcircuit that are harder to make and should benefit from the greater depth of field possible with X-rays. These logic chips and custom-designed 'smart' chips are the more profitable high-value microcircuits made in smaller quantities. By 1992 X-rays from Helios were being used to overcome a particular production problem on a specific prototype 64 megabit static random access memory (SRAM) chip for IBM (see Figs. 19.6 and 19.7). Helios was performing brilliantly, but masks were still proving very difficult to make. IBM had succeeded in cutting them successfully, as had NTT in Japan, but others were having difficulties. No companies seemed confident in planning expensive XRL installations until the mask problem had been fully solved, however costly the stretched optical fabrication systems were becoming.

Fig. 19.6. A silicon wafer full of microchips fabricated with X-ray lithography

For Oxford Instruments, in spite of producing the only really successful compact synchrotron in the world, it was still 'jam tomorrow', as the White Queen said to Alice. As soon as the success of Helios I had become clear to all, Alistair Smith had launched a new marketing campaign; Japan was the prime target. There were prestigious seminars for senior executives from semiconductor companies and academics, and there was lobbying of Japanese Government agencies. Peter Williams and Alistair Smith joined high-level trade missions to Japan, one hosted by John Major, the Prime Minister. But the world was then sinking deeper into recession; sales of computers and the chips needed for them were falling fast. Late in 1991 the Tokyo share index dropped dramatically, and the property bubble burst. Work on the next generation of chips slowed down, and companies pulled in their horns.

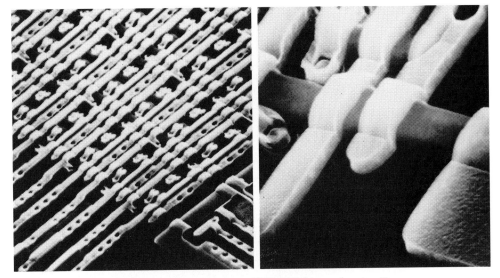

Fig. 19.7. High magnification of a prototype 64K SRAM chip made using X-ray lithography

In May 1997, after discussions over two years, Helios 2 was finally sold to the National University of Singapore. It was installed in 1999 in the new National Synchrotron Radiation Facility, to be used by the University and by the growing Singapore semiconductor industry. Oxford Instruments is itself to use this facility, especially for developments in the micromachining area (Fig. 19.8). As Peter Williams announced in his 1997 Chairman's review, 'A team of engineers and scientists within our newly established Singapore subsidiary will conduct a development programme on micro-machinery and nanotechnology, aimed at markets in instrumentation, healthcare and industrial measurement; for example to explore the potential of devices such as X-ray detectors and implantable medical sensors.' But this was an isolated sale. One swallow does not make a summer.

So, is Helios a failure? Surely not—it is a brilliant product ahead of anything similar in the world, and the British Design Council has designated it as one of the UK's Millennium Products. Is it ahead of its time? Here one must say yes. Although Helios 2 was sold, the closing years of the twentieth century were a difficult time for manufacturing industry, in particular for semiconductors. The market was still not ready for a compact synchrotron, either for XRL or for research. Oxford Instruments has sometimes been accused of being 'technology led' rather than 'market led'. As I have shown, in the case of Helios the early market pull appeared to be strong. In contrast to the MRI development, where the doctors and companies involved in diagnostics soon saw the benefits, the production engineers in chip factories hesitated as they discovered the complexity of the various technologies needed for XRL. But X-ray photons are about a

Fig. 19.8. Mould for a gear 0.8 mm in diameter—an example of successful micromaching using X-rays from Helios

thousand times more energetic than ultraviolet photons, and much of the development work on masks and other major problems is now complete.

L. Grant Lesoine and Jeffrey A. Leavey, after working for five years with Helios, wrote in the journal *Solid State Technology*:

In five years of industrial R&D operation, the advanced lithography facility (ALF) has proved the viability of synchrotron-radiation-based x-ray lithography for IC [integrated circuit] manufacturing. The Oxford Helios electron storage ring (ESR) has been shown to be reliable and easily controlled . . .

As feature sizes of ICs shrink, there will come a time when the paradigm must shift and the semiconductor industry must adopt new technology to move forward. Optical lithography has taken us far, but with exponentially increasing costs and technology challenges at each generation. The industry must soon decide what comes next. We believe that X-ray lithography—which we are using today for sub-150-nm development—is the best option for future manufacturing.   (Lesoine and Leavey 1998: 101)

If and when the laws of physics finally prevail, Oxford Instruments will be there to move the technology forward again. Meanwhile, a lot of research is in progress using the very powerful beams of X-rays from the growing number of large conventional synchrotrons round the world. Even if Helios is never used for the mainline production of chips, new and important commercial applications, like the micromachining that will be studied in Singapore, may yet need these powerful X-rays from Helios, the only synchrotron suitable for a factory environment.

20 Through the Long Recession

A changing world, and how to tackle the Company's many challenges

O how full of briars is this working-day world!

(Shakespeare, *As You Like It* (1599))

He that will not apply new remedies, must expect new evils; for time is the greatest innovator.

(Francis Bacon, *Of Innovations* (1625))

Going back to 1990, things looked good as Peter Williams wrote his Chief Executive's review for the annual accounts: 'Our profit recovery is clearly evidenced in this year's results, cash flow is strong and I look forward to the future with much enthusiasm.' He had every reason to be relieved by the successful resolution of the MRI magnet crisis and pleased with the way the Group was shaping. The older core companies had grown well over the past few years and all were making profits, while Helios 1 was soon to be tested. The acquisitions the Group had made in 1987, 1988, and 1989, to give the security of wider markets, had performed acceptably in the year, one very well, and they had added nearly £5 million to profits.

These 1990 accounts were complicated by the new joint venture in MRI magnets, and by the Link acquisition. Although the 49 per cent share of the OMT profit was included in Oxford Instruments' figures, as the minority shareholder the Group could no longer consolidate any of OMT's turnover. (This consolidation is now mandatory.) The Link sales figure following acquisition was not dissimilar to OMT's revenue before the new joint venture, leaving much the same Group total, at around £100 million. The operating profit was up at £13.8 million, and from the sale of the Eynsham factory to Siemens had come an exceptional profit of £5.8 million. But this had to be reduced by exceptional charges of £1.5 million, as a provision for the extra costs that would be necessary for completing the synchrotron development, which was running behind schedule. The bottom-line profit figure, much affected by the property profit, was a record £19.85 million.

For some time clouds had been gathering over the US economy, where a recession had long been predicted, but Europe remained unaffected, with Germany doing well—the pains of reunification were still in the future. In the Far East the so-called tiger economies were stretching their muscles and Japanese industry seemed to be growing at its usual high rate. Paul Brankin was busy looking into the best way of improving the Group's performance there.

But we were soon to be reminded how quickly the world situation can change. In the autumn of 1989 the Berlin Wall had been pulled down, piece by concrete piece, watched on television by a euphoric free world. I saw the memorable images on a silent screen, placed high up in the corner of a Japanese snack bar, as I waited for a train back to Tokyo; no commentary—just the unmistakable wall swarming with young people as the breaches grew larger. In truth we were seeing the beginning of the end of many old certainties. Businesses started looking forward to a more liberal future in the once solid Communist bloc, with growing free enterprise and opportunities for trade and joint ventures. There were not many voices warning of the disruption, confusion, crime, militant nationalism, and devastating economic collapse that can follow the break-up of a long-powerful and repressive empire.

As controls were relaxed in the early years of the 1990s, a surge of nationalism rocked the member states of the Soviet Union. Other ex-communist countries with multiple ethnic populations, such as Czechoslovakia and Yugoslavia, soon started to break apart. China was encouraging freer commercial enterprise, but keeping strict political control, as the world saw in Tiananmen Square in June 1989. In the Middle East, Iraq invaded Kuwait in 1990, and there were months of tension in the area, followed by the Gulf War early in 1991—again watched blow by blow on millions of television screens. That August reactionary members of the Government of the USSR, who had clashed with President Gorbachev over his new reforms, attempted a *coup*, which almost succeeded. The Soviet Union soon broke up, to be replaced by the Russian Federation and a number of independent states. In the USA the autumn of 1992 saw the election of Bill Clinton, the first Democratic President in the White House for eleven years. The US business world, largely Republican in its political leanings, forecast deteriorating economic conditions.

Beside these changes on the international scene, there were changes in the UK's economic and political status. In October 1990 the country entered the disciplined regime of the European Exchange Rate Mechanism (ERM). This was expected to bring stability and low inflation, but the DM 2.80 exchange rate at entry was high and was to damage the competitiveness of UK companies selling in Europe for the next two years. The US dollar was also fairly high against many currencies, but in spite of this the pound stood at $1.97 that October. A month later John Major became Prime Minister, succeeding Margaret Thatcher, who had been outvoted in a Conservative Party leadership contest; Norman Lamont became Chancellor of the Exchequer.

The recession started in the USA and soon spread across the Atlantic. It was deep and long, and characterized by several false dawns. It was also patchy, both geographically and in the sectors affected at different times. The semiconductor industry was already feeling the chill; by 1991 almost all the electrical and chip manufacturing firms in Japan were reporting much reduced profits—some were down by 50 per cent or more. In Oxford, PT found its order book declining in spite of new products introduced over the previous two years. These advanced systems were also suffering early production problems, and shipments and profits slumped rapidly.

Initially, some parts of the Oxford Instruments Group held up well. These included NMR Instruments and RI, which had relatively long-term contracts. Beside supplying wire for the evaluation phase of the planned Superconducting Supercollider in Texas, OST had won a $7 million contract as the sole supplier of 60 tons of fine-filament conductor for a new accelerator at the Brookhaven National Laboratory. OST was to go on growing throughout the recession. But, as many companies started cutting back on their planned investments in equipment, the analytical-instruments sector of industry began to suffer. MAG (formerly Link) found its orders shrinking, especially its OEM sales to electron microscope manufacturers. Like PT, it was also struggling with technical teething troubles in new products. In most of the businesses profit margins contracted as buyers forced prices down; there had to be redundancies, and even stricter controls on overheads. The spread of market sectors kept the Group well afloat through these years, unlike some companies confined to one sector, but its profit record suffered severe erosion.

The recent acquisitions had added to the number of factories, offices, and overseas agencies, leading to some overlap of facilities. One of the Group's first actions in the 1990s was to rationalize sales offices and consolidate its factories. The Ion Beam Group moved to join PT near Bristol—they both served the semiconductor sector and this integration was designed to save overheads. Manufacture of Sonicaid's foetal-monitoring products was moved from Chichester and amalgamated with the rest of Oxford Medical's production in Abingdon. In Oak Ridge the final integration of the two nuclear companies into one enlarged factory was underway. Elsewhere in the USA, the various sales offices in the Bay Area of California moved onto one site, and on the East Coast several were consolidated at Concord near Boston.

Sir Austin Pearce had been appointed as Chairman of the Group for three years, but agreed to extend his tenure for a further six months to 31 March 1991. In his Interim Statement in November 1990 he wrote, 'The Board has decided to appoint Peter Williams as Chairman and Chief Executive. I strongly support this appointment and with his scientific background and business experience I am confident Peter Williams will provide the Group with the leadership it needs in the future.'

The dual role of Chairman and Chief Executive was then quite common and widely accepted, but the Board decided it would be wise to recruit another non-executive director with long experience of the City, and of high office in a larger public company. In December 1990 Sir Alfred Shepherd, who had just retired as Chairman and Chief Executive of the pharmaceutical company Wellcome plc, agreed to join the Board; he was soon to chair a new Audit Committee. Sir John Fairclough had already become a non-executive director that October; Martin knew him well from various scientific advisory committees they had served on together. Sir John had recently retired as Chief Scientific Adviser to the Cabinet Office, and had previously been a director of the IBM Company in the UK and Chief Executive of its research laboratory. He brought to the Company experience of state-of-the-art manufacturing systems, and he was soon to be a very useful member of TAC.

During 1990 and 1991 the Board also made changes among the divisional executives. The recession was throwing up administrative and management weaknesses, especially in two of the new acquisitions, which were flagging and needed to perform better. The Board started to question again whether the Company's historical structure, with its emphasis on semi-autonomous business units, was the most suitable for its current size and stage of development; the Group was strongly decentralized. As in so many unexpected matters, there are fashions in company structure. I once read an article advocating alternate centralization and decentralization every few years, to invigorate a company through the rethinking process; this would be change almost for its own sake.

There were other relevant views of company structure. Christopher Lorenz wrote in the *Financial Times* (2 Nov. 1992):

One of the most powerful maxims of modern management is that companies are best organised by splitting themselves not only into divisions or businesses, but also into smaller 'strategic business units' focused on particular product markets.

The approach fosters efficiency, market responsiveness and managerial motivation . . . Yet there is also a groundswell of opinion that today's business units . . . provide far too narrow a set of boundaries from which to make appropriate decisions about all tomorrow's competitive opportunities and threats, especially those in new industries.

Such opportunities and threats may either cross boundaries between divisions or business units, or may fall somewhere in between. As a result they may be fought over, confronted inadequately, or just ignored. A much broader corporate perspective needs to be taken if the company is to act appropriately . . .

One way of trying to do this used to be to rely on senior 'functional' executives at head office, such as chief technology officers. Even in its heyday, this was generally inadequate. In today's streamlined matrix structures, where divisions or businesses generally hold most power, such functional executives lack muscle. Their perspective is also usually too narrow—for instance, most technology officers lack market experience and insight . . .

[Some companies have] since relied instead on less formal cross-company coordination. But sceptics have questioned whether this has been really effective . . . [for] exploit[ing] the industries of the future.

Mr Lorenz went on to talk of Dr Gary Hamel of the London Business School, who had recently excited participants at the conference of the International

Strategic Management Society by his presentation. His thesis had been that companies should be looking to compete for 'industry futures' rather than market position in a particular product. Dr Hamel wrote an expansion of this theme with C. K. Prahalad, in a book entitled *Competing for the Future*. As the preface states, attention is given to 'companies that have managed to escape the curse of success and have rebuilt industry leadership a second and third time' (Hamel and Prahalad 1994: xviii).

In his history of IBM, *Father, Son & Co.*, Thomas J. Watson Jr. wrote of his company's one-time inability to see beyond its present products:

My father initially thought the electronic computer would have no impact on the way IBM did business, because to him punch-card machines and giant computers belonged in totally separate realms . . . IBM was in the classic position of the company that gets tunnel vision because of its success. In that same period the movie industry was about to miss out on television because it thought it was the movie industry instead of the entertainment industry. The railroad industry was about to miss out on trucking and air freight because it thought it was in trains instead of transportation. Our business was data processing and not just punch cards—but nobody at IBM was smart enough to figure that out yet.   (Watson and Petre 1990: 189–90)

There are many examples of this blinkered and unimaginative approach to novel products for taking technologies a step forward. Established aircraft companies did not think retractable undercarriages had any future, and, later, companies making engines saw no advantages in the jet engine. When Baird called on the Marconi Company in 1925 he was told that they were not interested in television. Perhaps the most amusing line of all came in a speech by the British Secretary of State for War in 1910: 'We do not consider that aeroplanes will be of any possible use for war purposes.'

Oxford Instruments has always been a 'people' company. Its niche businesses covered a spread of markets, and the structure that evolved reflected this. In the early 1990s the Group Board had three executive directors, Peter Williams, John Woodgate, and Martin Lamaison, and normally had three or four non-executive directors. The management of the Company was basically in the hands of the important executive committee, which included the heads of the divisions and the Group executive directors, but not, at that time, all the heads of the strategic business units (SBUs).

Christopher Lorenz's article rang a bell, as there was, indeed, an element of competition between the SBUs. Sometime in the mid-1990s a manager was heard to remark that it was easier to cooperate with outside organizations than with other SBUs within the Group. With separate budgets for turnover and profit, no SBU wanted to risk being dragged down in a joint development with another unit. The small headquarters organization included an executive Director of Business Development, Ian McDougall. Working with TAC, he looked at each business in turn to review development projects, and he assessed the future technology needs of the SBUs and the Group. One or two projects had crossed unit boundaries successfully, but others had failed to gain the support and enthusiasm they needed. There were moves to improve communication between the units, and to promote the sharing of technology and information in Group-wide training sessions and technology seminars. But was this radical enough? The rest of the small headquarters staff catered for a very few general functions such as marketing support, and the vital financial and legal functions. Other group models were discussed from time to time, but the executives always felt some of the drive and excitement in the businesses would be lost by introducing a more centralized organization with joint facilities for functions such as manufacture, R & D, or human resources.

There was not much fun and excitement from the end of 1990. The managers of the divisions and SBUs were all feeling the weight of responsibility as they strove to keep their operations profitable through the recession. Some had to make many people redundant, and all had to cut back spending wherever possible without compromising future potential, or the essential drive for more sales. Sales costs indeed had to rise in order to win a larger share of the fewer orders available—the world is a tough and competitive place in a recession. OM suffered forced delays in shipments and deferred payments in the tense Gulf area; and it had trouble getting paid in Poland, itself facing curtailment of its trade with a disturbed USSR. In the USA and Western Europe OM now had to compete with scores of other companies, so it had been pushing sales of its products in these emerging markets. As a result its orders were holding up quite well, although the business in Eastern Europe was not very secure or profitable.

The Group's bottom-line result for the financial year to March 1991 was not as bad as it might have been, although one company had made a significant loss. The operating profit of wholly owned Group companies was down by 37 per cent to £8.7 million, but, ignoring the distorting exceptional profit of the previous year, the pre-tax profit figure was down by only 18 per cent to £12.8 million. This relief was due to the excellent performance of the flourishing joint venture with Siemens; Oxford Instruments' 49 per cent share of its profit was over £3.5 million that year. Throughout the long recession this profit share was to rise and to help sustain the Group.

This success in MRI was largely due to the popularity of the active-shield magnets. OMT had persevered in this development through the dark days of 1987 and 1988, after it had lost a large section of its market. The Group Board had continued to support this difficult and expensive project, and was at last reaping the reward for its tenacity. Through these magnets Siemens was gaining market share for its scanners and would soon be world number one in MRI. Picker was also buying more and more of these magnets, and other companies were ordering some for hospitals insisting on active shielding.

After Sir Austin Pearce handed over his responsibilities in March 1991, one of Peter Williams's first tasks was writing the Chairman's statement for the Annual Accounts. Peter paid a warm tribute to his predecessor: 'He guided us through three evolutionary years during which we made important corporate and technical advances.' These had been mainly the solution to the OMT problem with the Siemens agreement, and the Link acquisition. Peter also wrote tributes for the other two retiring directors. Sir Rex Richards had recently been appointed Chairman of the Royal Society of Chemistry, but would continue to serve on TAC. Dr David Ellis would be continuing his long association with the Company, dating back to 1970, as senior adviser on American matters. Peter expressed a hope that the recession had 'bottomed out', but there were to be two more depressing years ahead.

There were still some interesting developments and small acquisitions over those difficult years. In 1991 OM took an opportunity to strengthen its neurophysiology range when it bought Microtronics, a small company in Florida that made sleep monitoring equipment with interesting software. Apparently one in seven Americans has a chronic sleep disorder. Microtronics' equipment complemented OM's own sleep-analysis products, a niche area where there was less competition than in cardiology. Later, the neurology section would launch QUESTAR, a revolutionary technique for analysing sleep, developed in collaboration with Oxford University, and backed by the DTI.

As the recession deepened, many governments, as well as companies, cut back their support for research. At RI orders started falling, but some competitors suffered worse problems and one, Cryogenic Consultants, failed to survive. This company owned a small satellite group in Cambridge working on the science of thin film superconductivity and making related magnetic-field measuring devices. Known as SQUIDs (superconducting quantum interference devices), these tiny and very accurate sensors are based on the phenomenon known as 'Josephson tunnel junctions'. At the end of 1991 RI was able to buy this group from the receivers: 'Oxford buys Cambridge' read the headline in our local newspaper.

Dr John Lumley, who ran this Thin Films Group, had worked in Cambridge University's Cavendish Laboratory, a parallel institution to the Clarendon in Oxford. His small team of scientists was developing an interesting technology that could provide very accurate standards for many types of measurement (see Fig. 20.1). Much of the work was on development contracts for the European Space Agency, the National Physical Laboratory, and similar institutions, but the Group also sold the tiny sensors. These SQUIDs can measure one unit of magnetic flux, which is 10^{-15} T (a hundred thousand millionth of one gauss). The ultimate accuracy of this measurement has opened the way for equipment for use where exact standards are needed in the measurement of voltages, currents, X-ray energy, and temperatures, all of which can be determined through magnetic-field measurements. In the late 1990s this group started applying the technology in the exciting development of a 'gravity wave gradiometer' for mineral mapping by geophysical exploration from the air. This group later developed SQUIDs made with 'high temperature' superconductors that function in liquid nitrogen.

Fig. 20.1. Superconducting thin films are prepared under ultra-high vacuum conditions in the clean rooms at Oxford Instruments' thin-films operation in Cambridge

The business of NMR Instruments had held up well in the first couple of years of the recession, continuing to contribute substantial profits to the Group. But by 1992 its main customer, Bruker, was buying many fewer magnets. A new alliance was formed with Varian, which had been the first company to offer superconducting magnets for NMR, back in the mid-1960s. Varian is a long-established and very well-regarded instrument company based in California; its early years had, in some respects, been similar to those of Oxford Instruments. This Company had been making its own lower-field magnets, and had been buying more and more from Oxford for the higher fields. Soon Varian's spectrometers started winning market share.

For two or three years NMR Instruments had been working on the next generation of magnets. As the 1992 Chairman's statement announced, 'Significant progress has been made towards the development of a 750 MHz magnet to maintain our leadership in this field. Together with ICI and Oxford University we are participating in a major programme under the UK Government's "Link" scheme . . . aimed at problems in biomedical and pharmaceutical research.' There was something of a race on, as Bruker was also planning a 750 MHz system and had given a German Government laboratory a contract to develop the magnet. In the autumn of 1992 the Oxford magnet performed to specification, and, after more exhaustive tests, was installed at the recently established Oxford

University Centre for Molecular Studies. In January 1993 the Company announced this success in a scientific publication on magnetic resonance; in the very same issue Bruker also announced success with its magnet. The large magnet developed for Bruker was a 'pumped' system operating at 2.2 K—at lower temperatures higher fields can be achieved. The Oxford magnet reached its full specification at the normal liquid-helium temperature of 4.2K. Varian rapidly ordered three of the new 750 MHz systems, and it was not long before the Company also offered a pumped system for 800 MHz (see Fig. 20.2).

Several of the Group's businesses were facing problems through these difficult years, but the Board was becoming increasingly concerned over the poor performances of the two larger acquisitions, PT and the old Link Scientific Group, including the nuclear companies in Tennessee. The strategy, in place since 1983, had been to diversify by acquisition in order to reduce the risks inherent in the MRI market. Instead of supporting the Group, these acquisitions were going downhill, taking up a great deal of management time, and one of them had run into serious losses. There were other contributory reasons for this, apart from the recession; one company had been disrupted by the serious illness of its MD, and the other had suffered from rapid changes in ownership; both had launched sophisticated new products, which were not yet running smoothly through production. Were these factors enough to account for the fading performances? Had the 'due diligence' before acquisition been adequate? Had the Company underestimated the problems revealed then? Was the Group's strategy wrong? Was it being implemented in the wrong way? Should it have made rapid and fundamental changes in these companies? What should now be done about them?

There is no agreed textbook way of managing a new acquisition that is to run as a separate unit. Some advocate putting one's own management in on day one; others prefer to leave a good profitable company to continue as it is until it has settled down under its new owner. The first way may create resentment and poor morale in the new subsidiary, key people

Fig. 20.2. The 800 MHz NMR magnet system

may leave, and, with the sudden change in culture, everything that makes the operation 'tick' may be lost. We have seen acquisitions made by other companies fall apart when treated in this way. One, an instrument company we knew quite well, was bought for more than £200 million, but soon lost many of its key managers and scientists, the core of its previous value, and was sold off in bits some years later for a small fraction of its cost. Companies are not just balance sheets and profit and loss figures, they are organic communities; without sensitivity all the value may go. On the other hand, even if a company is left more or less as before, the new owners need to know what is going on, and to start planning any necessary changes for the future. Regular reports on performance against budget and other financial information are essential, and carrots and sticks should be in place early on.

At both PT and Link there had been committed entrepreneurial management with a record of success. Both had wanted to be taken over by Oxford Instruments. They were already aware of the problems shown up by the 'due diligence' and were intent on solving them, and the technologies of both companies needed the continuing commitment of the people at the top who had nurtured them. The culture of Oxford Instruments is very much one of trust and responsibility, and there were, in any case, no experienced senior managers to spare. So the two UK companies had been left under their existing managements, but, as financial and reporting weaknesses emerged, they were strengthened by the arrival of experienced accountants from elsewhere in the Group.

So what were the problems in these two acquisitions? In hindsight PT had been a little like Oxford Instruments in the 1960s, with a lot of enthusiasm, an R & D market, exciting technical achievements, interaction with universities, and all too many late development projects and difficult one-off systems delayed in production. It really needed someone like Barrie Marson, to lead the Company, to cut to the core of the problems, and to exert discipline. But the optimistic founders had been successful so far and had had no suspicion of serious trouble ahead. On acquisition PT had had a substantial and profitable share of the world market in equipment for R & D in semiconductor companies; soon afterwards it had launched unique new systems for short production runs. On top of debugging problems, these products had encountered a severe downturn in the industry.

The Group Executive demanded a performance closer to the budgets, which the PT directors themselves had signed off. One of the founders reacted against this pressure, became disillusioned, and left. During 1990 the other founder, John Ball, who had been striving to put things right, developed an illness that was sadly to prove terminal. It can take some time for a company to recover from blows like this. A temporary team was put in place, but the management also had to cope with the amalgamation with the Ion Beam Group, transferred from RI. In those difficult circumstances integration took much longer than expected, and too much time and effort from the depleted team. There were to be two more years of severe losses for this company and several more management changes before it came back to break-even point.

Link Analytical, in High Wycombe, by this time known as the Microanalysis Group (MAG), was a lot larger than PT and was used to the disciplines of a public listed company. At the time of the takeover it had been an exciting firm with a lot of potential, and it had recently developed world-leading products. But it had suffered a disruptive series of changes and the recession came before it had started to settle down. On top of the extra work needed to conform to Oxford Instruments' standards and reporting regime, and the developing pressures on prices, there were the technical difficulties in the production of the new products. These advanced systems had been welcomed by the market and it was a while before orders went down, but shipments got held up and profits started to fall.

One key problem was transatlantic. The world leading X-ray detectors MAG had developed for its new systems depended on ultra pure germanium crystals produced by NMG in Tennessee. This was a difficult technology in which only two other companies in the world were successful. At that time something went wrong with the process; NMG lost the art of growing crystals of the required specification (p type germanium), and for many months could not diagnose the problem. After some delay this material was bought in expensively from one of the other companies that *could* make it. NMG had suffered its own corporate trials and disturbances, with three changes of ownership within eighteen months, the death of an experienced vice-president shortly before Oxford Instruments came on the scene, and the final amalgamation of the two nuclear companies. The Group had certainly underestimated the effects of all these problems, which were exacerbated by the recession.

In a review of senior management resources the Board studied how it could support the underperforming divisions. The Group could give help with some problems, either through the limited central staff, or by bringing in consultants to assist with production systems, or IT, or quality. But the culture of the Group emphasized the devolution of responsibility, and SBU heads were all expected to overcome their own challenges; non-performers had to be replaced. As one director pronounced, 'when deficiencies are recognized change should be immediate'. This was accepted by all, but it is not always easy to implement. The cause of the problems may be hard to pinpoint, and it may take months to locate a suitable replacement.

J. B. McMasters, in his *History of the People of the United States* (1931–8), quoted President Thomas Jefferson as saying, 'No duty the Executive had to perform was so trying as to put the right man in the right place'. This is probably more important than any other executive action. Clearly, for a senior position the selection process must be deep and thorough. But it is not only *new* appointments that need careful attention; business units can outgrow their long-established top managers. This is not quite the same as the Peter Principle which states, 'In a hierarchy every employee tends to rise to his level of incompetence'

(Peter 1969: 7). In the sort of case I am describing the manager may have performed brilliantly as the unit was growing, but his or her management competence may since have been overtaken by events. It is easy to remember a manager's past achievements and put the problems down to outside factors alone, giving him or her more time to get everything under control again. But this may well be the wrong decision. Oxford Instruments, with its culture of reasonableness and trust, has occasionally left people in senior positions too long, doing a disservice to them, the unit, the Group, and the shareholders.

In the early 1990s several changes had to be made, mostly in the new acquisitions; and as the 1990s advanced there were other vacancies as a few long-serving managers retired. There were up-and-coming younger executives in the Group, but at that time of deep recession hardly any were yet ready to plunge into the deep end of top management. Executives were shuffled between businesses, and new managers were brought in from outside, introducing fresh ideas and experience to the Group. John Hearn, a British manager who had been working in the USA for many years, came in 1991 to tackle the problems at NMG, and later became head of the whole Instruments Division. He had earned his wide experience at Hewlett Packard and at the Finnigan Corporation, and he brought new techniques and questioning to educate and develop the managers under him. These included Andrew Mackintosh, who was to become Chief Executive of the Group.

The new troubleshooting directors and managers for the problem companies had a hard task ahead, and they were not always successful; PT had two new top managers in fairly quick succession. MAG had a long struggle before it could satisfy the Board; it would be 1994 before it began to contribute substantially to the profits of the Group. By 1996 MAG would be turning in the top profit performance of all the businesses and finally proving to have been a valuable acquisition, of central strategic importance. Looking even further ahead, in 1998, a year in which the superconductivity companies would face many problems and perform badly, MAG and PT would, together, account for over 40 per cent of the Group's profit before tax. These 'snakes-and-ladders' performances of the different operations again seemed to show the value of the Company's diversified structure and markets.

Soon after Peter Williams took the helm in April 1991, Oxford Instruments held its first formal International Management Conference. Peter wanted to forge a common goal to drive the Group forward in those difficult times. Eighty-five people attended from seven countries; all were heads of divisions, SBUs, sales companies, service units, engineering teams, or product groups. Peter's introductory talk, entitled 'Why We are Here', gave a thumbnail sketch of the Company's history and achievements, mainly for the newer managers. He told them of the 10,000 magnets of various shapes and sizes the Company had made since Martin's first small superconducting magnet in 1962, and how it had become the world leader in applied superconductivity. He described the major decisions facing the Group in 1982 with the opportunity to 'go for' the market for superconducting magnets for MRI; the decision to float the Group on the stock market to provide the necessary funds; and the euphoric days when we were the only Company able to make these magnets and our profits and share price were riding high.

This introduction was to remind everyone that, as a public company, Oxford Instruments was owned mainly by large institutional investors. The whole management needed to recognize that their prime responsibility was to look after the interests of the shareholders. At the same time they were not to forget their obligations to the employees and customers, to the suppliers who depended on the Company, and to the local community—the stakeholders as they came to be called.

Peter then told of the strategic re-evaluation in 1984, when the Board had already identified the risks ahead in MRI, as our customers 'backward integrated' by starting to develop their own magnets. He outlined the plan to 'hold the line' in MRI while the core businesses grew, the synchrotron was developed, and acquisitions added strength to the Company. He described how the plan had eventually worked out, but not within the intended time frame. The Siemens joint venture had come too late to avoid the 1987 cliff, with its profit warning, and the 1988 fall; the synchrotron market had not taken off as expected; and the opportunities to make acquisitions had come late in the day and only shortly before the recession struck. The share price had consequently retreated a long way. He went on:

We made some errors in that period—there is no point in denying it. I am, however, proud of much that we have achieved together—we have grown nearly tenfold in size since we were a private company, even excluding our share in OMT which we also founded and grew . . . But I am unhappy about what we have failed to achieve, in particular about our profits performance in recent years, which has been erratic . . . This must be rectified . . . recession, perestroika or Gulf wars do not exonerate us from blame. Let us all live up to our collective responsibility.

One purpose of the conference was to instil in all the managers the imperatives of quality, innovation, and training, and the need for risk-taking, matters that were to be central preoccupations in the competitive business climate of the 1990s. After many panel discussions and talks on finance and how the City saw the Company, Peter spoke again on his vision for the future and how to achieve the mission statement 'To become a major international company in scientific and medical instrumentation'. Quality, was his first subject. There were quality programmes in most companies in the Group, but Peter referred everyone to the role model of OMT, which had made such remarkable strides after embracing total quality management (TQM) in the late-1980s: 'output doubled in three years with fewer employees . . . The success of TQM around the whole Group would add millions to our bottom line profit figure.'

Innovation was Peter's second subject. The House of Lords had defined it as 'the commercial application of knowledge or techniques in new ways or for new ends'. Peter shortened this to 'the management of constructive change'. Innovatory change, he said, should apply to every activity within every business; but manufacturing processes, in some parts of the Group unchanged for many years, were an early candidate: 'With the real price of a colour television set one tenth of that in 1970, we live in an era where cost reduction is of paramount concern: our "high tech" products are not immune from this trend. Combating cost demands innovation . . . Let us just be sure we are not enemies of change.'

Peter's third point was training. As a relatively young company with a limited range of skills and experience available, Oxford Instruments would have to spend much more on training programmes, both external and internal. The existing internal courses, on financial awareness, and on technology, would be extended, and others would be added on marketing awareness, intellectual property rights, and industrial design.

On development, Peter reminded his audience that, in the short term, the evolution of existing products would take up most time and effort. Where we led the world, it was important not to let our position slip. On newer developments,

We won't always 'get it right' as there must be an element of risk-taking if we are to grow significantly . . .Could it be that we are developing a risk-averse culture around the Group? If this were to become true we are dead! Let us take risks and set demanding objectives and goals and then 'manage' as much of the risk out of our programmes as we can by professional attention to detail and quality.

Back in 1977 the Senior Management Committee had, for the first time, gathered for a two-day session away from their day-to-day preoccupations. Brainstorming exercises on strategic issues had produced useful guidelines. Among other matters, they had defined acceptable risks for any project; if successful it should be capable of adding significantly to the Group's profit, but if disastrous it should not be able to wipe out more than two-thirds of the profit. In those days Oxford Instruments was a private group. A public company, watched by analysts, the financial press, and the institutional shareholders, cannot contemplate this degree of risk. A reduction of profits by two-thirds would lead to such a drastic drop in the share price that survival as an independent company would be in doubt. But, if the same formula were to be applied to each *division* or SBU, it should still be acceptable, and might lead to brilliant future products.

There is a lot of inertia opposing innovation and risk-taking. I was reminded of a piece of paper that landed on my desk from some unknown source, entitled *How to Kill a Good Idea* (see Box 20.1). It contains twenty-six reasons for avoiding innovation.

In 1759, long before this list was concocted, Dr Johnson wrote, 'Nothing will ever be attempted if all possible objections must first be overcome.' And a great many centuries earlier Tacitus wrote, 'The desire for safety stands against every great and noble enterprise.' In the early 1990s there were several voices pleading for more creativity in the Company, for rule-breakers who could think

Box 20.1. *How to Kill a Good Idea*
Anonymous

Don't be ridiculous.
We tried that before.
It will cost too much.
It can't be done.
That's beyond our responsibility.
It's too radical a change.
We don't have the time.
That will make *x* obsolete.
We're too small for it.
That's not our problem.
We've never done it before.
Let's get back to reality.
Why change it? It still works all right.
You're two years ahead of your time.

We're not ready for that.
It is not in the budget.
Let's form a sub-committee.
If it was good we would be doing it
    already.
We'll be the laughing stock.
That doesn't apply to us.
We are doing the best we can.
We did all right without it.
Has anyone else tried it?
It won't work here.
Too hard to sell the idea.
Can't teach an old dog new tricks.

imaginatively. Some of the flair and sparkle seemed to have gone; was it the Group's increasing size, or was it the pressure for performance through the recession that was checking enterprise? The top priorities had to be orders and getting products out of the factory gate month by month. This often involved drafting in some R & D staff for troubleshooting, and this disruption was apt to blunt and delay the effort devoted to new developments; and no one had much energy left for 'blue-skies' thinking. Nevertheless, the Company *did* manage to keep up its investment in development through this period, and all parts of the Group launched new products during or just after the end of the recession.

In 1991 Peter's ambitious goal for his managers was a doubling of turnover by the middle years of the decade, with higher profitability. But as he spoke the recession was getting worse, not better. Most parts of the Group were affected by now: NMR Instruments, for long one of the top profit-earners, was getting fewer orders; NMG in Tennessee was on a four-day week; every successive Group reforecast of sales and profits was lower than the one before. Talk was of 'battening down the hatches'; conserving cash; drafting more engineers into selling; collecting debt; and moving technicians about the Group to where there was the highest workload. There had to be more redundancies and further changes in the management of the business units. The half-year results at September 1991 were the worst since the flotation. Had we reached the bottom? The press could still see no end to the recession.

The Board and the executives debated their best course through the continuing turbulence (see Fig. 20.3). Should they take a long-term view, keeping their options open, continuing to support all the businesses however badly they were performing, or should they look to short-term profits and cut and restructure the Group radically. With cash in the bank, the decision was to grit the teeth and ride out the storm, which could not last for ever. And judging by the large later contributions to profits from the companies then under potential threat, they probably made the right decision.

Fig. 20.3. The Board in 1992: back row left to right, John Woodgate, Sir Martin Wood, Sir John Fairclough, Sir Alfred Shepperd; front row, Martin Lamaison and Peter Williams

Although order intake was down, there was still some backlog in order books. The Company achieved sales of £103 million in the year to March 1992, down only £5 million, but, with poor margins, the operating profit was down nearly 65 per cent, to just over £3 million. That was before the welcome contribution of over £5 million from OMT, which still served a buoyant market for active-shield magnets, and had made record sales of £74 million. That year's pre-tax profit of £8.6 million was the lowest reached in the recession; the figure would have been £2 million higher but for the strong pound. The Company had cash reserves of £15 million, and there was little reaction from the City, who were seeing many companies reporting losses, and a string of business failures.

The pain went on unabated for the first half of our 1992/3 financial year. By this time the problems of reunification were damaging the German economy.

Fearing inflation, its Central Bank kept interest rates up at 8 per cent; the UK, linked through the ERM, had to keep its interest rates even higher in order to maintain the pound within its allotted limits against the other currencies. Industry was suffering rates of 10–12 per cent. In Japan the 'property bubble' had burst, leaving banks short of security for the huge loans they had made, and, with the yen still high, a lot of production was being moved to the developing countries of Eastern Asia. The semiconductor market was still in recession. With the dollar now lower, and declining interest rates in the USA, pundits were talking of an upturn, but even there the economy was stubbornly resisting recovery. With President Clinton in the White House there was uncertainty about the administration's intentions, especially on healthcare policies. It was proposing tough new restrictions that, if implemented, would affect the Group's two medical businesses, including the profitable MRI magnets.

By the autumn of 1992 the Company's moving annual total (MAT) of orders had sunk to £85 million—down nearly 20 per cent on 1990, and the three-month MAT was down at £72 million. This was our low point in the recession. All over Europe business confidence was faltering. But that September, after months of currency instability and a ruinous run on the pound, Britain abruptly left the ERM. The floating pound now found a lower level against other currencies, which gave exporters a more competitive position in world markets, and better profit margins. The main danger was renewed inflation as imports started to cost more. For the Company, the second half of the year was not much better than the first, but the interest on our cash reserves, a more favourable exchange rate, and an OMT profit contribution up at £6 million gave a pre-tax profit figure of £10.6 million at March 1993. Even more important, the order rate had started to turn up again, led by an unexpected surge over the Christmas period. There was light at the end of a very long tunnel.

21 Issues of the Nervous Nineties

Brief recovery, the debate on R & D, investors' time scales, risk, and sizeism

There is no natural end point or plateau in the development of the science-based industries. The performance that can be achieved can be increased almost without limit, provided the appropriate resources are devoted to the task in hand.

(Ieuan Maddock, *New Scientist* (1973))

A company must maintain a continuous pipeline of business-building initiatives. Only if it keeps the pipeline full will it have new growth engines ready when existing ones begin to falter.

(M. Baghai, S. Coley, and D. White, *The Alchemy of Growth* (1999))

The recession started to ease at the end of 1992. By the middle of 1993 reports were still talking of the 'halting recovery', but in August the *Financial Times* (*FT*) index of the top 100 shares on the London Stock Exchange—the FTSE 100—passed the 3,000 point. This was about double its low point after the 1987 crash, although during the recession it had gone down again, to something over 2,000.

In many ways the 1990s were to continue a nervous decade, with alarms, conflicts, and major changes, both nationally and internationally. There were more horrifying local wars—in Bosnia, Rwanda, Chechnya, Kosovo, and East Timor. Natural disasters disrupted many countries. Some resulted from the unusually strong El Nino Pacific Ocean currents in 1997; hurricanes, typhoons, floods, droughts, and famine were blamed on it, some in countries far removed from the Pacific. Volcanoes erupted, and severe earthquakes brought death, chaos, and fear to vulnerable parts of the world. Then in the 1990s the devastating AIDS epidemic spread through many developing countries, especially in Africa. In Britain 'mad cow disease' decimated beef farming, while arguments began to rage on genetically modified foods. Although there was famine in countries such as North Korea and Ethiopia, world overproduction of cereal crops, and changes in subsidy patterns, affected farming in several regions. Among many communities the ready supply of illegal drugs, and the crime wave that supports the resulting addictions, brought law and order to the top of the agenda.

In the economic arena the world became curiously divided. In the Far East the so-called tiger economies came crashing down, while Japan, whose large companies and financial institutions had invested huge sums in production in these countries, was dogged by bad debt and stuck in a deflationary dilemma. At the other side of the world, the US economy flourished, with millions of new jobs and a long running 'bull market', which took the Dow Jones index to unprecedented heights. There were plenty of prophets of doom in the last year or two of the decade but only occasional corrections, not the feared crash. In London the stock market followed New York. Parts of the economy were riding high, notably in the growing service sector, with financial services and IT, and the Internet, providing an ever-larger share of the action. Manufacture, and engineering in particular, suffered from the renewed strength of the pound, and from the many problems in export markets, especially in the Far East.

The British economy was out of phase with most of Europe. The countries lining up to join the Euro strove to meet the convergence requirements, which included more balanced budgets. Many suffered low growth, and in Germany, still affected by reunification, the economy remained sluggish and unemployment high for much of the decade. European currencies remained low, and the pound was seen as a safe repository for money, with relatively high interest rates, and little fear of inflation.

In the UK the economic recovery that followed the long recession was led by exports. After the traumatic exit from the ERM in 1992, the value of the pound dropped and manufacturing industry became more competitive. Among the population at large, confidence was slow to return. Many people who had bought houses during the boom of the late 1980s found the reduced value of their homes was now less than they owed their mortgage companies—the negative-equity phenomenon. Spending remained slow. It was not until 1996 or 1997 that what became known as the 'feel-good factor' started to return, but by then the pound was again strengthening, sucking in cheap imports and causing grief in manufacturing industry.

At Oxford Instruments, depending as it does on exports, the lower pound that came with the ERM exit brought relief. In the half-way results to September 1993 profit before tax was up 19 per cent on the previous year, and the 'value of new orders received showed strong signs of recovery'. When the full-year results for 1993/4 came out, the Group's sales had risen over 10 per cent to £112 million. Profits, at £12.8 million, were up 20 per cent in spite of a difficult market for medical equipment in the USA, which had reduced the OMT contribution by over £2 million from that of the previous year. The next year, 1994/5, was the first since 1986 in which every profit reforecast came in at a higher level than the previous one. That year the Company achieved sales of £125 million and profits of £18 million, considerably above market expectations. The three 'legs' of the Group, Superconductivity, Instrumentation (which then included OM), and OMT contributed roughly equally to profits that year. The shares soon went up to 400p.

In February 1995 Peter Williams wrote, 'We have seen all the effects of recovery we are likely to see post recession—now we have a hard slog for new business ahead.' Manufacture and quality needed attention—it can always be improved in any company at any stage. Inconvenient old buildings were to blame for some of the inefficiency, and, with recovery clearly established, new factories were on the drawing board. The rapidly expanding MAG operation in High Wycombe, RI, and IAG were all to have new factories over the next three years. The DM again started to weaken against the pound, but, in spite of structural problems in many financial institutions and the overhang of bad debts, the Japanese yen was then very high, averaging only 150 yen to the pound in 1995. (In 1998 it was to average over 200). This helped to provide a good market for some of our equipment. Through tremendous efforts, Oxford Instruments Japan would soon account for nearly 20 per cent of Group sales, and a considerably higher percentage of the sales of the magnet operations. In the UK, the Group was suffering from a shortage of scientists and engineers, and was advertising sixty vacancies, but, on the positive side, labour productivity was up by more than 10 per cent in the year.

Manufacturing companies and their City investors are not always agreed on the subject of R & D. During the recession the Group's businesses had worked hard to keep their 'technology edge'. Other overheads had been cut drastically, but spending on R & D, the vital seedcorn for the future, had been maintained; this was in spite of around a 30 per cent fall in sales from the Instruments Division, and 15 per cent from the superconductivity businesses over the years of recession. The new products launched over those years spearheaded the recovery. From time to time there have been criticisms of Oxford Instruments for spending more on development than on dividends. A company that serves the research and advanced industrial markets *must* keep up its development effort both for enhancing older lines and for achieving brand new cutting-edge products.

Major and minor updating of products is essential in a competitive world, and, in Oxford Instruments, this has been the responsibility of each business unit in the Group. They have not always achieved these improved products fast enough for the health of the Company. Many problems come from a failure to complete a new product development, and its production engineering, on schedule for a well-timed launch. If a new product is six months late, it can lose much of the initial premium profit margin, and, in our case, the Company can also lose the esteem of its scientific customers. If the market knows the new system is coming, customers will no longer buy the older ones, and competitors will sometimes get there first. And, in a fast-moving technology, obsolescence can knock at the door surprisingly soon.

Towards the end of the decade Diane Coyle wrote in the *Independent* newspaper (17 Mar. 1999), 'attempts to measure how much growth can be attributed to investment and how much to other causes has always found that by far the biggest chunk is the result of technological change'. The 1990s were a decade of galloping development in digital electronics and the software to serve it; of ever more powerful computing in ever smaller equipment; of the Internet; the mobile phone; and the geographical positioning system; communications were revolutionised by fibre optic technology. Technological developments changed our lives faster than ever before, and it was hard for many established companies to keep up to date in this fast-moving environment.

Through the mid-1990s most of Oxford Instruments' businesses managed to keep up to date with the latest computing developments. They upgraded products frequently as the successive versions of Windows software appeared, and used new processors and interactive facilities in their ever more user-friendly systems. Some of our businesses provided equipment with which other companies developed their own new products for this growing surge of digital technology. The research-based business, RI, makes equipment for the very basic research on materials, particularly on semiconductors, the starting point of many of these technologies. High magnetic fields and ultra-low temperatures are essential tools in much of this R & D. Through the 1990s MAG developed successive generations of equipment used with electron microscopes for studying the chemistry, mapping the elements, and looking at the crystal structures of semiconductors and other materials (see Plate 12). Among several applications these are used to analyse defects arising during the fabrication of chips. PT found a growing market in equipment for making microcircuits and other devices from compound semiconductors such as gallium arsenide and gallium nitride. These chips can be smaller, act faster, and use much less energy than the commoner silicon-based chips. They are used in devices such as mobile phones, which need to become smaller and smaller while their capacities get larger and larger (see Plate 14).

Most developments aim for effective improvements to existing types of equipment for familiar markets. But R & D can also lead to products that make fundamental changes to some walk of life. In the early 1980s the development of magnets that made MRI scanning possible was one of these, as it changed the way many medical conditions were diagnosed. The synchrotron development, half a decade later, was meant to change the way very dense circuit patterns were drawn on tiny chips, and so make further miniaturization of devices possible. It is hard to assess the potential of these revolutionary projects in advance. If successful, such developments can lead to a step function in sales and profits, as we saw with MRI. For these big new ideas there must be very careful selection of the project, with input from sales and marketing and other departments. There should be a timetable with milestones and possible cut-off points, and a clear focus towards a highly profitable future for the product, consistent with its development costs. This is easier said than done, but is vital.

For a really successful strategy a company needs to grow a *series* of these new-concept products (see Fig. 21.1). As one takes off, like MRI in the early 1980s, another should be in development, to be ready to take its premium-profit-earning place as the competition for the earlier product emerges, and its margins shrink. But unexpected difficulties often occur when developing these advanced products. The synchrotron was supposed to fill the gap as other companies inevitably learned how to make MRI magnets. But the widespread adoption of XRL for making chips was long delayed, although hopefully not superseded, by the stretching of the technology of optical lithography far past the previously believed limit. It was the MRI magnets that, almost two decades after they were first developed, continued to generate good profits, although there had been major new developments, and prices had come down by half. At the end of the decade OMT was probably still the world leading MRI magnet-maker thanks to its 'built-in' market, its innovative magnets, its quality, its reliability, its on-time delivery, and its perpetual work on cost reduction.

The UK ranks consistently high in fundamental research. The problem area is the transfer of the knowledge resulting from university research into industry, and its conversion into competitive products. In the early 1990s Peter Williams

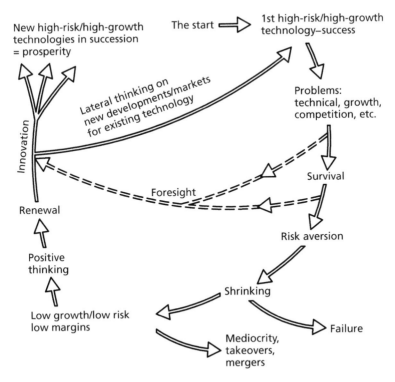

Fig. 21.1. Some routes high-technology companies may travel

chaired the Innovation Advisory Board of the DTI, and in 1991 he promoted an 'R & D scoreboard' for UK companies, which is now produced annually. Companies are ranked on their R & D spending by total sum, by amount per employee, by percentage of sales, by percentage of profit, and by percentage of dividend. The scoreboard also compares R & D in UK companies with that in other countries—and the UK still comes off badly in the comparison. In the period covered by the 1991 scoreboard, UK companies, on average, spent about 1.5 per cent of sales and under 20 per cent of profits on R & D. For the world's top 200 companies, the equivalents are about 4.5 per cent of sales and over 90 per cent of profits. The top UK spender did not come within the top fifty world-wide, raising questions on the future of British manufacturing industry in an increasingly competitive world. But, of course, the *sum* companies spend on R & D is not the only matter to watch; the *effectiveness* of the R & D depends on other factors. Here, a small focused team, like the Synchrotron Group, may be more cost-effective than a large corporate laboratory; it all depends on how the work is conceived, directed, and managed. Oxford Instruments was regularly spending some 7–8 per cent of sales on development, which, in 1992 and 1993, was almost double the depressed operating profit.

In 1993 the Company welcomed the Government's White Paper, *Realizing our Potential—A Strategy for Science, Engineering and Technology*. This recognized that Science and Technology were an 'immense national asset', mostly held in the 'science base' of university and government research laboratories. The paper recommended partnerships between academic or government science and industry in order to harness this knowledge for the creation of wealth for the nation. For Oxford Instruments this idea was far from novel; we had been creating partnerships and supporting developments in research laboratories throughout our history. Profitable partnership developments have included the continuous-flow cryostats, developed with the Oxford University Department of Engineering Science in the early 1970s; Medilog, developed with the Medical Research Council; NMR magnets developed with the guidance of the Physical Chemistry Department at Oxford. And there were, and are still, many other joint projects, some supported by government initiatives such as the Foresight Scheme and Link partnerships.

The City is often ambivalent about the amount of R & D in a company's profit and loss account. There was a time when many companies capitalized some of their R & D spending, regarding the resulting patents or product designs as intangible assets that would help to earn future profits. But in the 1970s the aero-engine company, Rolls-Royce, capitalized its development costs on the RB 211 engine to such an extent that no future sales of the product were likely to make any return from this 'investment', financed largely by high-interest loans. Rolls-Royce collapsed, and the capitalizsation of R & D was seen as imprudent and came into disfavour. There are still sporadic discussions on the investment value of R & D, and of intangible assets like goodwill, and how company accounts can best reflect the *real* profit and asset situation.

Undervaluing R & D may be a cause of short-termism among investors. In an article in the *Independent* newspaper (1 Mar. 1994) Roger Trapp discussed a report by David Collison, John Grinyer, and Alex Russell of Dundee University. The conclusion of these researchers was that the *perception* of City short-termism, rife among managers of engineering companies, often led them to cut investment in R & D. These managers were making investment decisions with an eye on the stock price, and they *believed* that the City valued their stocks primarily by reference to the current year's prospective earnings. Heavy R & D expenditure would reduce earnings, as it had to be written off straightaway against a firm's profits, but new *equipment* to enhance production and profits was a capital expense; it would be written off against the profit and loss account only as depreciation, over several years. The Dundee team saw the current rules of financial accounting as largely to blame for this threat to innovation, although they also felt managers should give City investors credit for more sophistication in their valuation methods. And companies themselves are often to blame for not keeping their shareholders properly informed about where all the R & D money goes and where it is leading.

The debate on pressures on public companies for short-term performance still goes on. Pressure certainly comes from some institutional shareholders, who want larger dividends, and all investors look for higher share prices. And one must not forget the pressures from executives—for high profits to provide good bonuses. An article by Andy Coghlan in the *New Scientist* was unequivocal—the title gave his views: 'City's Dash for Cash Crushes Research'. He quoted Lars Ramqvist, the Swedish President of Ericsson, who had recently given the annual UK Innovation Lecture: 'You need a society that encourages innovation. And in Britain, that's not the case . . . It is of utmost importance for innovation and development that financiers take a longer view on investments in R & D' (Coghlan 1994: 4).

Some UK companies, notably those in the pharmaceutical industry, seem to have investors who understand the need for high development spending. Here the cost of the necessary long incubation period for a new drug for a large market can, if the drug is successful, be matched later by very high profits protected by strong patents. But many companies and their investors do not look so far ahead. They want tangible returns quickly. As Peter Williams pointed out in a Stockton Lecture he gave on short-termism in February 1991, 'over the past five years, dividend payments to the shareholders of public companies have risen at a rate over five times greater than the rate of increase of R & D expenditure in the same companies'. The problem is real.

On 4 November 1999 Oxford University held a Science and Technology Day at the DTI, with a programme of talks on 'Oxford University in Business'. In a perceptive speech entitled 'Business Needs Science', Sir Robert Malpas spoke of the difficult task for managers in keeping a balance between long- and short-term needs. He said, 'Optimising the present to the exclusion of creating options for the future . . . will eventually lead to extinction. Investing too heavily in money and in people to create options for the future will severely affect current performance, to the point where there may be no future. It is a delicate and difficult balance.'

The balance must depend on the nature of the company, but some developments have to look far ahead. Warren Buffett, the ultra-successful US financier, looks for underlying value in a company, and is willing to wait a long time for his eggs to hatch. The Japanese think nothing of a ten- or fifteen-year development before any payback in a major new product. When Sir Austin Pearce, our previous Chairman, was head of British Aerospace (BAe), he found the times-pan for the development and marketing of a new aeroplane curiously at odds with City time scales. The plane had a ten-year development period, five years for delivery to airlines and a twenty-five-year airframe lifetime; against this, the pension-fund managers holding investments in BAe were judged and ranked by performance every three months! When the City talks of 'long term' it often means just five years. With a *stream* of new instruments or drugs or software programmes, many companies no doubt *do* make good profits consistently. But not all new developments succeed. Even with visionary forward planning, failure to realize their profit potential may result from all sorts of factors. Economic problems in major markets, internal company disruptions, or an uncompetitive currency can soon lead to unfulfilled sales budgets or lower margins and can cause cuts in R & D budgets and derailment of development timetables.

City perceptions of risk can have a major effect on a company. When substantial companies like BAe spend large sums developing a new product like an aeroplane, there is rarely any fear that the company will 'go under' through the project. The shares may go down if the market thinks the development costs will eat into the next few years' profits. But smaller companies with large R & D budgets for innovative new ideas are frequently seen by the City as extremely risky, the domain of venture capital. This risk may look less if an investor can understand the company and what it is trying to do. The young Oxford Instruments must have looked pretty fragile in the early 1970s, but David Ellis, of 3i, understood its potential, and our patient venture capital investor was rewarded, fifteen years later, with gains of more than 100 times the original investment.

According to some statistics in the late 1990s, venture capital in general was proving a less risky and better-rewarded sector than many others. In a press release in May 1998 the British Venture Capital Association reported that UK venture capital funds, some of them owned by the investment institutions, had shown 'significant and repeated out-performance of total UK pension fund assets'. The net returns of venture capital funds raised between 1980 and 1997 showed a 21 per cent overall return over one year, 29.3 per cent average over five years, and 14.6 per cent over ten years. Over three and five years the venture capital funds also outperformed the FTSE all share index, and the FTSE 100 index. (This was before the huge rises in top shares during the last two years of the 1990s.) Venture capital funds are invested in many kinds of company, and, in the 1990s, much of the money has supported management buy outs of established companies, or of divisions of larger groups. Nevertheless, the funds invested in young technology companies can do well. The most important factor for these funds is the *spread* of companies, as some are sure to fail, and others will never shine. The few that really succeed may do so spectacularly, more than making up for all the rest. In the USA there is considerable competition among venture capitalists for investment in new technology companies, and there the Nasdaq market caters largely for technology stocks.

Some venture capitalists insist on a planned 'exit route' before they will invest in a company, and, at an appropriate time, may press an investee company to seek a public flotation or a 'trade sale'. The London stock market in its present form is not an easy home for smaller companies. Some institutions cover the whole market, but there is simply too wide a gulf between the top and bottom of the range for the well-being of smaller companies. There may be a lot of interest at the time of a company's launch, and the share price can go very high. The venture capitalists and some of the managers may take part of their reward, but once the company ceases to be newsworthy the turnover in the shares may drop to a very low level, and the price will sink with it. Unless a company is seen as a takeover target or is in a fast-track sector, such as biotechnology in the mid-1990s or IT at the end of the decade, it is hard for it to keep up much interest from investors. Many of the shares listed in the FTSE SmallCap and Fledgling indices and on the Alternative Investment Market (AIM), are neglected by the City. Private investors may keep the market moving, tempted by the chance of high rewards, and guided by the few remaining private client stockbrokers or by specialist subscription publications. Smaller-company shares tend to be illiquid, and small deals may move the market price, either up or down, out of all proportion to the numbers traded. Perhaps Internet trading will change this situation.

Most of UK industry is now owned by the large institutions, the pension funds, insurance companies, investment trusts, and unit trusts. They all invest money that, ultimately, belongs to individual members of the public. During the 1990s, in parallel with major mergers in industry, there was a move towards a consolidation of City firms—some finished up with treble- or even quadruple-barrelled names to preserve their multiple origins. These organizations now have vast resources under management, and, with company takeover bids of many billion pounds to consider, most find it almost impossible to study and invest in smaller companies, capitalized at under £100 million, in which it is hard to buy a significant stake. Large companies are becoming larger by acquisition and merger. Philip Coggan reported (*Financial Times*, 12 Mar. 1999) that at the end of 1998 just twelve companies constituted the top 30 per cent of the London stock market by value, and the position must have got worse since. The institutions have, without doubt, been able to make good profits much more securely on substantial blocks of these shares than on smaller-company stocks. Many funds invest only in the top 100 companies. In the 1990s there was also a flowering of 'tracker' funds that shadow one or other index. These can distort share prices, as the funds *have* to acquire the right weighting of each new stock floated, or graduating to a higher index. Nowadays 'big is beautiful' in the City.

Even the Government is becoming worried by this understandable but unhealthy neglect of smaller firms. Companies do not start big. They become big by organic growth and by the amalgamation of smaller companies. To have good big companies there has to be a continual renewal of the pool of good small ones both in and out of the market. It will not be in the interests of the country or of investors if currently unfashionable and often undervalued smaller quoted companies—some small only by capitalization, not by industrial reality—disappear from the market. This is happening already. Some are seeking security in mergers, some are snapped up by overseas companies conscious of the unrecognized underlying value going begging, and some are 'going private' through venture capital-backed management buyouts. At the same time the queue of companies coming to the stock market for funds has been shrinking, except for the flood of IT or Internet hopefuls.

What can be done to help these smaller companies? The Government has helped many companies with net assets of less than £10 million by imaginative tax-relief schemes to encourage new investment in these 'risky' ventures. These schemes are envied and admired by entrepreneurs and investors in other countries. The rules cover most start-up or small private firms, and some companies listed on AIM or on the less regulated OFEX share-exchange system. Through venture capital trusts (VCTs) and enterprise investment schemes (EISs) many fledgling companies in the UK are now obtaining the modest equity investments they need. But in the middle there is this gap where companies hesitate to go to the market for the funds they need because their share prices are so low. In these days of 'sizeism' perhaps it is time for new imaginative schemes to help these companies in the middle. In the City there is no lack of money for investment, just a lack of time for the institutions to deal with them, and a lack of liquidity in their shares. Perhaps some tax incentive could be devised for institutions, such as the pension funds, to compensate them for the extra work involved in investing relatively small sums in sound companies with market capitalizations below a certain level. Perhaps we need a greater choice of market, maybe local markets, where smaller companies can receive better attention and more consistent support than in a London market dominated by giants. Oxford, with its growing stream of new companies, would be a prime location for a share exchange, perhaps modelled on OFEX. Local investors might be able to follow the progress of their companies more easily, and stay with them for the long term.

At the turn of the century a seismic shift seemed to hit the world of investment. Some so-called new-economy companies, in IT, innovative telecommunications, and, above all, the Internet, saw their share prices rocket upwards, sometimes by multiples of five or ten within a few months. These thrusting businesses became known rapidly in much of the world—through the Internet. Investors jumped onto the bandwagon. Many of these companies, while becoming large in terms of capitalization, had little or no track record of profits, and not much in the way of tangible assets. But a few of them were displacing solid profitable old-established companies in the MidCap Index (FTSE 250) and even in the FTSE 100. A growing chorus of voices has been warning of a shakeout, if not a crash. There has been some correction, with a little less confidence in the future of many of the 'dot.com' companies, and more attention given to investing for genuine value. But, as I write, it is still not clear whether the spectacular rise of the new technology sector will be a brief bubble or the pattern for the future—most likely somewhere in between. The Internet and other new technologies are already changing the world of business. No doubt there will be some consolidation among these companies, but only time will tell what is in store for most of them.

For exporting manufacturers in the UK, the recovery from recession was not to last many years. The Nervous Nineties continued a difficult decade for firms like Oxford Instruments. The Company had a second successful post-recession year in 1995/6, with the return on sales topping the 10 per cent level and profits exceeding City expectations. For most of the year exchange rates were favourable and inflation low. The turnover of the Group rose to £146 million (not counting OMT), and the pre-tax profit to £21.5 million, a record. In spite of health-cut fears and competitive pressures leading to static prices, OMT increased its market share and contributed £6.8 million of these profits. Its constant drive for efficiency and quality was delivering excellent results. In the rest of the Group total R & D costs rose to £10.4 million that year (see Fig. 21.2).

During 1996 the pound started to climb again against other currencies, bringing renewed pressure on profit margins in a company exporting nearly 90 per cent of its output. In the first half of the year OMT's profits were down 20 per cent, and the overseas sales companies, which had all made good profits in the previous year, found the going tougher. The remainder of the financial year to March 1997 was dominated by the strengthening pound and strong competition from the USA, and from Germany, where the DM was very weak. There was a widespread squeeze on healthcare spending, and by the year-end OMT's contribution to profits was down to £5.1 million. But the superconductivity businesses, with their long-term contracts and many customers funded by governments, did well that year, contributing £10 million to operating profits. The instrument businesses, hit sooner by the strengthening pound, contributed £6.5 million. The Company achieved total sales a little higher than in the previous year at £147 million, which would have been £155 million but for the movement of sterling, and Group profit before tax was again a record at £23 million.

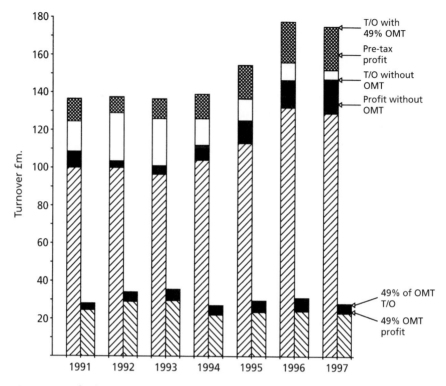

Fig. 21.2. Oxford Instruments' progress from 1991 to 1997, showing the effect of the OMT joint venture on the results

In May 1997 a divided Conservative Party was convincingly voted out of office after eighteen years of mixed fortunes. The Common European Currency, the proposed 'Euro', was the most notable cause of this rift. The potential benefits of a single European currency were clear to many people involved in industry, with international markets, but in other circles there was a strong wish to retain as much independence as possible, and, above all, the sovereignty of the UK Parliament. Apart from the Euro issue, there was, in general, a feeling in the country that it was 'time for a change', and 'New' Labour came to power under Tony Blair. The Government soon gave the Bank of England control over interest rates; it wanted to prevent the stop/go switchbacks in the economy seen in the past. The desire for electoral popularity had often dictated inflation-inducing reductions in interest rates, with public spending sprees, as an election approached. In the months following this change of government the Bank of England decreed four small upward steps in interest rates. The pound grew stronger and stronger as foreign money poured in to benefit from the higher returns, which could be enjoyed with little risk of higher inflation. In the summer of 1997 the pound went well over 3 DM, and remained high against the German currency through the two succeeding years.

With the strength of the pound eroding competitiveness, much of British manufacturing industry again started to suffer. This was especially so for the exporters, although cheaper imports also hurt those producing for the home market. Added to this there were the severe banking and industrial problems in Japan, the harsh falls in 'emerging-market' currencies, and the depression in recently prosperous 'tiger' economies. These adverse factors affected the Company and especially Oxford Instruments Japan. Having grown to account for nearly 20 per cent of Group turnover, it made valiant efforts to minimize the decline. Sales of research equipment there were reprieved to some extent by the Japanese Government's 'supplementary budgets', but the going was still hard, and profit margins went down in the attempt to fulfil the sales budgets. In January 1999 the Euro was launched and in its first year declined significantly against the pound.

The last months of the decade brought more hope. Manufacture in the UK was just starting to recover and the German and other European economies were looking up. The Japanese economy appeared to be turning the corner, but many believed it would not flourish until major structural reforms had been accepted. There was a little improvement in the Russian economy, and some of the ex-'tiger' countries were making progress. There was some fear of the millennium bug, but Western stock markets remained very high, and, although some people talked of the 'new paradigm' of endless growth without tears, others wrote articles with titles like 'Bubbles will Burst'.

On the last day of trading in the century, 30 December 1999, the FTSE 100 index reached an all-time high of 6,950.6. But early in the new year most markets suffered corrections and continued volatile and vulnerable to bad news. The dreaded millennium bug proved to be more like a wood worm—doing minor damage in a few neglected, but not vital, programs.

In this fluid world economy how can a small company have confidence in its long-term plans, or even in its annual budgets? With new ideas and new technology, international business is changing faster than ever before. In this evolving environment a company has to remain open to this wave of new ideas, which may be shaping the future. In the many recent articles and books on the philosophy and methodology of management, there are recycled old ideas, but there are also original and perceptive insights. Some innovative thinkers advocate replacing the rigidity of budgets, which emphasize past constraints, with targets that concentrate on future possibilities (see Hope and Fraser 1999). For others the patterns of growth and renewal are all important (see Handy 1994). Empowerment of managers, a human resources economy, and a knowledge economy are among other strong themes for the new millennium (see Mazarr 1999). The 'knowledge revolution' is a favourite topic, and the need for intelligent search engines for using the ever-widening data stores. As Frances Bacon said back in the sixteenth century, 'Knowledge itself is power.' But knowledge and power are not everything; companies need the human virtues and good leadership to maintain a commonality of purpose in the new millennium.

22 Towards the End of an Era

Developments in the Group and its businesses in the latter years of the century

They have their exits and their entrances;
And one man in his time plays many parts.

(Shakespeare, *As You Like It* (1600))

Applaud us when we run; console us when we fall; cheer us when we recover; but let us pass on—for God's sake let us pass on!

Edmund Burke, from a speech on 'Economical Reform' (1780))

The later 1990s were a time of ups and downs for the Group and its various businesses. After the efforts needed to retain profitability through the long recession had come the drive for new business in the few years of recovery. There were improvements in profits in 1995 and new records in 1996 and 1997. But further disruptions to the world economy and a rising pound lay ahead, and were to bring renewed frustration and a decline in profitability. A certain weariness seemed to settle on the Company as it struggled, yet again, with external pressures and internal problems. Was Oxford Instruments becoming middle-aged? Many of its managers were highly experienced and some had seen their businesses expand rapidly in the 1970s and 1980s. But, in the changing world of the late 1990s, were they still flexible enough in their thinking? Were they analysing their problems effectively and taking the necessary actions? Were they willing to take risks to reposition their operations? Old risks, taken on board in brighter days, were indeed one of the problems. A few projects, which proved far harder than anticipated, were starting to overhang the superconductor-based businesses. And disappointment over the interminable delay in the market for compact synchrotrons probably deterred the Company from taking other opportunities involving a similar degree of risk, with similar potential rewards.

Decentralization into semi-autonomous business units had served the Group well, at least up to the late 1980s. With growing competitive pressures, and the strength of the pound, costs needed to be kept to a minimum, and this structure involved some duplication of functions and facilities. The advantages of small flexible entrepreneurial units no longer seemed so apparent in the new business environment. Many manufacturing companies that depended on exports were having similar problems and started taking stock of their products, their markets, their structure, and their profitability.

By the time the record 1997 results were announced the pound had already begun to climb against other currencies, particularly the DM and yen. In the year to March 1998, turnover was actually up nearly £25 million to just short of £200 million on a restated £175 million in 1997. (With new accounting directives, this now had to include the Company's share of joint-venture turnover.) But the pre-tax profit was down from £23 million to £15.2 million. The strength of sterling had produced an adverse effect of £4 million. With such a high pound it was hard to retain sales in competitive markets at the same time as maintaining the margins of the previous year. This time round, while PT and MAG had continued to perform reasonably well, it was the superconducting magnet businesses that were having a difficult time—partly from those tricky projects. OMT had suffered the problems of the new product launch (see Chapter 17). In two businesses the moves to new factories had disrupted shipments, and some of the fall in profits was due to exceptional items. These included provisions for extra costs on one long-term project and reorganization expenses following an acquisition.

Through the rest of 1998 the insecure economic background persisted, with problems in several important economies in the Far East, South America, and Russia. After spectacular rises in Wall Street and other stock markets in the summer, there was a bank failure in Japan and the near collapse of a 'hedge' fund in the USA. The US Government twisted the arms of the banks to save the situation at Long Term Capital Management and, instead of a feared crash in the markets, there was only a 'correction' in the autumn. The shares of the US investment banks dropped steeply, but the shares of many other large companies soon recovered to go on up to more new records.

Relative to the service sector, manufacture in the UK has been in decline for many years. In the late 1990s it employed only some four million people and was actually in recession for a number of months in 1998 and 1999. Factory-gate prices were static or falling, and many small and medium-sized manufacturing companies were reporting lower profits, if not losses. In contrast, prices in booming services were rising sharply, with big jumps in salaries, especially in financial services and IT. This was threatening renewed inflation, and this persistent dichotomy in the economy presented the Bank of England with a tricky problem over whether to raise interest rates. Efforts to damp down the over-effervescent sectors by raising rates would strengthen the pound, and do further damage to manufacturing companies.

This was the fluctuating economic background for the second half of the 1990s. For much of this period the Board remained optimistic. From 1995 to 1997, currency levels were favourable, and later, when the pound was over-high, there were expectations of its falling to a more reasonable level. There was also confidence in near-term solutions to a few internal problems. This optimism was later to cause some loss of credibility in the City as the lower profits, after 1997, proved to be not just glitches but a steady decline. This was to lead to major changes after the 1999 results, when a new era was to begin. But back in 1995 the Group and its businesses were looking forward to a steady improvement.

So, what notable changes and technical developments had there been in the world of Oxford Instruments through this variable period? Changes of people were, perhaps, most significant for the 'old-stagers', as our companions through the long years of ups and downs started to retire. Many changes were positive, with new blood joining the management, new acquisitions, and new products coming out of development. There were always some internal problems to worry the management, but they did not appear worse than normal until the later years of the decade.

In 1995 John Woodgate retired as MD of OMT, the joint venture with Siemens, a post he had held for eight eventful years. He still remained on the OMT and main boards, and undertook the task of reviewing and monitoring manufacture throughout the Group. With his experience of the spectacular turnaround at OMT in the early 1990s, he was well placed to work with the other businesses to guide them towards achieving 'best practice'. He also helped to plan the new factories to promote better work-throughput, with higher quality and productivity. In February 1999 he retired from the Board due to ill health. He had been with the Group for thirty-three years in which he had made a truly outstanding contribution to its development. His many friends all round the world were saddened by his death in September 1999, but, in characteristic style, he had organized his own memorial celebrations. After a moving service there was a great party, which he had decreed instead of his sixtieth birthday party, cancelled because of his illness. Before his death he had asked that all his friends should be invited to his home to celebrate his life and achievements with his family, with food and wine, enlivened by a jazz band he had booked for the occasion.

In June 1995 Sir Alfred Shepherd retired and two new non-executive directors were appointed to the Group Board. Professor Mike Brady (see Fig. 22.1) had just finished his term as Head of Department in Engineering Science at Oxford University—a post rotating between several professors. He is of an entrepreneurial turn of mind, and has helped to start two or three small companies, as well as joining the Board of AEA Technology plc. He was already a member of TAC, and soon took over its chairmanship from Martin. Professor Alec Broers (now Sir Alec), head of the Engineering Department at Cambridge University (and later its Vice Chancellor), and Sir Rex Richards were also members of TAC along with Sir John Fairclough, who was to retire from both TAC and the Board in July 1998. The other new director who joined the Company in 1995 was Richard Wakeling (see Fig. 22.2), a barrister, who had been the Chief Executive of Johnson Matthey. He was also a director of Logica plc, Staveley Industries

Fig. 22.1. Mike Brady

plc, the Costain Group plc, and other companies, and brought very useful experience to the Board's deliberations.

On 1 April 1996, as the financial year ended, Andrew Mackintosh, who had been MD of IAG, was appointed a main board director and COO of the whole Group. He took responsibility for all operations worldwide, enabling Peter Williams to focus on strategic issues. Andrew had joined the Group in 1984 at the age of 28 after a spell in university research—in Cambridge and in Japan (he speaks fluent Japanese)—and an early career in Courtaulds and EMI. In his previous experience in Oxford Instruments he had been a project engineer and then Sales Director in the old magnet company, so had a deep knowledge of both the superconductivity-based businesses and industrial instruments. He was to become Chief Executive of the Group in February 1998.

RI was the embodiment of the original Oxford Instruments. Everyone who had joined the Group before 1974 had worked in it. From 1983 until his retirement in 1998 its MD was John Pilcher, who had joined the Company back in 1964, in Middle Way days. He was a dynamic, warm, and caring person, held by his staff in great affection—sometimes tinged with awe. He could motivate his team, and had an extraordinary ability to get many difficult projects finished, tested, and out of the factory door in time for the financial year-end. Although not in the mould of a management whiz kid, he knew every facet of the operation and often performed miracles. In all, he worked for the Company for thirty-three years, and was one of the greatest contributors to its success for most of its history.

After John retired Steve McQuillan took on the job of MD. An electronic and electrical engineer of 38, he had been working as Global Sales Director for Marconi Instruments in the USA. His excellent track record included work in engineering, sales management, and strategic marketing with Mars Electronics

Fig. 22.2. Richard Wakeling

International, and later as MD of Sodeco in Switzerland. He brought increasing professionalism to the magnet business, promoting more use of carefully documented technical, cost, and marketing information in the choice of standard new products. While recognizing the value of know-how and 'green fingers' in the demanding project business, he instituted stricter criteria, including better risk analysis, for the design, quotation, and contract negotiation for difficult custom-made superconducting magnets.

In 1997 RI moved from its 50,000-square-foot factory in Eynsham, where it was bursting at the seams, to brand new 80,000-square-foot premises in Tubney Wood, a few miles west of Oxford. This had been a small wartime industrial site, where Bofors guns were made and tested in a series of very long windowless buildings with thick bombproof roofs. Modern bulldozers soon reduced them to rubble. The new factory, in its beautiful woodland setting (see Plate 4), gave more scope for this group to overhaul its management procedures and production methods with a view to greater efficiency. There was soon some improvement, although there was a backlog of difficult long-term projects that was to hold the operation back for several years.

Project work is still an important part of the business. Throughout its history the Company has designed and supplied very advanced and specialized superconducting magnets and cryogenic systems to the leading research laboratories of the world (see Plate 5). Nobel Prize winners have used our equipment for their seminal experiments. Stretching the technology in this way has not always proved profitable, but the results obtained by scientists using the systems sometimes open up new avenues of research. This can lead to new markets for standard commercial products developed from the original equipment made for a particular experiment. This process has resulted in ranges of products, like the Teslatron family of magnet systems with fields of up to 20 T. Others include the Cryofree superconducting magnet systems, which are cooled by mechanical refrigerators instead of liquid helium, and are making products based on superconductivity more acceptable to industry. And there are the MagLab systems for the characterization of materials—an essential stage in the hunt for new and improved semiconducting materials. Still newer are the non-superconducting compact Pulselab systems, which have been developed in conjunction with the Clarendon Laboratory. These can achieve peak fields of over 45 T for short periods of time, and have important applications both in academic research and in industry. In 1998, with improved materials coming from OST, the Company achieved fields of nearly 22 T (see Plate 9).

Many research scientists require low temperatures in conjunction with high magnetic fields. The Company's ultra-low temperature systems have been refined over many years (see Plate 9). Among the nine standard Kelvinox dilution refrigerators, three models, developed in the late 1990s, use an entirely different pumping system, while still providing temperatures down to 10 mK or lower. Instead of conventional oil and mechanical rotary vacuum pumps, connected to the system by large-diameter pumping lines, the cooling cycle in these

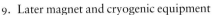

9. Later magnet and cryogenic equipment

Above: a 21 T magnet for the Clarendon Laboratory, shown without its case
Right: core components of two models of dilution refrigerator for achieving ultra-low temperatures
Below: a 'Telstar' system developed by Stereotaxis. Oxford Instruments provides the 'helmet' containing three magnets in their cryostat. These guide a catheter, tipped with a permanent magnet, through the blood vessels to the heart or brain for diagnostic, drug delivery, or surgical procedures. A model of a heart shows a catheter being guided into the heart chambers

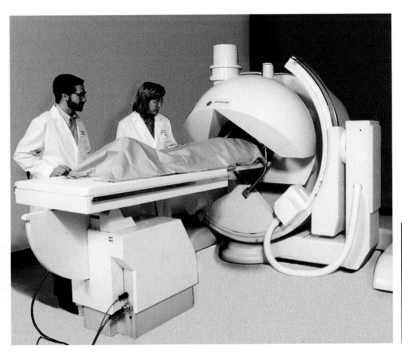

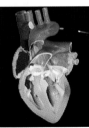

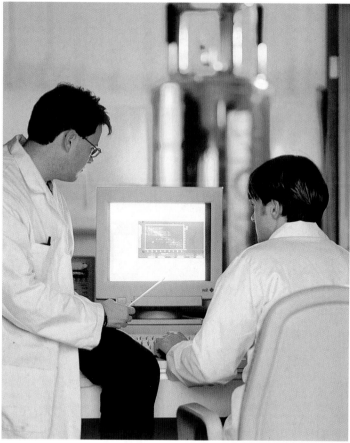

An NMR magnet system

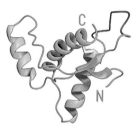

BSE prion

10. Research using an NMR spectrometer

Used in the analysis of the structures of large organic molecules, these systems are essential tools for work on viruses, genes, and other proteins in the body, and for drug development

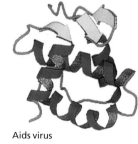

Aids virus

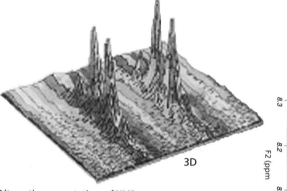

3D

Alternative presentations of NMR spectra

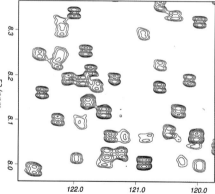

2D

11. Equipment for use with a particle accelerator

Top: the CLAS toroidal particle-detector magnet installed at CEBAF, Virginia, USA, in 1996
Above: a unique magnet/refrigerator combination, for polarized particle experiments, designed to fit inside the CLAS magnet, shown being installed in 1998, and before delivery

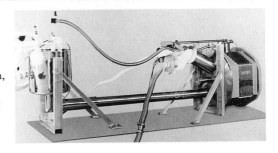

12. A microanalysis system in use

Below: an example of the kind of chemical information available

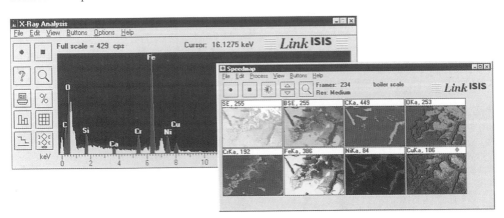

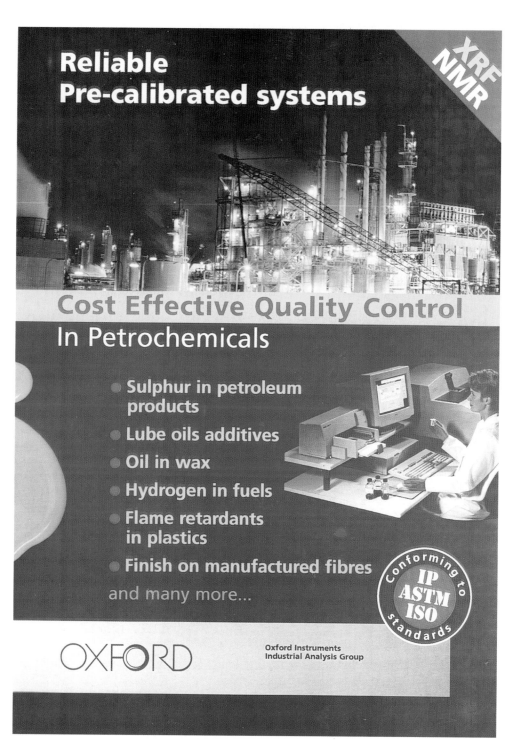

13. An advertisement for analytical instruments for industry

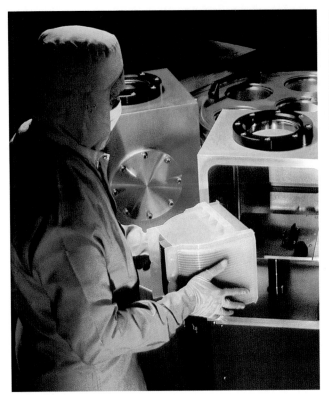

Versatile 'Plasmalab' System

Micromachined silicon motor

Integrated circuit interconnects exposed during 'failure analysis' etching

A gallium arsenide 'via hole' feature used in circuits for mobile phones

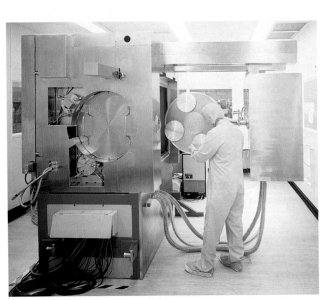

'Ionfab' system for optics applications being assembled

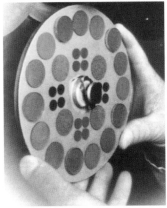

Optical filters coated in an 'Ionfab' system

14. Plasma and ion beam equipment for research and industry
Above right: some examples of the precision etching and coating that can be achieved with these systems

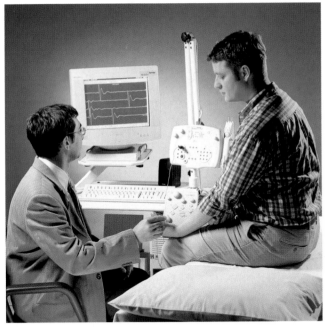
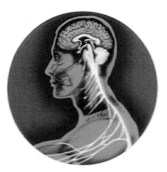

15. Medical equipment used in diagnostics

Above: testing nerve conduction and muscle response with a 'Synergy' system
Below left: the simplest of several systems for foetal and obstetric monitoring
Below right: analysis in progress for a sleep problem

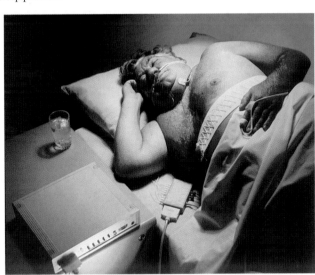
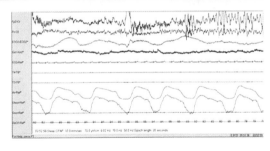

John Woodgate, Sir Ashley Ponsonby, then Lord Lieutenant of Oxfordshire, and David Hawksworth MD, in the magnet test area following the presentation of a Queen's Award for Industry

16. At Oxford Magnet Technology, and an MRI system in use

Above: the factory, outside and in, and work in progress on a magnet
Below: an MRI scanner in use, and images of a head (with teeth), a knee, a spine, and kidneys

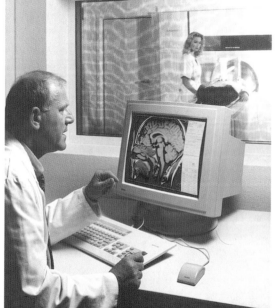

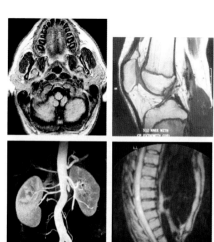

systems uses sorption pumps within the cryostat itself. This innovation has led to much smaller systems, which can be inserted into various different cryostats, and even into liquid-helium storage vessels. These refrigerators also have the advantage of very low intrinsic vibration, which is important when they are used in conjunction with NMR magnets and other sensitive equipment.

Magnet systems for commercial purposes have been growing in significance since the 1980s. Over twenty of the large superconducting magnets for magnetic separation had been delivered by the end of the century, some to very remote areas (see Fig 22.3). There is an expanding market for magnets for crystal-growing furnaces, and at least thirty large gyrotron magnets were produced in the 1990s. Gyrotrons are used for high-power microwave generation at extremely high frequencies, and some are designed for heating up plasmas in fusion research.

Fig. 22.3. Large superconducting magnet separation system in a mineral processing plant

In the medical arena, in 1998, RI started working with Stereotaxis, a US company, in the development of complex systems for guiding a catheter through the body. RI developed a prototype magnet system that would steer a tiny permanent magnet attached to a catheter through blood vessels to an exact position in the brain or the heart. This was to be used for diagnosis, for delivering drugs, or for other medical procedures (see Plate 9). The potential market for these systems may be in the hundreds. One article commented, 'Of course the question for LTS [low-temperature superconductivity]-magnet manufacturers is whether

MSS [magnetic surgery system] can open up a new, widespread application for superconductivity in medical markets, perhaps even on the scale of MRI?' (Bitterman 1990: 14). There also seems to be a renaissance of other possible new medical and commercial applications deriving from recent research.

Over the years, national institutions for research in high magnetic fields have been important customers. The French national research centre in Grenoble bought its first high-field magnet from Oxford Instruments in 1963. In the intervening years many magnets and cryogenic systems have gone to this laboratory, and in 1998 RI started collaborating with the team there on a very innovative and difficult project for a new type of hybrid magnet. In 1961 the Frances Bitter National Magnet Laboratory at MIT in Cambridge, Massachusetts, promoted the first International Conference on High Magnetic Fields—which set us on the road to superconducting magnets. This laboratory made great pioneering strides in magnet technology, and has been a long-term customer for research equipment. In the 1990s it bought NMR magnets, as it was then becoming an important centre for NMR research. Centres for magnetic field research in Japan, the Netherlands, and Germany have also been good customers, as has the new National High Magnetic Field Laboratory at Tallahassee in Florida. Established in the early 1990s, this has grown rapidly to become a major centre of high-field research. The Company has supplied several 20 T magnets for work there, and NMR Instruments have made for them both standard NMR magnets and a unique 9.4 T Fourier transform ion cyclotron resonance system. This is used for ultra-high resolution mass measurements of complex molecular structures. The system has revealed fine structures known to exist in protein molecules, which had not previously been observed. This work is proving so interesting that a second system is being supplied with a higher field and larger bore.

Like many of the other businesses in the Group, RI publishes its own newsletter for its customers. 'Research Matters' is full of information on what is going on in the world of high magnetic fields and cryogenics—both in the Company and elsewhere.

In 1998 NMR Instruments moved into the Eynsham factory vacated by RI, and Alan Street moved from ATG to become its MD. Alan had previously worked in RI as well as ATG, and realized the importance of better interaction between all the superconductivity-based operations. He instituted regular interbusiness meetings on technical matters, and promoted widespread informal contacts both inside and outside the Group.

In the thirty-five years since work started on Oxford Instruments' first NMR magnet around 5,000 have been delivered and, in the late 1990s, the Group was producing about 230 of these systems a year. Several new models were introduced in the 1990s, some actively shielded like the imaging ones, which makes their siting in laboratories easier. NMR spectrometers have become vital tools for determining the structure of complex organic molecules (see Plate 10).

They are used everywhere in biochemistry, molecular biology, biotechnology, and pharmacology laboratories, and have assisted in the development of most modern therapeutic drugs. In this respect they have probably been of more direct and practical benefit to the 'man in the street' than any of our other products, although few people recognize the role they play. In 1996, in the Wellcome Witness Seminar 'Making the Human Body Transparent', Sir Rex Richards made a comment that shows the continuing importance of NMR for the future of medicine—and of our NMR business:

NMR continues to be a source of a tremendous amount of information and it's worth asking why. The reason is the huge information content that there is in almost any NMR spectrum. There are five or six independently measurable parameters for every distinguishable nucleus in a sample. There is a huge amount of potential information in the NMR spectrum and I don't myself believe that we have exploited all of that, by any means. (The Wellcome Trust 1998: 17)

A new initiative in the USA showed the importance accorded to the technology there. In January 1998, at the suggestion of the US National Science Foundation, a conference of NMR spectroscopists was held in Washington, DC, entitled 'High Field NMR: A New Millennium Resource'. The report following this meeting began:

Nuclear magnetic resonance (NMR) spectroscopy has profoundly influenced science over a diversity of disciplines. Today, advancements in NMR technology coupled with computational and connectivity power are making NMR an essential tool for the nation's biological, chemical, and materials science communities; a tool that can advance the frontiers of science and help address the national scientific agenda including:
  • Beyond the Genome . . .
  • Gene Regulation . . .
  • Neuroscience . . .
  • New Materials . . .   (US High Field NMR Committee 1998: 3)

The report on the conference went on to say that the Japanese were investing $500 million in a new NMR centre, and to recommend a cost-effective US solution in 'a National Magnetic Resonance Collaboratorium (NMRC), a virtual laboratory . . . that would network collaborations among sites and the nation's scientific communities'. It was envisioned that approximately ten leading NMR sites, linked by 'Next-Generation Internet', would serve as 'sectors' building on existing investments (US High Field NMR Committee, 1998: 3).

In the late 1990s NMR Instruments' development section started working on an important contract for the Batelle Memorial National Institute in the USA, for building a wide-bore 900 MHz NMR magnet system. This is for a major research programme concerned with the environmental problems in a large area of highly contaminated land, the site of nuclear research and bomb manufacture during the Second World War and the cold war. With a field of 21 T, and a bore of 65 mm, the magnet will be at least twice the size of the standard 800 MHz magnets, and will hold much more stored energy when it is running. The forces involved are huge. It is a very big technological step forward. Another team started developing a smaller-bore 900 MHz magnet system that is a little less demanding and will become a standard product. This team, led by Martin Townsend, reached its major goal in August 2000—just in time to be included here. The unique magnet was quickly mated with its 900 MHz Varian spectrometer that soon produced remarkable spectra from samples rushed to Oxford from distant NMR scientists. Several were shown by speakers at an international NMR conference in Florence two weeks later, to general acclaim. Ray Shaw of Varian said the alliance had taken pole position in NMR once more. In the US High Field NMR Committee's report, the forecast budget suggested $32.5 million for the development of five narrow-bore 900 MHz spectrometers and probes, and $55 million for five wide-bore systems. The next stage was to be 1 GHz (1000 MHz) systems. These are exciting times.

OST continues to supply NbTi wire in large quantities to OMT and to other Group operations (see Plate 8). It also supplies materials to other magnet companies and laboratories, and, in 1988, it received a new contract to supply another major MRI company. But the wire business is now quite competitive and, taking all the suppliers around the world, there is over-capacity.

In 1997 OST acquired a new 'Modified Jelly Roll' NbSn product. This method of making multifilamentary NbSn conductor needed further development, but the material is now being used for the highest-field NMR and RI magnets. Development of high-temperature superconductors (HTS) still goes on. In a collaborative magnet project with the Tallahassee Laboratory, OST provided some of its latest form of the superconductor BSCCO 2212 for a central coil, which was fitted inside a 19 T superconducting magnet. Together these magnets achieved 22 T. Still higher fields will be achieved with these remarkable HTS materials, but their high cost still prevents their use in standard magnets. Low-temperature superconductors remain much cheaper.

OST also services an installed base of around 3,000 MRI magnet systems in the USA and Canada, produced by the Group's imaging magnet operations over the 1980s and 1990s, and growing steadily in number.

In 1998 ATG expanded into the next-door building on Osney Mead that NMR Instruments had left. Alistair Smith had been in overall charge of the Semiconductor Systems Division, including PT, and, after Alan Street transferred to NMR Instruments, he took direct control of ATG as acting MD.

The work on the Helios synchrotrons and other advanced equipment resulted in a team with strong capabilities in precision motion and systems integration. ATG was finding more opportunities for designing and making the complicated experimental end stations needed at the large-scale synchrotron research facilities built over the past decade or so. The first complete 'turn key' integrated X-ray beam line was installed in 1998, and at the end of the century there were twelve systems under manufacture for this growing market (see Fig 22.4). Eight

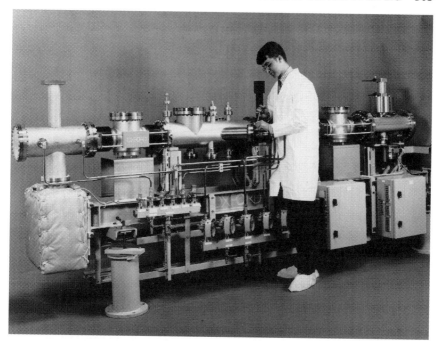

Fig. 22.4. Part of a beam line: the 'front-end' module for the European Synchrotron Radiation Facility in Grenoble

of these systems are designed for the new German synchrotron facility at Karlsruhe. Early in 2000 it was announced that a new large synchrotron is to be built at RAL, near Oxford. The experience built up at ATG should enable it to win orders for engineering projects at the new facility. ATG continued to address the market for larger custom-designed magnets and cryogenic projects, such as the manufacture of magnet test stations for the CERN Large Hadron Collider programme. The delivery of Helios 2, for the National University of Singapore, was delayed while its building was completed, but it was shipped late in 1999, and commissioning started soon after.

MAG was the largest of the instrumentation businesses. By the mid-1990s its old premises at High Wycombe had become inadequate, and a new factory was built on the same site, in stages. In 1998 Professor Alec Broers came to open the fine new building; he was then the Vice Chancellor of Cambridge University, and he flew in by helicopter, between other commitments, to open the factory. The assembled visitors were treated to fascinating displays showing how micro-analysis is used in forensic investigations, in the assessment of jet-engine wear, and in other interesting applications (see Plate 12).

Back in 1992 MAG had launched ISIS, a brilliant product that spearheaded its recovery from the recession as business started to pick up. In 1995 a second generation of ISIS instruments was launched to make use of Microsoft Windows 95. That year MAG acquired a small company, the Microspec Corporation, based in California. This brought to the Company very sensitive wavelength-dispersive X-ray (WDX) spectroscopy products, which are complementary to the energy-dispersive X-ray (EDX) products MAG was already producing. In July 1998 ISIS was replaced by INCA, a versatile, industry-leading, system, in which both EDX and WDX techniques are combined.

The Link OPAL product, developed in the mid-1990s, is an electron diffraction system that enables the user to perform crystallography on a micro-scale, using a scanning electron microscope. This micro-crystallography can be integrated with the electron imaging provided by ISIS or INCA to form a powerful analytical tool.

MAG also launched lower-cost variants of existing products for the competitive markets that prevailed in many countries in the late 1990s. Semiconductor defect analysis was one of MAG's chief markets and the severe recession in that industry in the last few years of the century cut its order intake. As 1999 closed there was hope of a revival in the semiconductor industry. The development team is always on the lookout for potential new applications for these analytical techniques. Thin film composition and surface characterization are important areas as devices become smaller and smaller, and the molecular structures more relevant.

For MAG it is important to train users to operate their systems properly and interpret the results correctly. A team there developed a brilliant CD ROM training package, *Principles and Practice of X-ray Microanalysis*, which won an 'Oscar' from the British Interactive Multimedia Association. As part of this training programme the learner watches an interactive cartoon-style movie called *Mike Roscoe, Private Investigator (Microscopy)*. This Philip Marlow-like character talks the watcher through all the stages of microanalysis on a strange small specimen he finds on his desk. The graphic maps of elements and corresponding analysis results come on screen, and, in the end, the specimen turns out to be a tooth filling he has just lost.

Software is always of major importance, and the analytical-instrument businesses have increased their work on 'applications-specific' software for particular tasks they have identified. This spreads the market for their core technologies. They can also diagnose difficulties with systems at a distance, and they provide worldwide support.

IAG in Abingdon found the last few years of the 1990s difficult as its client manufacturing and processing companies cut back on capital spending. IAG's strategy lies in concentrating on a few important industries where it can achieve an understanding of the processes involved, develop appropriate measurement capability, and deliver true value to the manufacturers in proportion to the cost of the new instruments. The industries addressed in the last years of the 1990s were fuels and lubricants in the petroleum industry (see Plate 13), polymers, synthetic fibres, paper coatings, minerals and cement, and environmental waste analysis. IAG focuses on solving a range of problems in each of its chosen industries, using its existing technologies of low-resolution NMR and XRF, and adding others to increase this range.

Many countries have started legislating for more safety in the environment and in the workplace, and new opportunities lie ahead in sampling for certain quality control criteria. In 1994 the US Government imposed a standard for sulphur in fuels. IAG's QX product was the easiest to use and the lowest-cost instrument available to meet the regulatory needs, and took 50 per cent of the market until all the refineries and distribution depots had acquired instruments. With the pattern of these regulations, IAG has to keep on its toes to meet the next directive somewhere in the world, and satisfy what is likely to be a temporary market there. For sulphur levels in oil this will be in Europe, where the maximum comes down from 500 ppm (parts per million) to 350 by the year 2000 and 50 by 2005. The enhanced product, MDX, which replaced QX, can cope with this, but IAG has also developed a dedicated version of Lab-X 3000, the most successful product it has ever made, as a still lower cost solution.

In the late 1990s IAG was working with the cement industry, for which it developed a successful and popular on-stream 'raw meal' cement analyser called OSCA, based on the Lab-X 3000 analysis head. The mix of materials going into the kiln is all-important for the quality of the cement, and is normally analysed by sampling every hour. The new system adds value by automatic control of the mix from the online sampling, which both saves technician time and improves quality.

In 1999 the Company bought, for $4.5 million, the business of Auburn International Inc., which makes online instruments, based on NMR, for industrial process control. The instruments can measure key process parameters such as density and crystallinity, which are important in the manufacture of polyethylene and polypropylene. Auburn has also developed a range of sophisticated devices that can measure the flow of solids in a gas or liquid carrier. These may have applications far beyond the polymer industry. This operation has been amalgamated with IAG in Abingdon, and has given the Company a wider range of products in its chosen fields.

After many changes and difficulties, PT started running smoothly, and its sales and profits began to grow. Since 1993 it has been headed by Alan Goodbrand. He joined Oxford Instruments in 1992 after a successful twelve-year engineering, sales, and marketing career in the company Electrotech, including service on the Group Board. During the 1990s ion beam processing equipment was made compatible with standard wafer-handling equipment. This enabled different processing systems to be 'clustered' together for robotic transfer of wafers in an automated chain of processes (see Fig. 22.5).

The ever-shrinking digital devices of the 1990s needed smaller and smaller components. Hard disk drives for PCs were reduced in size to fit laptop, or

Fig. 22.5. The 'Plasmalab System 100' hexagonal robotic wafer-handler at the heart of a versatile cluster tool

palmtop, computers. But there was a parallel demand for ever more processing power and storage capacity. For packing information more tightly on these smaller discs, micro-miniaturized 'thin film magnetic read–write heads' (TFMH) were needed. PT developed equipment for IBM and other companies that they needed for the manufacture of these heads.

Microcircuits made from compound semiconductors, such as gallium arsenide, have become more and more important as devices like mobile phones have gone on shrinking. They need much less power and are faster in operation than silicon-based chips. By the end of 1999 around 60 per cent of PT's sales were in equipment for working with compound semiconductors, and these sales were growing at 25 per cent a year (see Plate 14). As the worldwide mobile-phone market exploded, this became a significant opportunity for PT, although the market is not without competition.

The small operation, X-Ray Technology Group (XTG), in California, produced a number of innovative X-ray tubes during the 1990s. XTG makes all the tubes used in IAG's spectroscopy products, like the Lab-X 3000, as well as supplying tubes for purposes such as thickness gauging, particle analysis, density measurement, and industrial radiography. For one of several collaborations, XTG started developing X-ray tubes with focal spots at or below ten microns, about one-seventh the diameter of a human hair. These are for inspection systems for examining the tiny arrays in the smallest semiconductor devices. In the medical field XTG has long-term contracts to supply X-ray tubes for real-time bone imaging products for use during surgical procedures, and in the diagnosis of osteoporosis.

NMG, acquired as part of the Link Analytical Group, had performed quite well in its early years with Oxford Instruments. At that time there was a substantial programme in the USA aimed at cleaning up the nuclear pollution still remaining after the end of the cold war. But in the mid-1990s, moves towards a balanced budget cut back some of the work of the US Environment Department. In spite of some strong product lines, the business was unable to compete profitably in a static and consolidating market. In 1999, after years of underperformance, the business was sold for $9.5 million, with continuing royalties on one unique product.

For OM, the most important event in the second half of the 1990s resulted in a doubling of its size. This event was the purchase from Vickers Medical, late in 1997, of its 'Medilec' and 'Teca' neurology businesses, for about £10 million. The Vickers businesses were based at Old Woking, and OM's small existing neurology section soon moved there.

With this major acquisition came a reassessment of the priorities for the Medical Division. OM started life in the early 1970s with the then revolutionary cassette-tape physiological recorders, which were very successful through several generations of instruments. Growing competition, and cost-consciousness among doctors, steadily cut profit margins until the cardiac recorders became almost commodity products. The Company continued to offer a range of these recorders and analysers, but other products overtook them in profitability. The Obstetrics section, which produces a variety of antenatal and labour and delivery monitors, remained very significant for the future growth of OM (see Fig. 22.6 and Plate 15). At the end of the 1990s this section developed a new range of obstetric monitors. One system was launched in December 1999, and was well received by the market. This range is promising for future business.

The most important products in the Medilec range were for electromyography (EMG), for testing nerve conduction and muscle responses. In this business they were leaders in the market and in the technology, which included the use of modern Windows software. There was also an accessories business that every year made millions of the complicated electrode needles, needed for EMG, and for other purposes, as well as producing many other types of electrode used in diagnosis and for therapy. These mainly disposable products were sold widely, many to competitors. In other areas of neurology Medilec had a strong position in digital EEG, and in equipment for sleep studies. These products had been

Fig. 22.6. Antenatal
monitoring is possible
at home or at a local
clinic: a communica-
tions pack transmits
the unborn baby's
heartbeat to the
hospital for analysis

developed in different directions from the Oxford systems, which had concentrated on the portable or ambulatory area of diagnosis.

This amalgamation of medical businesses formed the largest operation in the
Group. The acquisition also brought to OM an excellent distribution network,
which complemented OM's existing outlets. One new instrument launched
after the acquisition was 'Synergy', an EMG system used in the diagnosis of diseases such as muscular dystrophy and multiple sclerosis (see Plate 15).

Apart from the hiccup in 1997/8, caused by the unexpected problems in a
newly launched magnet product (see Chapter 17), OMT continued to flourish.
These magnets, the OR 24 for 1 T and OR 70 for 1.5 T, which were identical on
the outside, proved very successful (see Plate 16). They are shorter than previous magnets and this gives easier access to patients during scanning operations,
and leads to less claustrophobia. They are also relatively light, and produce
smaller stray fields than most actively shielded magnets, making them easy to
site in hospitals. Their success led to record sales of £121 million in the year to
March 1999, with operating profits of £14.6 million for the joint venture.

As new applications for MRI continued to enter clinical use, and demand
increased, the Company started applying a programme aimed at doubling the
capacity of the existing facility. OMT has continued to work hard on cost
reduction, and this is broadening the range of hospitals and clinics that can
afford to offer this prime diagnostic facility.

A collaborative development between OMT, RI, and Siemens Corporate Technology resulted in a prototype magnet made with tapes of high-temperature superconductor. The magnet was designed to integrate with the Siemens C-shaped Magnetom Open scanner, which normally incorporates a resistive copper-wound magnet. The HTS magnet achieves a field of 0.2 T. Although the project demonstrated the use of HTS tape in a large-scale MRI application, commercialization will not be possible until the cost of the materials comes down a long way.

In the last few years of the century links between OMT and the other super-conductivity-based businesses of the Group were strengthened and formalized. A team from across these businesses developed joint strategies to handle certain emerging product lines and started exploring better ways of coordinating technology development programmes. This started to lead to benefits for all these operations.

In the press release following the disappointing results for 1998 Andrew Mackintosh commented, 'Oxford Instruments remains a world leader in our core technologies with strong market positions in chosen sectors. We have continued with investments on several fronts over the past 12 months and I am confident that these will contribute significantly to results in the near future.'

'Near' is not a precise word. As I have already indicated, the year to March 1999 was to prove even worse. With the persistently strong pound and a poor half-way performance, City expectations for our 1998/9 profit were not high, and a trading statement in April made everyone aware that a poor result was likely. Turnover for wholly owned Group companies was down by £4.4 million to £167.8 million, but when our 49 per cent of the OMT turnover was included it was up from £199.3 million to £217 million. The pre-tax profit was down from £15.8 million to £11.5 million, again strongly supported by OMT, where the popular new magnet was running smoothly through production after the problems of the previous year.

During 1998 and 1999 a lot of thought had gone into the Group's difficulties. Consultants had started working with the superconductivity-based businesses to improve their operational performance, including the organization of manufacture, quality, on-time delivery, and new product introduction. In the instrument businesses very poor markets, especially in Europe and Japan, and increased competition, had led to lower orders, and overheads had been cut back. Although the profit was weak, and the share price with it, the Group remained basically sound, with net assets of around £100 million, and net cash of £6 million. The businesses were still organized, as they had been for some years, into ten semi-autonomous operations (see appendix to this chapter). But, with the help of the consultants PriceWaterhouseCoopers (PwC), a study was in progress to assess the benefits of joint procurement. These consultants were also involved in discussions on the possibility of introducing more-fundamental structural changes and on other new initiatives. These were aimed at coping

with external problems, such as weak markets, and overcoming internal inefficiencies in R & D and in manufacture. Change was in the air. An era was coming to an end.

As the 1999 Report and Accounts stated:

Oxford Instruments is a global leader in advanced instrumentation. Our products are
used by people all over the world for scientific research, industrial chemical analysis and
quality control, healthcare and semiconductor processing. Our customers include most
of the world's major companies as well as the leading research institutes. . .

We are proud of our tradition of innovation through joint development with our customers and are committed to using our strengths to create value for our shareholders.

The new era started in July 1999 with a decision to reorganize the Company. But
it was expected to take a couple of years for major new initiatives to work
through to much better profitability and better value for the shareholders.

Appendix The Structure of Oxford Instruments at 31 March 1999:

The End of an Era

The Board of Directors and directors of operational functions

Chairman: Sir Peter Williams CBE FRS FREng
Deputy Chairman (non exec.): Sir Martin Wood OBE FRS FREng. DL
Chief Executive: Andrew J. Mackintosh Ph.D.
Finance Director and Company Secretary: Martin Lamaison FCA
Professor Michael Brady FRS FREng. (non exec.)
Nigel J. Keen FCA (non exec.)
Richard K. A. Wakeling, Barrister FCT (non exec.)

Director of Business Development: Ian McDougall Ph.D.
Director of Corporate Development: Roger Humm MBA FCA BSc.
Director of Information Systems: John Lewis-Crosby FCA BA

The Divisions and Strategic Business Units and their key staff

Superconductivity

Research Instruments

MD: Steve McQuillan
Senior Executives: Ken Bailey, Nick Kerley, Jim Hutchins, Mike McEvoy, Steve Vale

NMR Instruments

MD: Alan Street
Senior Executives: Kevin Hole, Neil Killoran, Andy Taylor, Richard Thomson,
   Nick Wilkins, Claire Goulding

Oxford Superconducting Technology

President: Jim Worth
Senior Executives: Seung Hong, Scott Reiman, David Andrews, Rowdy Teague,
   Bill Patterson, La-Verne Tyler

Accelerator Technology

MD: Alistair Smith
Senior Executives: Nigel Boulding, Tracy Cuthbert, Dave Fisher, Vince Kempson, Peter
   Penfold, Kevin Smith

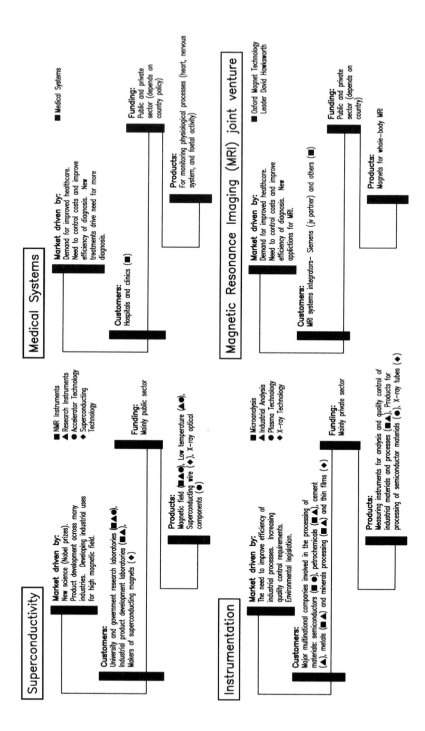

Superconductivity

Market driven by:
New science (Nobel prizes).
Product development across many
industries. Developing industrial uses
for high magnetic field.

Customers:
University and government research laboratories (■▲●),
Industrial product development laboratories (■▲),
Makers of superconducting magnets (♦)

Funding:
Mainly public sector

Products:
Magnetic field (■▲●), Low temperature (▲●),
Superconducting wire (♦), X-ray optical
components (●)

■ NMR Instruments
▲ Research Instruments
● Accelerator Technology
♦ Superconducting
   Technology

Instrumentation

Market driven by:
The need to improve efficiency of
industrial processes. Increasing
quality control requirements.
Environmental legislation.

Customers:
Major multinational companies involved in the processing of
materials: semiconductors (■ ●), petrochemicals (■▲), cement
(▲), metals (■▲) and minerals processing (■▲) and thin films (♦)

Funding:
Mainly private sector

Products:
Measuring instruments for analysis and quality control of
industrial materials and processes (■▲), Products for
processing of semiconductor materials (●), X-ray tubes (♦)

■ Microanalysis
▲ Industrial Analysis
● Plasma Technology
♦ X-ray Technology

Medical Systems

Market driven by:
Demand for improved healthcare.
Need to control costs and improve
efficiency of diagnosis. New
treatments drive need for more
diagnosis.

Customers:
Hospitals and clinics (■)

Funding:
Public and private
sector (depends on
country policy)

Products:
For monitoring physiological processes (heart, nervous
system, and foetal activity)

■ Medical Systems

Magnetic Resonance Imaging (MRI) joint venture

Market driven by:
Demand for improved healthcare.
Need to control costs and improve
efficiency of diagnosis. New
applications for MRI.

Customers:
MRI systems integrators– Siemens (jv partner) and others (■)

Funding:
Public and private
sector (depends on
country)

Products:
Magnets for whole–body MRI

■ Oxford Magnet Technology
   Leader David Hawksworth

Fig. 22.7. Oxford Instruments' structure at 31 March 1999

Table 1. Statistics for the divisions from 1997 to 1999

Division and year	Turnover (£ m.)	Operating profit (£ m.)	Operating assets (£ m.)	Employees
Superconductivity Division				
1997	74.2	10.0	42.4	797
1998	90.0	5.2	38.7	815
1999	74.7	0.1	34.6	809
Instrumentation Division				
1997	54.4	7.2	24.1	515
1998	59.3	7.8	35.8	586
1999	54.0	3.6	35.6	556
Medical Systems Division				
1997	18.5	0.7	5.4	223
1998	22.9	(0.1)	18.4	450
1999	39.1	0.3	21.5	423
OMT Joint Venture				
1997	72.8	9.0	10.6	390
1988	75.7	5.7	13.1	426
1999	121.0	14.6	17.4	525

Instrumentation

Microanalysis Group

MD: Ron Jones
Senior Executives: Dan Varnam, Dave Scott, John Comfort, Peter Statham, Peter Roberts

Industrial Analysis Group

MD: Simon Bennett
Senior Executives: Brian Mitchell, Andy Ellis, Colin Cleaton, Damian Campbell

X-Ray Technology

President: Bradley Boyer
Senior Executives: Scott Snyder, Brian Burk

Plasma Technology

MD: Allan Goodbrand
Senior Executives: John Field, Frazer Anderson, Jim Williams

Medical Systems

Oxford Medical Systems

MD: Paul Brankin
Senior Executives: Mike Russell, Victor Regoczy, Robin Higgons, Peter Roberts,
   Michelle Lucas, Mike Pattinson

Magnetic Resonance Imaging (joint venture)

Oxford Magnet Technology

MD: David Hawksworth
Senior Executives: Graham Gillgrass, Frank Davies, David Grime, Rob Owen,
   Terry Allen, Bob Graham

Overseas Operations and Federal Offices

Oxford Instruments Japan

President: Jiro Kitaura
Senior Executives: Toa Hayasaka, Yasunori Yoda, Masahiko Nishiyama, Tony Ford,
   Kinya Matsutani, Yuichiro Shima (total staff of 52)

Oxford Instruments Federal Offices

USA (staff of 162)
France (staff of 35)
Germany (staff of 67)
Latin America (staff of 2)
China (staff of 6)
Singapore (staff of 6)
Belgium (staff of 11)
Spain (staff of 21)
Italy (staff of 24)
Holland (staff of 5)

23 The Beginning of a New Era

Restructuring and other remedies for a better future

> We dare not lengthen this book much more, lest it be out of moderation and should stir up men's antipathy because of its size.
>
> (Aelfric, teacher, of Cerne Abbas, later Abbot of Eynsham (*c*.995–1020))

> Cost cutting, downsizing, re-engineering, are all important and necessary activities, providing the efficiencies and hence profit to survive today and to finance the future. It is, however, new and improved products, new and improved manufacturing processes to make them, and new markets to be won, that produce growth.
>
> (Sir Robert Malpas in a speech to the 'Oxford University in Business' seminar (4 November 1999))

With the June announcement of the 1999 results came news that Peter Williams would be retiring as Chairman after nearly seventeen years with the Company. He had always been interested in higher education and the promotion of research and innovation on a national scale, and his appointment as the new Master of St Catherine's College, Oxford, was announced that July. Earlier in 1999 he had been elected a fellow of the Royal Society, and he was to become President of the British Physical Society. In his years as Group Managing Director, then Chief Executive, and finally Chairman, Peter had seen through several cycles, both in the country and in the performance of the Company. He had succeeded in raising the scientific profile of the Group and in his time there were many other achievements. Together with John Woodgate, he had been responsible for securing the very successful joint venture in imaging magnets with Siemens, which had contributed strongly to the profits of the Group. He had negotiated the acquisition of the Link Scientific Group, which was the largest of the instrumentation operations and, in the late 1990s, was the most profitable of the businesses. We were sad that his departure came at the bottom of a cycle in the Company's performance, although the seeds of better things had already started to germinate.

Departures create new opportunities. The newly elected Chairman, who took office at the Annual General Meeting, was Nigel Keen (see Fig. 23.1). Appointed a non-executive director in February 1999, he had extensive operational experience of investment and management in engineering and high-technology companies. He was then 52, and had both engineering and accountancy qualifications. He was Chairman of the Cygnus Group of companies, Deputy Chairman of the substantial Laird Group plc, Chairman of Shield Diagnostics plc, and Chairman of Deltex Medical Holdings Ltd. His leadership qualities, his range of experience, his pragmatic attitude to what needed to be done, and his ability to push it through soon proved very valuable to the Group. Nigel became Chairman at a difficult time for Oxford Instruments, when its fortunes were at a low ebb and much needed to be done to enable it to regain its former position and go on to further successful growth.

Fig. 23.1. Nigel Keen

Why had the Company suffered this decline in profitability? Some of the reasons have already been outlined, but there were other underlying, as well as specific, causes. It is quite hard to get the recent past into perspective and see just where and when the problems began. Like economies, companies have cycles, and the two are not always in phase. As I have said, in the economic boom of the late 1990s the service companies prospered; the financial institutions, the software and Internet companies, and the telecommunication providers. Meanwhile manufacturers, especially the exporters, faced a hard and often depressing grind. But company cycles depend on many other factors as well as the general environment for the country or the sector. These can reinforce or temper the effects of national cycles.

Any company that, like Oxford Instruments, has survived independently for forty years knows about these ups and downs. Some of ours were due to external influences and some to our own successes or problems. Oxford Instruments grew very rapidly, and quite profitably, for four years after we embarked on the new technology of superconductors in the early 1960s. It suffered a setback as technical problems and a shortage of working capital hit us later that decade. Under new management, in the difficult 1970s, it clawed its way back to prosperity through the success of NMR magnets and Medilog. In the 1980s, through recession and boom, it achieved a spectacular cycle of profitable growth through imaging magnets. It suffered the cliff in MRI in 1987, and it was just recovering from this through the joint-venture solution of 1989 when the long and painful recession of the 1990s dragged it down again. From 1994 to 1997 it won a brief recovery when it reached new profit records, only to be hit by the strengthening pound and other external factors such as the downturn in the semiconductor industry, as well as suffering from certain internal factors.

One of these specific internal factors was yet another episode of superconductor problems. When operating at the boundary of the possible, a science-based company cannot avoid problems with materials. Magnets for the highest achievable field need the latest high-performance superconductors. Problems, either in the materials themselves, or in how to get the best out of them, are often hard to diagnose and to cure. The most difficult projects may get held up for many months. The metallurgists in New Jersey and the scientists in the other superconductor-based operations worked hard to remedy these latest problems, and by late 1999 had found at least partial solutions. But, for at least a year, some of the highest-field magnets suffered from this problem, which caused delays and extra costs and lower profits. Oxford Instruments is world leader in applied superconductivity precisely *because* it has in the past managed to overcome these problems in advance of its competitors. But setbacks with the new superconductor were certainly one of the causes of the latest 'down'.

Another specific factor was in the IT area. The Group was one of the first companies in the UK to install integrated business information systems, back in the 1980s, but the technology had since moved on. The providers of the old ManMan system were not prepared to update it into the new century. From the mid-1990s there were discussions on what would be needed in the future and investigations of possible systems. John Lewis-Crosby, the Group's Director of Information Systems, started planning a new programme that would need new hardware. A decision was taken to use software based on 'One World' from J. D. Edwards. In 1998 work started on the installation of the new system. This was to take place sequentially, business by business, using a small team that would also train the users at each site, including many of the production workers. This system is designed to bring greater transparency to all the processes of running a business. On the shop floor the system offers real-time information on all the stages between receiving an order and delivering the product. The installation process was speeded up in 1999 and was expected to be running smoothly in all the UK businesses by the time of the 2000 audit. A change of system inevitably causes disruption and gaps in the flow of information to management, and it costs many man-hours in training. And such a novel system always turns up anomalies; there had to be rewriting of bits of software at J. D. Edwards and modifications to new large computers. Although this innovative IT system promises better management communication, more efficiency, and considerable savings in the future, its introduction has been an expensive process.

I have mentioned concerns over new-product introduction more than once. New developments have often been rather slow to reach the marketplace. With nine or ten fairly autonomous business units, some in related technologies or markets, it has not always been easy to get new ideas developed rapidly and efficiently, especially when progress has depended on more than one unit. The Director of Business Development and TAC have worked with the businesses to encourage them to try new ideas over and above their normal updating of products. Ideas have often come from interaction with university customers using the equipment for their experiments. These scientists have sometimes thought of new ways of reaching higher specifications—of interest to all the magnet businesses; or they have conceived brilliant new ideas for making some measurement they need for their work—research is all about measurement. In the mid-1990s the Company invited the business units to bid for additional R & D funding for special new development projects. There were several new initiatives, especially on industrial applications of NMR spectroscopy, mostly in the oil and gas exploration field in partnership with larger companies. But the special-bids scheme tended to be treated too much as a top-up of the normal R & D budget. Some of the projects were not given a high-enough priority, and were often starved of labour, while cross-unit developments ran into other problems.

In the spring of 1999 the Company introduced a new, and much better researched scheme, aimed at tapping the creativeness of employees in the hunt for innovative ideas for new business. A New Ventures Board was instituted, chaired by the Chief Executive and supported by a strong team from across the Group. Structured with the assistance of consultants, who had seen it work well before, the scheme provides more focus for new commercial opportunities. It generates concentrated and directed developments, spurred on by project champions, with a personal interest in their success, working towards important product goals. The Company has great hopes for this scheme for future growth.

The strong pound was a very real cause of lower sales and profits, but it must not be used as an excuse for everything. For many years German manufacturing industry prospered although its currency remained high. A better performance is possible here too. OMT has striven over many years to win international recognition for quality and reliability, and has at the same time achieved good profits and amazing reductions in prices. But not all parts of the Group have followed its example and used the techniques it has found so successful. Why has the Company been unable to reach the same level of organizational success in all the businesses? What has been missing? At the Group's International Management Conference in 1991, Peter Williams laid out the things that needed to be done (see Chapter 20), and did not accept the strong pound and other outside events as an excuse. But with the recession, and the many other problems faced through the 1990s, the focus on creating value was sometimes lost, and achievement was patchy. OMT was indeed lucky in having a captive, if exacting, customer in Siemens and, unlike some of our markets, magnetic imaging continued to grow through most of the 1990s. Some of this growth came from cost reductions, which enabled more hospitals and clinics to buy scanners.

With the assistance of the consultants PwC, a detailed study of the problems in the Group had been going on for many months. In the late summer of 1999, after wide-ranging discussions, the Board came to the conclusion that increased centralization was necessary and that far-reaching changes should be made to the Group. The new Chairman, Nigel Keen, was deeply involved in the work, and the retirement of two or three senior managers eased a few of the problems of a major reorganization. A fundamental restructuring programme was put into action, code-named 'TopFlight'. On 16 September 1999 a press release announced a streamlining of the organization, with a medium-term cost reduction target of about £8 million per annum. Most of the UK operations were to be merged into three new businesses: for magnets and cryogenics, for industrial and analytical instruments, and for medical equipment. Their MDs would be newly appointed, and would report to Andrew Mackintosh, the Group's Chief Executive (see Fig. 23.2). Combined with a series of other moves, this was designed 'to lever more effectively the many strengths of the Group and to drive the return to shareholders back to acceptable levels', as Nigel Keen put it in the press release.

This ambitious programme was very carefully thought through. It came on top of the new IT system introduction, which was having teething problems, and it was bound to produce yet more disruption to production. But the Board viewed it as essential for the future prosperity of the Group, even if the year 1999/2000 were to be badly affected.

As the resulting businesses were in effect new, all existing office holders, and others not in 'ring-fenced' jobs, were invited to apply for the various positions

Fig. 23.2. Andrew Mackintosh

in each of the three new companies. Interviews took place in stages. By 21 September the three new MDs had been appointed. They would have the difficult task of amalgamating existing units into effective companies. By 21 October the next tier of management, the functional directors of the three businesses, had been appointed. Those who had held similar positions in one or other of the old operations would now have wider responsibilities. Within another month the management teams were complete. The TopFlight team was assisted through the whole difficult process by the consultants PwC.

The new magnet business incorporated RI, NMR Instruments, and ATG. Its new name became Oxford Instruments Superconductivity (OIS). Its newly appointed MD was Steve McQuillan, who had been running RI for the previous year. Alan Street became his deputy MD, a necessary position as OIS became a large and complex company. The wire operation in New Jersey is a different type of business, and it remained separate. The MRI joint venture with Siemens was, of course, not included in the reorganization.

MAG and IAG were to be merged in the new instrument business, which became Oxford Instruments Analytical (OIA), while PT and X-Ray Technology, while remaining separate, came under the same new organization. The new MD was Charles Holroyd, who had come from United Industries plc, where he had been the Divisional Director responsible for the Engineering Plastics Division. He had also been Chairman of the Electronics Division of B. Elliot plc, and a Managing Director at Bowthorpe plc. He had a strong track record in the management of change, which was to stand him in good stead in this new appointment.

Paul Brankin had recently retired from OM. The new medical business, which continued to offer the Company's full range of medical monitoring equipment, became Oxford Instruments Medical (OIM). Its newly appointed MD was Alan Cousens. Alan was then 42, and had joined the Company in August 1999 from Smiths Industries, where he had been MD of Medical Systems Europe. Before that he had been MD of Graseby Medical (acquired by Smiths Industries in 1997), where he was responsible for the Company's strong organic growth and rapidly increasing profitability. He had also successfully managed the acquisition and integration of the 3M Infusion Therapy business.

The new initiatives did not stop with the newly integrated companies. With the help of PwC, four new process improvement programmes were launched. One was centralized purchasing, which was introduced in order to leverage more effectively the Group's buying power—some £95 million a year was spent on materials. The central team for this new Group function was to report to Martin Lamaison, the Group Finance Director. Another new senior appointment was in market development, to improve worldwide sales effectiveness for the whole Group. Ron Jones, who had run MAG successfully for many years, moved to this position. The third project was in new-product introduction, an area where OMT had succeeded in reducing the 'time to market' phenomenally. Alan Street, who had spent some time with PwC sorting out new-product development problems at NMR Instruments, was to coordinate this team. The fourth initiative was aimed at improving manufacturing yield, and the team for this vital project was to report to Andrew Mackintosh. These four process improvement programmes were designed to deliver major cross-business benefits. Each of these key areas was to have a best-practice project team, selected from the new companies, who would work to increase the pace of change and stop the businesses falling back into old ways.

Reorganization and cost cutting mean redundancies. This was the sad side of the renewal programme. The original announcement mentioned 120 redundancies. It was no comfort to those who had to go that many other UK companies were doing the same thing, and that these painful times were necessary for the recovery of the Group's prosperity. The Company said goodbye to a lot of loyal people who had given their best, some over very many years.

With the consolidation within the Group, and more subcontracting of production, fewer factories would be needed. The OM factory at Abingdon was to close at the end of the financial year, and remaining employees who were willing to go were to transfer to the factory at Woking. Consolidation was also expected in OIA, either in Abingdon or in High Wycombe.

In December 1999 two new non-executive directors were appointed to the Group. Peter Morgan had had a distinguished business career in IBM UK, where had had been involved in sales and marketing and had been a main Board director. He had been Chairman of SWALEC and the Director General of the Institute of Directors. At the time of appointment to the Oxford Instruments Board he was Chairman of Pace Micro Technology plc and the National Provident Institution as well as the software company KSCL, and Deputy Chairman of Baltimore Technologies plc.

The other new director, Peter Hill, was President of Invensys Air Systems. He had been Head of Strategic Planning for BTR plc and, as Director of Merger Integration, he had been responsible for the successful merger of BTR with Siebe plc to form Invensys plc, in which he had then taken charge of an operating division. He had previously been an executive director of Costain Group plc. These two appointments extended the range of senior business experience on the Oxford Instruments' Board.

The TopFlight reorganization was expected to cost the Company about £6.5 million, which would come on top of poor trading in the first few months of the financial year. There were also other disruptions, in particular from the introduction of the new IT system. There would be overhead savings, but it would be very difficult to predict the performance for the year until quite near its end. The Board gave strong support to the TopFlight programmes, regarding 1999/2000 as a transition year, and hoping for recovery in 2000/2001. The Company had a strong balance sheet. It would be able to ride out the problems involved in making a new start, and the directors looked to the future for major gains in performance and shareholder value.

I am sometimes asked what I think are the most important factors for success in a company like Oxford Instruments. In first place I would put people. Van Herpen's law, which I found in a small humorous booklet called Murphy's Law, states 'The solving of a problem lies in finding the solver.' If a company can find exactly the right person for an important job, nothing should be allowed to stand in the way of recruiting him or her, particularly when the position involves leadership. And the reverse is also true.

In second place, for a science-based company, I would put keeping ahead in the technology. This is really again about people, the creative people who absorb information from outside, who think laterally, and who come up with brand-new ideas for leading-edge products. Other innovators can cut through accepted practices to find better and cheaper ways of making existing products. In 1871 Charles Darwin wrote to his son Horace: 'I have been speculating last night what makes a man a discoverer of undiscovered things; and a most perplexing problem it is. Many men who are very clever—much cleverer than the discoverers—never originate anything. As far as I can conjecture, the art consists in habitually searching for the causes and meaning of everything which occurs' (Cattermole and Wolfe 1987: 5).

As important as the technology is the marketing; the technologists must work with those who have a good insight into existing and potential markets. And, of course, without successful sales operations the new products may not get a proper chance to show their paces.

Then, again about people, a company needs its really efficient production engineers and managers, administrators and accountants, who can make a complicated organization work smoothly and profitably. So people are all-important. I believe that good management is, first and foremost, about finding and enthusing committed people; keeping them in the right positions at the right time; harnessing a good team with a united agenda for the success of the company; and developing a community of *all* the employees in an inclusive culture.

With this perfect team, which must always be learning and changing and moving ahead, a science-based company needs to develop a series of unique and desirable new technical products, preferably with patent protection, but at least ones that will take competitors several years to match. With these it can establish a very strong position in its ever-changing markets. Looking back, Oxford Instruments has had six or seven such successes, which produced the 'ups' in our cycles, but they have not yet occurred frequently enough to eliminate the 'downs' due to national cycles or other circumstances. In April 1999 Martin gave a paper to the second Cambridge Enterprise Conference. In this he said, 'I have come to see them [downs] as almost inevitable, situations to be borne—situations which . . . when properly handled, can become the inspiration for new starts, new ideas, new developments, sometimes higher growth than before.'

I asked Nigel Keen what had tempted him to become a director, and then Chairman, of Oxford Instruments. After a little thought, he wrote of how he saw technology companies, and the opportunities ahead for Oxford Instruments.

I have spent the last twenty years working with young high-tech companies. The strengths that these companies must enjoy if they are to compete with larger, more established and better-financed companies is that they must 'own' a piece of the world. This ownership gives the young company a window in time, which allows it to exploit its technology or product without intense competition from the world at large. This window, when coupled with the fleetness of foot and the speed of response of the small company, gives the opportunity for the young high-tech company to prosper in today's super-competitive commercial markets. One of the principal disadvantages that these small young companies have is that they have to reinvent the wheel in every part of their organization, and this all at the same time. They have to introduce business disciplines, they have to perfect product introduction, they have to establish manufacturing, and of course they have to raise the capital they need to finance their endeavours.

Oxford Instruments owns its world in several of its businesses. It has the operations and finance that it needs to exploit the technologies and products that come from this ownership. With the changes that have been made to streamline and improve the operations there is a great opportunity for Oxford Instruments to move ahead strongly in the years to come.

24 Bridges, Networks, and Nurseries: The Oxford Trust

A foundation for promoting technology transfer and guiding new science-based companies

No man is an island, entire of itself; every man is a piece of Continent, a part of the main.

(John Donne, *Devotions upon Emergent Occasions* (1624))

The great end of life is not knowledge but action.

(T. H. Huxley, 'Technical Education', in *Science and Culture* (1877))

In 1985, two years after the flotation of Oxford Instruments, Martin and I set up and endowed a new charitable foundation, The Oxford Trust (TOT). Its mission was 'to encourage the study and application of science and technology'. This was to be achieved:

- by promoting interaction between various groups including schools, institutes of higher education and research, and scientists in industry;
- by the organization of relevant courses and meetings;
- by supporting certain research projects in educational establishments;
- by letting surplus trust premises to science-based companies applying new ideas and the results of research;
- by using a small percentage of its endowment to provide seedcorn capital to selected small companies.

Ten years on, the mission statement remained the same, but the means to its achievement had been reduced to:

- by exciting and enthusing young people in science, engineering and technology;
- by encouraging economic growth through innovation and technology transfer.

These remain the Trust's two main groups of activities, but, in this account, I shall have little room to deal with the first, which may, however, be the more important for the long-term future of our economy.

The Oxford Trust has its origins in the history of Oxford Instruments, so I do not apologize for including it in an account of the latter. In her paper 'Science-Based Enterprise: Threat or Opportunity?' Elizabeth Garnsey (1997: 17) wrote, 'Few science based companies have been as successful as Oxford Instruments.' In earlier chapters I have explored the reasons for this success, as well as telling of our mistakes and missed opportunities. The life of Oxford Instruments goes on, with its ups and downs, its new developments, and the changes every organization must embrace if it is to prosper. But the Company is part of a wider movement, an upsurge in the knowledge-based economy of Oxfordshire and of the UK. The Oxford Trust is playing its part in promoting this movement, and this chapter tells of these developments and how the Trust has sought to drive them forward.

What motivated us to set up this Trust? In 1990, when interviewed about the Company and why we had started TOT, Martin made a remark he has repeated on other occasions: 'As time went on . . . I was amazed that there were not many more companies like Oxford Instruments coming out of the rich research community in Oxfordshire' (The Oxford Trust 1990*b*: 1). For years we had both taken an intellectual interest in the process of starting science-based businesses, and finding out how young firms can succeed on several planes. We believe it is important to integrate the staff into a company community as well as producing clever products and developing good management techniques. From the 1970s onwards a trickle of scientific entrepreneurs had seen us as role models and come for advice. We had invested in a couple of new firms, and in the 1980s joined with others in forming a seed-capital company. This was to help fill the funding gap at the very bottom end—investing much smaller sums than most venture capitalists are willing to contemplate.

After Oxford Instruments had become a public company, its success gave us the means to extend our interest and gear up our influence through a new organization. Although the environment for industry in general had improved, for small new spin-off companies there were still many problems: premises continued to be difficult; there was limited basic management training available for busy entrepreneurs; securing finance was still a minefield; and legal and administrative rules and regulations were often unfathomable and always expensive. Apart from these obstacles, there was a persistent gulf between academia and industry that formed a barrier to the cross-fertilization of ideas and the exploitation of the results of research. Science education was also failing the future economy; companies like Oxford Instruments were finding the shortage of well-qualified scientists and technologists a brake on growth. As a secondary-school governor, I was disturbed by the diminishing choice when appointing science teachers, and by the falling numbers of children wanting to study science and technology at school and university. Many people had written about these problems, but in Oxfordshire, in 1985, there was not much action. When we see something that needs to be done we like to act on it if we can. The Trust was our way to get things done.

Since 1985 The Oxford Trust has acted in many contexts. It has been in the lead in some developments, and has made a contribution towards changing the small-business environment and the attitudes to economic development in Oxfordshire and in the country. It would take another book to do justice to all its many-faceted activities, and to the people who have worked in it to drive its projects forward, both in its relations with industry and in its programmes with schools. It is not alone in these activities; there is now a tide of new bodies, schemes, and grants for helping companies to create wealth by innovation. Many other organizations partner the Trust in its various projects, many generous sponsors contribute to the costs, and a great number of individuals have given freely of their time.

In the early 1980s the most pressing problem for potential spin-off companies was still the dearth of small premises for short-term occupation within easy reach of Oxford University. A scientific entrepreneur, wanting to see if his ideas were practical, could be faced with the millstone of a long lease for an unnecessarily large and smart business unit—that is, if he or she could find premises at all. We had experience of a better system; we already had a small 'science nursery' at No. 3 Middle Way. After Oxford Instruments had moved to Osney Mead in 1965, we had retained the informal unwritten lease from Mr Lindsey, the butcher, and the valuable, if vague, planning permission originally attached to activities involving Martin personally.

Between 1965 and 1985 four companies in turn had grown to a viable size in the old stables and slaughterhouse, and had graduated to larger premises. The first, Orbit Precision Machining, started by two technicians from the Nuclear Physics Department, had moved on to an old factory in Wantage. The second was the small electronics company that first worked on Medilog, and was absorbed into Oxford Instruments in 1971. The third, Analysis Automation, had been started by a chemist, and had filled every corner of the stable yard until 1982, when it had finally discovered a new place in Eynsham. This company was followed by Oxford Lasers, which had been founded by Professor Colin Webb and others from the Clarendon Laboratory. It later moved to a factory in a business estate on the edge of Abingdon. In 1985 the current tenant was Exitech, started by Dr Phil Rumsby and Dr Malcolm Gower from the RAL. In their spare time they were using a specialized laser for micro-machining tiny parts under contract, mostly for research. At the end of the century they had a substantial factory on the Long Hanborough Business Park, fifty employees, their own pulsed laser products, which they sell for micro-engineering, offices in the USA and Japan, and a profitable turnover of £6 million.

John Lindsey had a *laissez-faire* attitude to his old building and was pleased that it was serving such a positive purpose. Occasionally I would suggest that we ought to pay him a little more rent, and we would raise the modest charges to our current tenant company. This informal arrangement ran smoothly for nearly twenty years with little administrative work needed from me. Middle Way never had to wait for new tenants, and, in the entrepreneurial climate of the early 1980s, we knew of several more scientific start-ups. We felt the need to multiply Middle Way. A more substantial nursery or innovation centre was what Oxford required for its inexperienced entrepreneurs, with management support and training, and help with administration—things that had not been possible for a single company at Middle Way. There was then no Oxford Science Park, although there were proposals for this kind of development, which usually fell down on planning concerns.

Cambridge was ahead of Oxford in science-park development. As far back as 1969 a University subcommittee, chaired by Sir Neville Mott, had demonstrated strong links between the University and science-based industry in the area (see University of Cambridge (Mott Report) 1969). Cambridge had a higher ratio of science to arts activities than Oxford, and, unlike Oxford, had long accepted applied science as a proper subject for university study. In 1875 Cambridge had established a new Chair of Mechanisms, and the Professor's tiny workshop became the forerunner of the large Engineering Department, which later developed a strong electronics section. A Government-funded Computer Aided Design Centre—one of the first—was managed by the Department and attracted many able scientists who forged ahead developing computer applications. Small companies started to germinate from this rich Cambridge seedbed, many of them developing software, and often with a large consultancy element. This kind of company, which can often start from home, is much easier to get going than, say, a firm embarking on the development of some complex chemical or one needing to manufacture early on.

In 1970, in response to the 1969 Mott Report, the Cambridge authorities relaxed the tough planning restraints that had kept IBM's proposed European Research Laboratory from setting up in the area, and had prevented Metals Research even from expanding near its own base. Trinity College obtained planning permission on its land on the northern edge of the City, and that year, through the efforts of its Bursar, Dr John Bradfield, established the Cambridge Science Park. It grew slowly at first—most tiny computer and software companies stuck to their spare bedrooms, garages, or ex-shops.

In 1979 Matthew Bullock, working as an assistant manager in a Cambridge branch of Barclays Bank, identified forty-one companies based on computers. He was instrumental in getting them together into a Computer Club, which later became the Cambridge Technology Association. He saw the difficulties some of these companies met when looking for funding, and later travelled in the USA studying the situation there. In 1983 he reported on academic enterprise and the financing of small new high-technology companies in the USA. These had been starting up near Boston and in California in increasing numbers ever since the war (see Bullock 1983). The USA was at least two decades ahead of any such movement in the UK. Matthew played an important role in making the Cambridge establishment aware of this trend towards scientific entrepreneurism.

The escalating number of small high-technology companies became known as the 'Cambridge Phenomenon', named thus in a well-researched and widely read report on the situation published by Segal Quince & Partners in 1985. The Science Park became the focus of this 'phenomenon', although the vast majority of these new companies, put at about 300 in the 1985 report, used tiny premises elsewhere. In 1983 the Park boasted twenty-three companies on site, and in 1984 gave space to sixteen new arrivals. It catered for companies engaged in R & D and associated light industrial production, and for professional support firms. It was not for mainline industry or for start-ups and really small companies. Other universities, like Aston in Birmingham, were developing their own parks for spin-offs and similar types of company, often aided with funds from the local authority or the private sector; the Technopark on the South Bank in London was financed by an insurance company. There had been some movement in Oxford too, although more sympathetic planning rules were not adopted by the County Council until 1984, and it would be 1990 before Magdalen College founded the Oxford Science Park to the south of the City— twenty years after Cambridge.

Late in 1984 Oxford Instruments decided to get rid of the old builders' yard on Osney Mead, which had recently been the Packing and Despatch department for the nearby operation. On the road frontage of the half-acre site stood a solid single-storey brick building with several offices and a large room. Behind it were some 3,000 square feet of older wooden buildings and newer temporary 'portacabins'. This would be an excellent place for the multi-cloning of Middle Way, with room for other desirable activities at the front. The Company agreed to sell it to us, for this purpose, as just over the highest outside offer, and we planned to go ahead and buy it.

The sudden availability of a base stimulated us to think through our rather vague ideas on the problems of small high-technology companies in Oxfordshire. We wanted to form a new organization as a charity, and for this we needed clear objectives that would be accepted as charitable. There were precedents here, but still some confusion, so we found Michael Macfadyen, a solicitor in London, who specialized in charity work; he took us to see a Charity Commissioner to explain our intentions. With approval for our plans, the Trust was set up as a charity and a company limited by guarantee, and endowed with shares in Oxford Instruments. We started with three trustees, including Jonathan Welfare, who remained the Chairman for ten years.

We now had clearer objectives and some ideas on how to fulfil them, but these needed formulating and carrying through to actions. The Trust needed a really good director. Just at that time Paul Bradstock, an Oxford physicist, came to get advice from Martin in his search for a managing directorship in a small scientific company. He had previously worked in one of the Oxford Instrument Group's operations, and had a range of experience in technology-based companies, especially in development and marketing. We persuaded him to get the Trust up and running, even if he might go off to manage another company six months later. From then on Paul was so busy he had few moments even to see if any jobs were going. At the end of the century he was still the Director of the Trust, and it is through his leadership, good judgement, and hard work, and the outstanding staff he has been able to attract, that it has grown into an active, multifaceted, internationally recognized organization with a strong and widespread influence in its field.

Paul set to work with frequent planning meetings. All the buildings needed refurbishing but there were corners where he could put up a table. The first tenants, the Thames Business Advice Centre (TBAC), actually came into offices in the main building, which they undertook to renovate instead of paying rent. TBAC's first director was seconded from Marks & Spencer and most of the furnishings came from a makeover at a local store; the whole building, including the big meeting room, benefited from large quantities of redundant carpeting. TBAC was a new agency that had been set up to advise prospective entrepreneurs and small companies, mostly in low-technology areas, in what they had to do to make their businesses successful. The Trust would be offering similar help to its tenant companies and to other science-based businesses. As they served different strata, the two bodies, in adjacent offices, could easily pass clients to the appropriate source for advice.

Among the first wave of builders, carpenters, and boutique hopefuls approaching TBAC, there were two people setting up a software company. They were looking for a base, and soon moved into the first incubation unit to be finished—and the smallest. It was more like a garden shed, with room only for a bench for the computer, two chairs and a few shelves, but the rent was low and it served its purpose. The other units averaged a more comfortable 300 square feet. When up and running, with all ten units complete and in use, the place was quite a warren, and some affectionately named it our 'Science Slum' (see Fig. 24.1).

The Science and Technology Enterprise Project, or STEP Centre as we called it, was essentially for very early stage companies trying out their ideas. Rents had to be commercial to satisfy our charitable status, but we used flexible licences rather than binding leases, and the tenants could start with little overhead investment as they were able to use secretarial and telephone answering services, photocopier, fax, and the conference room, mostly at cost price. Very soon the tenants developed their own informal network and ways of helping each other. If a science-based company is going to flourish, it normally proves the worth of its new product or service within two or three years, and starts to grow. There was no room in the STEP centre for static 'lifestyle' companies, or for people whose interest seemed to be in the research itself rather than its appli-

Fig. 24.1. The STEP Centre on Osney Mead

cations. Most of the STEP tenants grew out of their units within this period and moved out to larger premises, in Oxford or around the county, although, in some areas, these were still quite hard to find in the late 1980s. The few tenants who stayed on and on found that their rent increased year by year, until the advantages of the centre were outweighed by the cost.

Among the successful tenants who used the STEP Centre in its early years, and then moved on, were: Update Software in computer applications; Ridgeway Systems in microelectronics; Voltech Instruments in voltage measurement and control for the power industry; Otoz, which was testing the optical quality of materials for compact disks; Apex Organics, which was offering esoteric organic compounds; and Oxford Glycosystems. The last of these was a biotechnology spinout from the University, which used a unit for eight months while it planned its future operation. It moved out to Abingdon and, renamed Oxford Glycosciences, became a thriving public company listed in the Pharmaceuticals Sector.

High on the action list for the new Trust was building bridges between research establishments and industry, and encouraging networking between the people in these organizations and teachers, accountants, solicitors, and other business facilitators. To some academic scientists industry was a foreign world. Most had no concept of the intellectual stimulation and excitement of starting or running a science-based company. We needed to mix people up. We needed a forum for discussion, lubricated by food and wine, where a network could replace the isolated boxes many seemed to work in. We looked for a really prestigious occasion for launching the Trust. Through his years with one foot in academia and one in industry, Martin had made many friends and had his own wide network of contacts. For the inaugural talk and meeting he invited Sir Robin Nicholson, then Chief Scientist to the Cabinet Office, whom he had known when serving on the Government's Advisory Council on Science and Technology (ACOST). His letter to Sir Robin presents our thinking so well that I will quote a few passages from it:

Dear Robin,
May I ask you to focus your eyes down from the broad horizons of Whitehall to a small local problem? . . .

With the deadlock over various Science Park proposals still almost complete, we decided about six months ago to start our own project. Albeit it is on a microscale, I think it may, by example, actually break down the log jams in peoples' minds and show them the advantages of encouraging the right sort of industry . . .

Six months ago, Audrey and I thought we would set up this trust, and endow it with Oxford Instrument shares to recycle the locally made capital gains. We've bought a half-acre site with 5,000 square feet of office and workshop space on it, appointed a Director, laid out a large conference room, divided the workshop space up into ten little units, got out the postcard PR blurb and we're off . . .

We are currently emphasizing the educational side of the work. This is a matter of getting people together, people from schools, research labs, the professions, the local authority, and budding entrepreneurs. We had a preparatory meeting last month and it was actually rather exciting, with representatives from most of the research laboratories within the county—the University, the Poly, Harwell, Culham, the Hydraulics Institute, British Non-ferrous Metals Institute, etc., and senior people from several professional bodies. I think we will have a forum which may be highly productive. When the workshop units are filled, I believe we'll have the example needed to get other bigger projects moving.

We are preparing an autumn programme of three meetings at which we shall explore the scope of the research done in Oxfordshire, with senior people from these organizations coming and saying their say. We have sponsored a survey of Oxfordshire industry and we hope to use the output of this in further meetings in the winter when we shall lay out the industrial side.

Nothing may come of all this. After this initial flurry no one may come to our seminars, and no one may crawl out of the university to set up his business. In that case, eventually, we'll accept there's nothing we can do to change the downward trends that have been operating for the last century, flog the premises and go on just doing our own thing—but at least we'll have tried.

Anyway, we want to have a good inaugural meeting in the autumn and I wonder if you'd come and set the scene by talking quite briefly about the need to create wealth by exploiting the results of research . . . *The Times* has said it would like to do a feature on the Trust to coincide with this meeting. I want to have a good send-off, and if the cause is seen to have the backing of the Vice Chancellor and the top brass from all the local laboratories and the City authorities, I think we may start a bandwagon . . . Any chance?

The inaugural meeting was all that we had hoped for. Paul described the Trust's activities, Martin explained its purpose, and introduced Sir Robin, who gave just the right speech. The gathering was graced by the Vice Chancellor, Sir Patrick Neill, and by many senior scientists, Oxfordshire dignitaries, headmasters, science teachers, leaders from the professions, and successful entrepreneurs. This meeting, the first of many, was certainly a great send-off, and the Trust became regarded, almost instantly, as a reputable local institution, and one that the 'great and the good' were happy to address. As Peta Levi wrote in her article in *The Times* (14 Sept. 1985), 'Sir Robin told me "The Oxford Trust could be as important a catalyst to the high-technology scene as Matthew Bullock . . . has been in developing the Cambridge Phenomenon". . . Oxford has had no similar high-tech focal point and no one has known how many high-tech firms exist in the Oxford area. The trust is now sponsoring research into this.'

The Segal Quince & Partners report, *The Cambridge Phenomenon*, had made an attempt to compare the Oxford situation with that in Cambridge: 'It must at once be recognised that there *are* high technology firms, some of them university spin-outs, in Oxford and the surrounding villages and towns—Oxford Instruments is the best-known example, but one estimate we have been given is that there are perhaps 50 such firms in the area' (Segal Quince & Partners 1985: 63). This was compared with some 300 they had identified in Cambridge. We were pretty sure this was a gross underestimate. In order to develop suitable programmes the Trust needed information on the economy of the county. Early on we had decided to use our limited resources to focus on Oxfordshire, and we needed a sound database of what was going on already.

A few months after the Trust had been formed, it sponsored Helen Lawton-Smith to survey the industry in the County, define 'high technology', and study the companies covered by this definition. She embarked on this work under the auspices of the Geography Department of Oxford University, using the research for a doctoral degree. She charted the distribution and provenance of these companies and their links with university and government research laboratories. She studied the birth and growth of new start-ups, which she found had been accelerating from the 1960s onwards. In 1986, long before her thesis was submitted, she produced an extremely useful initial report for the Trust. On a stricter definition of 'high technology' than that used in the Cambridge analysis, she had identified 186 such companies in the county. This work became the core of a valuable database, which the Trust was to use widely in its own work and in helping others needing information.

From its early days the Trust was active in promoting science parks in Oxfordshire, and innovation centres within the city itself as well as the county. There was understandable local resistance to the possibility of another industrial estate, maybe in the green belt. There were already a number of these in the county, notably one on the western edge of Abingdon, where Oxford Medical was situated, and Milton Park, which was losing its wartime depot image with new factories and landscaping. Some of the more substantial science-based companies had settled happily in these estates. In 1987 the Trust issued a discussion paper, gathering together the experiences of other science parks and innovation centres, listing the current local proposals, and giving case studies showing company development patterns.

Paul organized a public meeting in St Catherine's College to promote a more informed debate. Lord Bullock, whose son had done so much for the computer industry in Cambridge, chaired this discussion and went on to chair a new Advisory Council on Central Oxfordshire Science Parks (ACCOSP). This council involved all the major science-based institutions in the area, including the health authorities, the local councils, and other interested bodies; the Trust provided support, and Paul became the secretary. The Council was to consider ways of fostering technology transfer from the rich science base in Oxfordshire, and how this could be aided by encouraging appropriate centres or science parks. While ACCOSP was mulling over these matters in its regular meetings, and forming useful ground rules and criteria for acceptable companies, one or two developments were already underway, including the Oxford Science Park. Prompted by earlier discussions with the Trust, and with the enthusiastic drive of Keith Wills, the Estates Bursar, Magdalen College with its partner, the Prudential Insurance Company, was already putting up its first science-park buildings and was planning a 50,000-square-foot innovation centre just south of the City.

In 1991 ACCOSP produced a report, financed by the Oxfordshire County Council, and published with The Oxford Trust, entitled *Oxford & Technology Transfer: The Role of Science Parks*. As Lord Bullock put it in his introduction, 'There is no other area in Britain outside London, and few in Europe, with as large a concentration of scientists working in close proximity to each other as in central and south Oxfordshire' (ACCOSP 1991: 5). There was the University, with its wide range of faculties; there were the Oxford hospitals, the Medical School and the strong medical research departments; there was the Oxford Polytechnic, soon to become Oxford Brookes University; there were the large UKAEA research sites at Harwell and Culham, and near the latter the Joint European Torus, working on nuclear fusion; next door to the Harwell site there was RAL, and, within its precinct, the laboratory of the National Radiological Protection Board. There were a further eighteen scientific research institutes, or company R & D laboratories, and a substantial R & D effort in about ten smaller high-technology companies. These completed the picture of a thriving scientific community. But companies springing from this fertile soil still needed much more space to get going and to grow. By that time the Trust's database had reached 400. Cambridge could no longer hold quite such a pre-eminent position in this kind of technology transfer, although, to this day, the media usually look in that direction when discussing scientific enterprise.

At The Oxford Trust, meanwhile, the 'networking' was gathering pace. In 1986 a series of lectures on 'The Two Aspects' attracted many participants. Two scientists, one from industry and one from an academic or government research laboratory, gave talks on the same subject as seen from their own viewpoints.

Matters such as artificial intelligence, robotics, biotechnology, and lasers were presented, debated after the talks, and taken further over food and wine. To round up this series, in December, Matthew Bullock came to present the Cambridge scene, while Helen Lawton-Smith and Paul Bradstock could now answer for Oxford. New links were forming at these meetings, technical problems received the benefit of lateral thinking, and people from small companies were getting informal advice from others in similar positions on administrative matters, such as suitable accountants, patent agents, or venture-capital firms.

An early objective had been the provision of the sort of practical business training that had been so hard to find in the early days of Oxford Instruments. Small companies form a special sector, and managers who have spent all their working lives in large companies are not always the best at explaining financial and legal matters to busy small-firm managers. Soon appropriate talks and training sessions were arranged. In 1989, the year the newsletter *Scintilla* was launched, Paul talked of the Trust's 'virtual science park'. Firms and individuals throughout the county could, by then, attend quite a range of meetings. There was the Innovation Club, with business games and short topical talks and networking over a glass of wine. Then they could sign up for tutor-led workshops, lasting several sessions, on subjects such as marketing and planning for growth, with case studies drawn from a restricted number of participant companies. They could attend technology update lectures, or go to half-day seminars on topics with titles such as 'Intellectual Property', 'Innovation through Strategic Alliances', and 'The Effective Use of Consultancy'. Many outside experts gave freely of their time in this valuable educational process. Some of these sessions were organized in partnership with outside firms—Grant Thornton, the accountants, regularly gave a seminar on 'Starting a Science-Based Business'.

Since 1988 the Trust has held an annual SMART workshop to help innovative small companies and aspiring entrepreneurs to apply for these important and competitive Small Firm Merit Awards for Research and Technology. The scheme is part of the DTI's Enterprise Initiative package. The quite substantial Stage 1 awards are intended to help a company to carry out feasibility studies; the larger Stage 2 awards are for companies clearing the first hurdle, and are designed to lead to market launch, with additional finance expected from the private sector. These awards would have made a great difference to Oxford Instruments during the days in the 1960s when we were developing superconductor and cryogenic technology on a shoestring. Each year several local firms benefit from the awards, and the presentations usually take place in the Trust's premises.

In 1989 Dr Janet Efstathiou came to work at the Trust for a period, and helped to launch the four-monthly newsletter, *Scintilla*, which she edited for its first year. Sent to a large mailing list, it contained high quality-information on technology transfer and innovation, as well as featuring the Trust's many activities with schools. There were short articles and reports on technical matters and reviews of recent lectures; there was news of the companies in the STEP centre, and other success stories covering technology as well as business; and there was a programme of future events, both at the Trust and elsewhere in the county. From its first number *Scintilla* has been widely read; the twenty-eighth issue was sent out in September 1999.

By 1990 there was something of a bandwagon in encouraging small enterprise in general. The Trust always looked to its mission and, as other agencies started appearing on the scene with similar objectives, it sought to work with them rather than being competitive. The endowment, which secures the salaries of the core staff, has always been a great benefit, in that it enables the Trust to remain independent and adventurous in its thinking; it can act in a pioneering and catalytic way. The security of its position also led to a general feeling that it was objective and was to be trusted, in fact as well as in name. But we could not have carried out half our successful projects over the fifteen years since 1985 without the generous sponsorship, and support through partnership, given by many organisations. These have included the DTI, the Heart of England Training and Enterprise Council (TEC), Business Link, the Royal Society's Council on the Public Understanding of Science (COPUS), the Royal Institution, the County, City, and District councils, the Leverhulme, Gatsby, and other trusts, several banks, accounting, and legal firms, and many established companies, including Oxford Instruments. The list of people who have given their time to serve as trustees and on the Advisory Council includes many well-known names from the various institutions and companies in the county. Dr Mick Lomer, who was once the Director the UKAEA's fusion research laboratory at Culham, served as Chairman for some years, and Judith Iredale, from the world of school science, took over from him in 1998.

Dr David Kingham, who had a strong academic and scientific-business background, arrived in July 1990 to be Assistant Director with responsibility for technology transfer. He soon took over as editor of *Scintilla* and in 1991 he developed the Associate Membership Scheme. Without wanting to disturb the successful informal networking service, he felt corporate and individual memberships would improve communication and benefit the Trust's 'clients' more effectively. A *Scintilla* supplement was launched specifically for the associate members, and aimed especially at promoting technology transfer to small companies.

What is technology transfer? This convenient but loose term covers a wide range of channels for information to travel from research organization to industry or from company to company (see Fig. 24.2). There is the transfer of ideas; the transfer of a proven concept; the transfer of a working prototype, or of a fully engineered product; the transfer of the formula for a drug or other chemical; the transfer of people with knowledge and skills from one organization to another; the transfer of a small research group to another university, or of a small company, together with its products and people, to a larger company. Finally, there is the spin-out from a larger company, either assisted or resisted, of a group organizing itself into a new small company, perhaps to manufacture some recently developed product, which the larger company has rejected as having too small a market. Many people think of technology transfer as coming only from an academic research environment to industry. This may be the most important route, but there can be a cultural difficulty. Academics may want to publish their achievements as soon as possible to enhance their reputations and prospects, while companies may want to retain a competitive advantage through secrecy. Increasing mutual respect and understanding, and the involvement of more academics in potentially valuable start-ups, are reducing the severity of this problem.

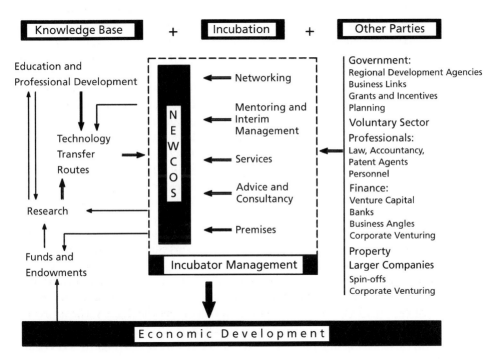

Fig. 24.2. Technology transfer and fledgling company incubation
Source: Paul Bradstock and The Oxford Trust

Most universities had, for some time, employed industrial liaison officers; the bulk of their work was in technology transfer through licensing agreements and sponsored research contracts from larger companies. In the early 1980s new rules permitted universities to set up their own systems for dealing with inventions arising from government-funded work in their laboratories. The intellectual property rights (IPR) had previously been channelled through the then-publicly owned British Technology Group, which, as a result, held valuable patents on important drugs and inventions such as cephalosporins and magnetic resonance imaging. Different universities decided on different systems. Cambridge always had a more relaxed view on this matter than Oxford, and left the IPR largely in the hands of the inventors. In the late 1980s Oxford University set up its own company, Isis Innovation, under the founding Chairmanship of Sir Peter Hirsch, to deal with the licensing side of technology transfer, and also to offer help to any academics wanting to start their own companies. In this second activity Isis had a slow start. It was not until the 1990s that the successful flotations of Oxford Molecular, Oxford Asymmetry, and Oxford Glycosciences, in which the University held substantial stakes, convinced the academic authorities that spin-off companies could become a valuable source of funds in the future. Tim Cook, who once led one of Oxford Instruments' subsidiaries, became its MD in 1997, and by 1999 the staff had grown to eleven. Peter Williams became its Chairman soon after Tim settled in. In 1998 Isis filed nearly 200 patents, negotiated thirty licences, and organized the formation of five university spin-off companies, and the pace has since been increasing.

A less direct form of technology transfer is through intermediate research and development institutes, between university and industry, which usually serve larger companies. A successful network of Fraunhofer Institutes in Germany carries out contract research for industry, and works on new technologies at the precompetitive stage. Heriot-Watt, in Edinburgh, has long followed this route. It is predominately a science and engineering university, and is traditionally in close touch with local industry. In the 1960s the academic authorities there did not want teaching and research staff to spend a lot of time on outside work for industry. As Ian Dalton (1989: 2) put it, 'Technology transfer involves at least one, and often several orders of magnitude more effort than ideas transfer.' Next to the new Heriot-Watt campus the University set up a Research Park, of which Mr Dalton became the Director. On this it established Technology Transfer Institutes for each of its strongest departments. Originally financed by the University, they then had to pay their way, so most of their work is in contract research. There are strong interactions with the academic faculties as well as with local industry, with periodic transfers of people who carry their expertise with them. As in the Fraunhofer Institutes, there are also more-speculative developments, which can lead to new spin-off companies, encouraged by the institutes.

In 1990 the Prince of Wales, who has long been interested in innovation, asked Sir John Fairclough to get together a working group to stimulate action. This eminent group published an interim report in 1992, in which it proposed 'Faraday Centres' on a similar line to the Fraunhofer Institutes, for this kind of intermediate role (see Working Group on Innovation 1992). This group advocated more of a *local* focus to the Government's various initiatives, like the Link projects and Teaching Company scheme, set up to foster innovation.

The Oxford Trust has always been most concerned with the various factors influencing technology-based small and medium-sized enterprises (SMEs).

These companies could be the seedcorn for future economic prosperity through good technology and management. In 1991 the Leverhulme Trust provided support for a two-year study of the various routes for technology transfer in Oxfordshire. David Kingham supervised the project, and Dr Elizabeth Bell was given leave from the Economic and Social Research Council to research the subject. Technology transfer was defined widely as 'The process of promoting technical innovation through the transfer of ideas, knowledge, devices and artefacts from leading edge companies, R & D organisations and academic research to more general application in industry and commerce'. The report went on to demarcate invention from innovation, which are often loosely used:

Invention, the process of creating new technology . . . is frequently confused with innovation which describes the process by which the successful application and exploitation of a new technology is achieved in practice. Innovation is . . . a complex, vulnerable and often poorly defined process, frequently requiring a wide range of technical and business problems to be solved. Technology transfer can be a crucial part of innovation.   (Bell *et al.* 1994: 3)

This study contained an 'action' component. It aimed to *stimulate* technology transfer, as well as researching it, by carrying out a pilot-scale technology audit of a science department at Oxford Brookes University. This was to lead the Trust to undertake more technology audits. For a couple of years from 1992 the cost of an outside consultant for these audits was shared between the DTI and a university. The pilot audit helped the DTI to formulate its ideas in this area, and it soon recognized the importance of more activity at the *local* level for promoting innovation. The Trust's team carrying out the audit at Oxford Brooke's School of Biological and Molecular Sciences not only looked for promising areas for exploitation; it aimed to make the department aware of additional sources of funding. These included research grants, training and teaching opportunities such as short courses, and the possibilities for consultancy and for contract research. From all these areas thirty-four potentially exploitable opportunities were identified in the work of five research teams. Two of these needed patents—which were rapidly organized; the largest number, eleven, were for consultancy, seven were for short courses, and the rest in the other areas.

The main findings from the Trust's study of fifty-four companies were: that technology transfer was important to most of them; that there were far more informal links than formal ones, and most informal links were local; that companies were unclear about how to approach university departments for help; and that technology audits were valuable for identifying opportunities for technology transfer.

When Matthew Bullock visited the USA in 1981 to study academic enterprise, he found round Boston, and near Stanford University in California, an active culture in which university research scientists were on the *look-out* for industrial links. Many small companies there began life on the low-entry-barrier route of consultancy; he called these 'soft' companies. Some then evolved through intermediate stages into 'harder' companies, making products. Few university spin-off firms started life as 'hard' companies aiming, from the beginning, to make and sell standard products. But 'harder' firms would often splinter off from larger companies to offer a product after little further development. The large number of small consulting companies, and more established new-device manufacturers, were widely used by larger companies as expert consultants, or for new components to help solve a design problem. Matthew Bullock called this process 'component technology transfer'. He wrote, 'Only a very few small research-based companies grow from acorns into full industrial oaks. Their impact on the rest of the economy tends to occur in the undergrowth of technical, industrial markets' (Bullock 1983: 2). He identified a strong technology transfer route in the USA, in the acquisition of many of these technically clever, but small, companies by larger firms. This was an accepted life cycle for these companies. The 'growing oaks' had the means to exploit these clever developments, while the original entrepreneurs often used the proceeds to start again on another idea once the technology had been absorbed by the new parent. With developing management skills some became successful serial entrepreneurs.

The climate in the UK is not so kind; a liquidation makes it quite hard for an entrepreneur to try again, in contrast to sentiment in the USA. Even selling one's company is often seen as a failure here, with the result that some small companies, which might do better under the wing of a big brother, struggle on until they are in such a mess that larger companies shy away from taking them over. The success or failure of a 'hard' company may depend crucially on its type of market. As Peter Marsh (1987: 48) pointed out in the New Scientist, 'Most of Britain's successful small companies have developed specialised products for small markets. In this way, they can grow without attracting competition from too many big international firms.' Among the successful companies serving niche markets he mentioned Eurotherm and VG Instruments, which have both been taken over, as well as Oxford Instruments.

Innovative companies looking towards mass markets could be very successful at first, but often suffered severely as the 'big boys' caught them up. Acorn, Apricot, and Sinclair Research, all in computers, fell over a cliff when IBM introduced its PC in 1982, and have never regained their early sparkle. In contrast, the Oxford company, Research Machines (now RM), which was also in computers, specialized in one small area of the market, in schools, and is now a flourishing publicly quoted company. In Oxford Instruments the ambulatory monitoring markets grew enough to attract many other companies, large and small; later computer developments have made it a highly competitive 'low-entry-barrier' business, and Oxford Medical looked outside for less competitive products. In the late 1980s the market for our whole-body imaging magnets dropped suddenly as one of our few big customers started to make its own magnets. Here our solution was to find a strategic partner, and, after a difficult hiatus, we formed the successful joint venture with Siemens. Some form of alliance is probably the best solution for small biotechnology companies developing drugs of universal importance, and for other small companies with unique technology that might have wide applications. Large companies have the means to provide the necessary development resources, and they are usually much better than small companies at sales and marketing. But many larger companies are increasingly willing to 'outsource' development and to form alliances with small firms holding important patents or knowhow. Few science-based firms in the UK have grown to be substantial independent companies of international importance. They do not have the large home markets of the USA, and, on the whole, the UK is behind on management skills. It also lacks a large pool of experienced entrepreneurial executives.

Some people ask why anyone should want to start a 'high-risk, high-tech' enterprise. In fact the little research that has been done seems to show that they may be less risky than most small firms, especially when provided with 'big brother' support, as in an innovation centre. From the time the first STEP Centre opened to late 1999, 124 companies had occupied innovation centre premises under the wing of the Trust or of its subsidiary, Oxford Innovation. Only twelve of these companies have gone out of business. This compares with the national statistic of about a third of all start-ups failing in their first three years. There *are* some risks in new technology, and a firm building up on one or two new ideas is clearly more vulnerable than a large established technology company with a spread of products and markets. But these risks from the technology may not be worse than, say, the risks to a food company from the possibility of a contamination scare. Then, those who start high-technology firms are often very intelligent and committed people; they want to see their unique ideas become a reality of use to many others. They are often flexible and able to change course, taking on temporary jobs like programming if their plans get delayed.

In a science-based company the other side of the 'high-risk' coin reads 'high reward'; but most scientist entrepreneurs are not in the game *entirely* for the money. When The Oxford Trust was five years old, in 1990, the third issue of *Scintilla* started with a short article called 'Before The Oxford Trust'. I was interviewed about the early years at Oxford Instruments, and I am quoted as saying, 'Few people realise how stimulating and exciting our sort of company can be . . . You need courage to take risks and the resilience to work all hours and take the knocks. You need to be willing to take advice and to bring in experienced managers when the time comes. And it helps to have more than a little luck. But it's a great challenge' (TOT 1990*b*: 1). Elizabeth Garnsey (1997: 15) has expanded on the same point:

Still today, in the 'winner takes all' ethos, idealism is viewed as ineffectual sentiment and competitiveness rules. In my experience of high tech firms, here and in the US, science-based enterprise is not part of that stream. Rather there is a strong strand of idealism in many such companies. Professional pride is focused onto solving problems effectively for users, or producing a product or service that enables users to do things better or to do something new . . . Making money is what happens when you solve problems that people are prepared to pay for.

Now, more than ever, the ethos of helping the customer is vital. In Oxford Instruments some of our most supportive customers are those who have experienced a problem with one of our products, which has received immediate attention, and has been put right at the earliest opportunity.

By 1990 so much practical experience had been built up, on matters like the technology audits and how to run innovation centres, that Oxford Innovation Ltd. (OI), the Trust's wholly owned subsidiary, formed a few years earlier, took over the commercial applications that were beginning to fall outside the charitable field. These remained compatible with the Trust's objectives, and, through covenants, it could now receive the profits from this valuable stock of information and skill, and plough them back into new projects. David Kingham became this Company's MD and Martin its Chairman. OI works with a growing staff and with a panel of associates to provide consultancy services and to run innovation centres; it has grown in strength, profitability, and influence. In its consultancy work, which it prefers to call 'innovation engineering', it promotes technology transfer, and carries out technology audits. This work has spread well beyond the confines of Oxfordshire. OI has fulfilled contracts from the European Commission to work with SMEs; it has undertaken projects for the Czech Republic, and has even been called to the USA, where it has advised the US Navy and the US National Technology Transfer Centre.

OI has planned, promoted, and now manages innovation centres at Milton Park, Upper Heyford, Rissington, Banbury, and Begbroke. At the end of the century three more were under negotiation, at the AERE Harwell, at RAL, and near Bicester. As the new century started, The Oxford Trust was still running the thirty incubation units in its second STEP centre, using the young companies there to study the processes of incubation. Altogether there were then ninety-seven units available for these fledgling companies, all with access to help and support, and the number was expected to double before long. In fact we are beginning to talk of a *cluster* of innovation centres.

Oxfordshire boasts several other clusters of companies, including those in older industries such as publishing and medical instruments. There has long been a small cryogenic cluster, sometimes referred to as Cryogenic Valley. Some twenty companies can be traced back in one way or another to Oxford Instruments. But Oxfordshire is also home to an important cluster of world-class companies involved in motor sport, mostly near Banbury in the north of the county. The Heart of England TEC commissioned a study from Oxford Innovation, in which Treve Willis found that these companies provided about 4,000 direct jobs (see Willis *et al.* 1996). He also found that they were desperately short of key engineers. The longer-term solution was to work with

Banbury School's Design and Technology department to devise a competition for 13 and 14 year olds to raise the numbers wanting to study engineering. Teams of eight children, with help from advisers in these companies, made Formula S (for schools) 1 : 8 scale remote control racing cars, and raced them for a prize. The next year ten schools participated, and a regional competition was planned for 2000. Oxford Brookes University now runs a degree course on automotive engineering, and the course is oversubscribed.

Oxford has a long tradition in the fields of medicine and chemistry. In the 1930s William Morris (later Lord Nuffield), who started the large motorcar manufacturing business in Oxford, gave £2 million to the University for a medical research establishment, a lot of money at that time. Now the county is home to a growing number of biotechnology companies. There were forty when OI surveyed them in 1996, and by 1997 the cluster accounted for over 2,000 jobs, while a continuing trickle of new companies was starting up or arriving from elsewhere. With the help of other sponsors, David Baghurst of the Trust took on the task of organizing some of these small companies to form the Oxfordshire BiotechNet, which became a company chaired by Dr John Gordon. There is an associated Biolink, run by the Heart of England Business Link, which assists networking among the longer established biotechnology companies, and now there are specialist local degree courses, which aim to provide the manpower these companies need. In 1998 BiotechNet secured one of the Government's 'Biotechnology Means Business' mentoring and incubation awards, worth £400,000, for which there was locally matched funding, some of it in promised services. A new prefabricated Biotech incubator, with laboratory provision, has been erected on the Yamanouchi Research Institute's site at Littlemore, near the Oxford Science Park. The first phase filled up rapidly with eight small companies, and, at the end of 1999, the buildings for the second phase were on order. This BioInnovation Centre aimed to house eighteen start-ups over its first three years. The longer-term goal is for a larger centre, a Bio-Medical Park, near the Headington hospitals and the new Medical Research Institutes.

Companies in clusters seem to benefit from association with others in the same type of business, especially when some mutual organization is developed, as in the automotive and biotechnology clusters. Other local clusters at an earlier stage of organization are in multimedia, centred on Banbury, IT and software, which are now widespread in many regions, and opto-electronics and advanced materials. The newly established Begbroke Innovation Centre is inside the new complex for Oxford University's Materials Science laboratories, and there are other companies working in this area based near AERE Harwell and RAL.

By the early 1990s the Trust was growing out of its premises on Osney Mead. Our 'networks' brought us early news of RM's plans to move from its excellent building not far from the STEP Centre. This gave us time to organize its purchase in partnership with Lansdown Estates, owners of Milton Park, and its forward-looking directors, Ian Lang and Nick Cross, who built the innovation centre at Milton Park and are great supporters of the Trust. This building, once a Pickfords Repository, can be seen beyond the churchyard as the train from London approaches Oxford Railway Station. This became the Oxford Centre for Innovation, and Mike O'Regan, one of RM's founders, became a trustee. Apart from the extra space for the Trust's own activities and more room for OI, the new incubation centre created there can house up to thirty tenant companies, with larger, commercially let, units above. This new building was opened formally by the Prince of Wales in 1996, in front of an audience of 150 of the county's business and academic leaders and policy-makers (see Fig. 24.3). The innovation centre filled up in a year or two, and there has usually been a list of would-be tenants waiting for companies to graduate to larger premises elsewhere.

Fig. 24.3. HRH The Prince of Wales at the opening of The Oxford Centre for Innovation. Mike Denis and Gillian Pearson help the children explain their experiment to him while Paul Bradstock waits behind him

One of the early objectives of the Trust had been to use a small percentage of its funds to provide seedcorn capital to a few small companies. In the early years we invested small sums in two companies, but very soon needed all available funds for the Trust's own work. For a while the Trust acted informally as a matchmaker, introducing would-be investors, often known as 'business angels', to promising companies needing equity capital. It catalysed investments in several local companies, including Oxford Glycosystems, Oxford Asymmetry, and Voltech Instruments. In 1994, after a feasibility study financed by the DTI's Innovation Unit, the Trust, and the Heart of England TEC set up the Oxford Investment Opportunity Network (OION). Inevitably called Onion, this is a non-profit-making company.

OION was officially launched early in 1995 in front of eighty-four representatives from local banks, accounting and legal firms, innovative companies, and potential investors. At regular presentation meetings technology-based companies needing investment talk about their business plans to registered OION investors, and answer questions. In its first two years the network hosted twelve meetings, and advised at least seventy-five companies seeking finance, of which eleven achieved the investments they needed. In 1996 the solicitors Manches & Co. and accountants KPMG drew up investors' agreement and 'due-diligence' packs to help new companies and business angels to keep down the cost of investment. At its launch Peter Angel, of Manches, explained how he had had to overcome all his legal training to write in normal English. By March 1999 over £3 million had been invested in eighteen companies, and more deals were under negotiation. Often these business angels become non-executive directors, or help the investee companies in other ways from their previous experience in business. The OION formula has now been copied elsewhere and, combined with the favourable UK tax treatment of early stage investment, is helping many small companies to find the seed capital they need. When well established, the running of OION was handed over to OI.

The Oxford Trust was an organization set up in the right place at the right time. Its work with small companies and in promoting technology transfer has been widely recognized. The Advisory Council, on which serve some twenty highly experienced people from different walks of life, holds biennial two-day meetings that have become a forum for innovative ideas and plans for the future, both on the economic side and in the education programme. There is little room in this book for describing the activities in this, the other side of the Trust's work, in 'exciting and enthusing young people in science, engineering and technology'. But a little must be said, as this side of the Trust's work is so important for the future of our increasingly knowledge-based economy. To get children interested in science and technology, it is important to 'catch them young'.

Soon after the Trust had been launched, Paul Bradstock started talking to the local education authority and to science and technology teachers to explore how children could be enticed into studying the sciences and into contact with local industry. So few teachers know much about industry—many move from school to university, to training college, and back to teach in schools. The advice coming back was that money would be the answer. Schools were so short of funding they would jump at ways to enrich the curriculum. In a school a small sum goes a long way; the Trust started a Schools/Industry Grant scheme to tempt older children to carry out technical projects under the guidance of a mentor in a company. The tiny Trust team was soon joined by Gillian Pearson, who became Assistant Director for Education. A graduate chemist who had taught in secondary schools, she had also been production manager in a cosmetics factory. She was exactly the right person for this job, which she continued to hold into the new millennium, although she has been joined by several others in the education section.

Gillian was soon familiar with every school in the county and could, from the Trust's extensive database, help students to find suitable companies for partnership; she was also very persuasive. When applying for the grants, schools had to give an outline of the proposed project and estimate the costs, which usually covered things like transport, materials, and photocopying. Each year Gillian visited all applicants and assessed the projects, and a committee at the Trust decided on the grants, which totalled £3,000–£5,000 a year. It was an extraordinarily cost-effective way of making talented children think through an idea and carry it forward to a prototype, and on the way they saw something of the best of local industry. Companies were very generous, but also felt they gained from the school links, and one or two projects turned out to be of *real* use to the company. One inventive young man, after a very successful relationship, which included working for his mentor company in his holidays, finished up coming back to the Trust several years later for help to start his own company. There was an annual exhibition of this work, usually hosted by RAL, where several assessors chose the prize-winners in each category—and the prizes were substantial by school standards. The Trust's annual Summer Lecture, given by a well-known scientist and aimed at GCSE and A-level students, usually formed part of the programme for the day.

There were many other projects. In one the Trust supported six teachers who were paired with science-based companies to develop teaching material to bring local industrial relevance into GCSE science lessons. Another project was for two-day innovation workshops, the first held in partnership with Oxford Instruments. Fifteen-year-olds were involved in an interactive assessment of a product, possible improvements, and matters such as cost reduction and marketing. For A-level students there were business seminars, with case studies to help them appreciate the commercial aspects of technology. Fifteen-year-olds were involved in a series of 'Sciencability' workshops, the first a three-day project supported by British Gas and the Royal Institution. With input from apprentices and other helpers, each team of children had to design a 'hands-on' science exhibit to illustrate a scientific principle. The best of these exhibits were later used in CuriOXity, the Trust's interactive science centre for younger children.

CuriOXity, now run by Mike Denis, was opened in 1990 in two small upstairs rooms in the old fire station in the centre of Oxford. Sarah Farley wrote in the *Times Education Supplement* (30 Apr. 1993), 'It must be the smallest interactive centre in the country, but like the Alice in Wonderland world with which it identifies, the smallness doesn't seem at all odd once you get inside.' In fact teachers welcome the small size and limited number of exhibits, as they can more easily control a class and try to instil a limited number of scientific principles. For primary school children there are visits from a travelling 'Science of Magic' show, which was developed by Mike Denis. While being great fun, the tricks make children think *why* certain odd-looking things happen. And there is the 'Experiment Experience' for schools or for family days at the Trust. When the Prince of Wales opened the Oxford Centre for Innovation he took great interest, both in the older children's projects and in the younger children who were busy finding things out by experiment.

Among other projects is the annual Festival of Science and Innovation, which has been growing in numbers of events and in geographical cover ever since 1992 (see Fig. 24.4). In 1999 it stretched over three weeks, with prestigious lectures, master classes, science theatre shows, and schools workshops, as well as the normal range of activities. In 1998 Nobel Prize winner Sir Harry Kroto gave a fascinating lecture on his discovery of the strange carbon molecules known as 'buckyballs'. But the afternoon before he had insisted on taking a class of eager children at a small village primary school. Every year several thousand people participate in these events. We all hope that the public of Oxfordshire will soon have a better understanding of science and technology and its importance for economic prosperity.

In June 1999 the Trust played a leading role in the very successful two-day 'Venturefest Oxford 1999', held in St Catherine's College. This is to be an annual event, to be held in Oxford Brookes University in 2000. As the next number of *Scintilla* (September 1999) reported, 'Venturefest . . . [was] the first event of its kind to showcase Oxford's particular strengths as an international centre for innovation and technology. It brought together technology companies, entrepreneurs and investors, academics and commerce.' The schoolchildren were also at the event, demonstrating their projects at the Trust's Innovation Awards Exhibition. With such a wide spread of participants Paul Bradstock talked of the event as 'networking the networks'.

Martin gave a keynote speech at Venturefest under the title 'Oxfordshire Connections—Partnerships for a World-Leading Community' (22 June 1999). He started his talk:

I have two jobs to do here today . . . first to explain why Oxfordshire has become such a fertile area for the development of new high technology companies. Secondly, as an entrepreneur myself, I'd like to transmit some of the excitement and satisfaction that I've felt over the years as a result of being part of the rise in the profile of the high-technology sector.

I hope this book has addressed both of these aims.

(The Oxford Trust can be contacted at The Oxford Centre for Innovation, Mill Street, Oxford OX2 0JX. UK. Telephone: (0)1865 728953 Facsimile: (0)1865 793165. E-mail: admin@oxtrust.org.uk Web: www.oxtrust.org.uk.)

Fig. 24.4. A child operating a working model of Boyle's air pump during the 1992 Oxford Festival of Science—The Oxford Trust aims to catch them young

Postscript: Oxford Instruments in the Context of Post-War British Industry

Richard Coopey

The story of Oxford Instruments is engaging in and of itself—an absorbing tale of scientific and commercial endeavour in which the reader shares sometimes the sense of achievement, sometimes a sense of exhaustion. There are crises over technology, the race to keep abreast of new technological developments, crossing new frontiers and creating new areas of business, outrunning finance and agonizing over merger decisions. There is also, of course, the fascinating account, which only an insider could relate, of the rise from a garden shed start-up to a globally recognized firm, market leader in a series of advanced technology products, some of which revolutionize key medical techniques.

In addition to being this gripping tale, however, the history of Oxford Instruments throws a rare illumination on a series of debates and ideas about the state of post-war British science, technology, and industry. Many of the details outlined by Audrey Wood confound, or call for a re-evaluation of, popular notions of the state of British entrepreneurship, the relationship between industry and the academic sector, banks and industrial finance in Britain, or the role of the state in aiding economic modernization. Other elements of the story re-emphasize the problems that face growth companies, particularly in fields of high technology, and indeed point to ways in which the ride could have been made significantly less bumpy—for Oxford Instruments or for other growing firms. The British economy has been deemed to be in a state of relative decline since the late nineteenth century. Debate rages over the periodization and the extent of this decline, and also over which factors contributed. Briefly, the banks and the City have been held to be unsympathetic to the needs of manufacturing; science (and education) have been similarly preoccupied with other aspirations or disconnected from the needs of enterprise; governments have struggled to formulate corrective policies, or been distracted by other targets ranging from defence to welfare; management and trade-union cultures have variously been held to account for entrenched and restricting modes of working; Britain has been hampered by more general legacies and inertias—the penalty of being the 'first industrial nation', an impossible position to sustain, or of the overarching influence of class and cultural formations antipathetic to industry. As stated, the history of Oxford Instruments supports some of this, but demands a re-evaluation of much of it.

Banks and the City

One of the key issues facing a growing company such as Oxford Instruments is the recurrent need for new rounds of finance—to fund new research and development, to launch new products, or to expand operations to meet new market opportunities. This is in addition to the support needed in the short or medium term in the form of banking assistance to meet the normal cycles of trade, fluctuations in the marketplace or business environment. For a small to medium-sized firm, the line between these two forms of finance may often be blurred. Unforeseen difficulties in production or the market can turn overdraft facilities into the equivalent of long-term debt. An understanding relationship between bank and enterprise is vital here, often dependent on the experience and autonomy of local bank managers.

The difficulty connected with the cost and availability of finance to small and medium-sized firms, whether for the short-term trade cycle, or for long-term investment purposes, has been viewed as a flaw in the British banking and finance system since at least the 1930s. A 'Macmillan gap' was identified in 1931 by the Macmillan Committee hearings. These hearings were prompted by the economic depression, and the popular accusation of a 'banker's ramp'. The 'gap' they found, though not the central finding of the committee, was seen as the product of Britain's dual system of banking and finance. Economists following Gershenkron have lauded the continental 'universal' banking system, where banking and investment functions take place within the same institution. German banks in the post-war period are similarly praised for their commitment to long-term industrial finance and a close working relationship with manufacturing firms. Britain's dual system comprised, on the one hand, a banking sector overconcerned with the security and liquidity of its deposits, and, on the other hand, a London-based capital market overconcerned with short-term profit indicators and with no particular affinity for investment in British manufacturing. There are those who detract from criticism of the banks and the City, pointing out the rationality of such a strategy, and indeed the longevity and stability of the system (Capie and Collins 1992). Critics continue to stress the distance generated between industry and finance in such a system (Ingham 1984; Kennedy 1987; Hutton 1995).

The Macmillan gap had occurred because the clearing banks did not traditionally see it as their duty to provide long-term investment capital, with all the attendant risks of commitment, to industry—a sector about which they knew less and less as they consolidated into the London-based big five after the First World War. The capital market, itself monopolized by London from around the same time, offered long-term funding in the form, for example, of a public issue of shares. The problem lay in the scale of investment needed. Medium-sized firms could neither afford the attendant costs of prospectuses, underwriters' fees, and so on, or issue enough shares to ensure an adequate secondary market. This gap, unearthed by the Macmillan Committee, has been perennially highlighted. The Radcliffe Report of 1959, the Bolton Report of 1972, the Wilson Reports of 1977–9, and the recent Cruikshank Report all restate the same case. The prohibitive cost and uncertain availability of finance for small and medium-sized firms continues to be a serious problem for the British economy. This is all the more of a problem since the SME sector has frequently been identified as the most dynamic of any advanced economy, with the exception of periodic fads towards giantism, such as in the 1960s (Birch 1979; Bannock 1981; Stanworth and Gray 1991). The German *Mittelstand*, the large pool of SMEs that provide the core Japanese manufacturers with outsourced components, the still expanding clusters of dynamic spun-off advanced technology firms in silicon valleys, fens, and glens—all provide strong support for the Schumpeterian view that entrepreneurship and innovation, sited in growth firms like Oxford Instruments, is a vital part of an advanced economy.

Firms like Oxford Instruments lie at the centre of this Schumpeterian world—high-technology innovators, creating new products and generating and expanding markets, at home and abroad. The firm's history illustrates very well the difficulties inherent in this process and the importance of a good relationship with the local bank at various stages of growth, or when unforeseen difficulties occur. In the case of Oxford Instruments this is illustrated time and again—as, for example, when material supplies, at the leading edge of development in superconductivity, begin to fail for unexplained reasons, or when market forecasts for semiconductor manufacturing technologies fail to emerge in the case of X-ray lithography.

In the initial stages, when a firm is being established, finance usually revolves around personal issues—mortgages, special relationships with the technology of 'patrons' (such as the connection with the Clarendon Laboratory and Harwell), and limited capital resources. Overheads remain low, expediency determines that premises, for example, will often be Spartan, and a make-do attitude will suffice. Oxford Instruments is the classic garden-shed start-up, and we get a real sense of the pioneering, firm-building process as we follow the subsequent expansion of the firm through converted stables, offices in a caravan, a disused laundry, and an old boathouse. Capital requirements of a growing firm will eventually call for substantial outside help. There are notable cases of expansion funded through profits alone, which stretch from the early years of Ford through to the Wordperfect Corporation, but these are generally the exception. Oxford Instruments, like so many small and medium-sized firms in post-war Britain, turned first to its bank for overdraft finance. Here again the story is illustrative of a general complaint by SMEs about the structure and culture of banking in Britain. Unease exhibited by its London bank led the firm to transfer to its Oxford high-street branch. Here it experienced, for some considerable time, a supportive relationship, still based on overdraft finance, but nevertheless extended to suit circumstances and fluctuating business cycles. This relationship touches on two debates. Banking historians have long debated the role of the overdraft in long-term finance of firms—that overdraft finance should not be ignored when assessing the role of the banks in the capital provisions of British industry throughout the twentieth century (Capie and Collins 1992). In the case of Oxford Instruments, for example, the purchase of a substantial piece of capital equipment—the helium liquefier—was funded by overdraft. Secondly, the relationship between the local Midland Bank and Oxford Instruments demonstrates the continuing importance of local networks—local trust and knowledge by regional bank managers—as a vital resource in lowering information asymmetries, enabling banks better to assess risk. This local connection was a feature of nineteenth-century banking in Britain, but was held to have faded with the concentration of the major banks for much of the twentieth century (Pressnell 1956; Newton 1996).

There comes a point, however, when overdraft finance, however flexible or accommodating, cannot serve larger ambitions. This moment came for the firm in the mid-1960s. Though the nascent venture capital sector was already burgeoning in the USA, British industry had little similar provision. Oxford Instruments did find a suitable investor, however, in the form of the Industrial and Commercial Finance Corporation (ICFC). Throughout the 1950s and 1960s ICFC went against the grain of investment banking in Britain. Though a private-sector institution, owned predominantly by the major clearing banks (though not experiencing an altogether happy relationship with its owners), ICFC had its origins in political arguments surrounding the reconstruction of British industry after the war and the need to close the Macmillan gap. Led by Lord Piercy, John Kinross, and later Larry Tindale, ICFC successfully invested in the SME sector using a mixture of networked branches, managerial and technical assessment, and equity investment (Coopey and Clarke 1996). In many ways ICFC represented the prototype for venture capital methods. Indeed, it was later commonly referred to as the 'university' of venture capital in Britain. ICFC's Technical Development Capital initiative of the 1960s, under the auspices of which Oxford Instruments received support, reflected the corporation's continuing sensitivity to political initiatives, coming as it did in the midst of the Labour government's Ministry of Technology initiatives under Harold Wilson, and resistance to the perceived 'American challenge'.

Entrepreneurship, Ownership, and Control

In many ways Oxford Instruments was the classic ICFC investment. The involvement of ICFC brought long-term funding through successive rounds of financing, some of which followed difficult periods of trading. It also bought advice and, if not 'hands-on' control, then certainly 'eyes-on' mentoring in issues such as management consultancy and help with recruitment. The local connection was maintained through the Reading office—indeed the closeness of the relationship can be judged by the recruitment to the board, years later, of the man who had been ICFC's local manager during the company's most critical period. ICFC's gain from the relationship was the value of its equity holding.

Founders of growing SMEs are notoriously reluctant to surrender equity, and, as they often see it, a degree of control in their firm. For investors such as ICFC, however, equity is a vital component of the relationship, ensuring a proportionate share in the success of well-chosen investments (and essential in covering the costs of failures). Equity enables risk, and, as ICFC demonstrated in the case of Oxford Instruments, can work successfully as part of a long-term relationship. The group eventually became one of ICFC's (later renamed 3i) most successful investments. Unfortunately, too much emphasis has been placed on shorter-term business plans and exit strategies in the venture capital industry that followed ICFC into the 1980s. Concentration on high-technology SMEs, in the classic American venture capital model, also faded somewhat in Britain during this decade, as the 'softer' and considerably more risk-free option of the management buy-out market emerged (Bygrave and Timmons 1992).

As the Oxford group grew, questions of control and ownership inevitably spread beyond the family, workforce, and ICFC, and the prospect of an increased shareholder base emerged. The firm's encounter with the City, though not as painful as some, raised familiar questions and problems. The accusation of short-termism and the City is now commonplace. The dominance of the stock markets by fund managers investing the vast assets of pension funds and insurance companies has come under periodic critical scrutiny from the 1960s onwards. The notion that these managers, sensitive to quarterly returns and short-term indicators of stock values, will be extremely reluctant to wait patiently for the fruits of long-term investment programmes is a powerful one. Thus the general trend in the British capital market is seen to militate against investment in R & D, for example—an immediate cost set against future prosperity. This is usually seen to be in contrast to the relationship banking and less volatile systems of investment in Japan and Germany. (Venture capital funds were, during the 1980s, seen as a way in which these institutional funds could be unlocked and directed towards long-term technology investment, but, as noted above, they have failed to live up to this 'classic' image.) Academic studies are divided on this issue, some maintaining that the effect has been exaggerated. For a group like Oxford Instruments this effect could prove extremely problematic. R & D clearly formed a foundation of the business from the outset. It is an essential part of the firm's business culture, and the very basis of the successful strategy that saw off a series of larger rivals throughout the firm's history. Another critique of the City is that of the herd mentality and sensitivity to rumour and peripheral factors (Ingham 1984; Kindleberger 1996). Ironically this can be seen working both for and against a firm like Oxford Instruments, floated on a boom in technology stocks, but later seeing its share price affected by press opinion concerning the US health-care markets, or the general state of the electronics sector.

The decision to widen the ownership of the firm with the flotation, in the 1980s, not only represented the beginning of a new phase of expansion and funding but also raised concomitant questions of control. Here is another theme in the history of enterprise in Britain that is informed by the history of Oxford Instruments. As the scale of any enterprise grows, it becomes inevitable that decision-making structures—management and organization—be more formally established. Yet there is a strong theme of personal control, or control devolved to the workforce, that emerges throughout the Oxford story. At various times, formal managerial control of the enterprise has been presented as the optimum for efficiency. Or, more recently, firms have increasingly sought ways of formalizing systems that incorporate both control and a 'voice' at all levels of the hierarchy. The business historian Alfred Chandler is perhaps the most notable proponent of big business and an accompanying rise of a cadre of professional managers, which he held to be the foundation of American business success in the twentieth century. Britain, by contrast, was held to be hampered by the longevity of family-owned enterprise. Historians have continued to debate both the extent and the effect of prolonged personal capitalism. One of the features that marks Oxford Instruments off from many small, family businesses was the willingness to bring in professional management when necessary in the early 1970s. The family lost majority voting power with the BOC deal in 1973, yet, despite this formal dilution of control, the enterprise retained, with the support of sympathetic outside shareholders, an independence that would avoid the dilution of 'neat Oxford', and, importantly, an independence that would 'keep the business alive in a world of Goliaths'. The ways in which a democratic element was introduced into this process at an early stage is seen in the decision to reject the approach of Bruker in the NMR business. This move, supported by the staff, provides a fascinating insight into the possibilities of real influence through a consultative process among the workforce. The early commitment both to shared ownership—now part of a fashionable 'stakeholder' ethos—and to a voice in key decisions again emphasizes the perhaps radical, certainly insightful, managerial culture that grew at Oxford Instruments.

Time and again in the story we see the balance being made between independence and control, and the need for partnership or increased capitalization. At several points the company has the advantage of being in advance of larger competitors in technology, but is faced with the fact that many of these are customers, who may, indeed do, attempt to develop key rival technologies of their own. Mergers and strategic sales become inevitable, as does the search for acquisitions to broaden the base of the enterprise when single successful products threaten to unbalance the group and leave it vulnerable to market changes. This story is one of propitious moments. The fear of takeover, and loss of identity, emerged early on with resistance to the overtures of Air Products, later with the rejection of the proposed merger with UEI, and yet later with the acceptance of the sale of a majority shareholding of Oxford Magnet Technology to Siemens. Again we see the fleetfootedness of the enterprise in the subsequent acquisition of Link Analytical Instruments to eliminate the cash 'hillock' created by the Siemens deal.

Science and Industry, Networks and Clusters

The theme of independence and control—of attempting to manage the destiny of the business, and preserve its culture of operation—seems to exist throughout the history of Oxford Instruments. This process bears the strong imprint of the founders, and Martin Wood's transition from engineer and scientist to entrepreneur forms one of the most fascinating elements of the story. Another common theme in the British 'declinist' literature, which seeks to explain the relative eclipse of the British economy, is that engineering and science have been neglected, or, more precisely, detached from enterprise. Arguments range from the specific—for example, the level of state support for science education, or overemphasis on military or prestige science and engineering—to the cultural explanations pointing to 'two cultures', and beyond, or the triumph of an 'anti-industrial spirit' (Snow 1959; Weiner 1985; Elbaum and Lazonick 1986; Edgerton 1991; Rubinstein 1994). A variant of this theme emphasizes the lack of scientists and engineers in the British boardroom. A predominance of accountants in control of British companies is compared unfavourably with the number (and the status) of scientists and engineers, in boardrooms, yet again, in Japan and Germany.

Again the experience of Oxford Instruments partly confounds, and partly confirms, this schema. Martin Wood could be portrayed as a rare combination of entrepreneur and scientist. Combined with Audrey Wood's flair for management, this ensured that the firm remained innovatory at the levels of corporate strategy or human resource management, while at the same time keeping abreast of, and often defining, technological change and market opportunities. As with all entrepreneurial histories, this is one of vision, risks taken, and fortuitous interconnections of circumstances. Entrepreneurial and scientific skill blend in examples such as stealing a march on BOC with the delivery service of liquid helium, and keeping one step ahead of a firm the size of Westinghouse in superconducting magnet development. The way in which the first, comparatively expensive, batch of superconductor worked, in comparison to later problems, could be seen, on the other hand, as a fortunate episode. The contexts in which the firm grew—the nexus in which the Woods operated—are also very important. One of the legacies of science and technology in Britain and its geopolitical history in the post-war period was the large effort directed towards the aircraft and nuclear sectors, as part of the quest for international prestige and cold-war participation (Smith and Smith 1985; Edgerton 1991; MacKenzie and Spinardi 1993). Large-scale research facilities such as those of the Atomic Energy Authority at Harwell or the Royal Radar Establishment at Malvern were a product of this history. It is interesting to observe the almost symbiotic relationship that emerges at points in the story as the establishments take advantage of Oxford Instruments' technological advances, or in turn place development contracts that boost the firm's R & D programme.

The success of Oxford Instruments is clearly several stories interwoven. The Woods' undoubted flair for technology and enterprise is set against a background of British scientific and technological legacies and the emergence of new market opportunities, eventually on a global scale. More immediate contexts are also important, however. Historians have, over the past few years, become increasingly interested in networks and clusters, whether virtual or geographic, in explaining scientific, technological, industrial, or economic change. Again the story of Oxford Instruments is very instructive. As the firm evolves, it can be seen as an integrated part of a series of communities. Prominent amongst these are the scientific and university communities. The firm's links with the Clarendon Laboratory were crucial in the formative phases. The advice and support of Nicholas Kurti (and his 'continental' ideas of collaboration); the use of equipment out of hours; the loan of the special winding machine; the help of retired university technicians; the use of the university computer in calculations on the first superconducting magnet; provision of liquid helium and nitrogen for the early tests; all these indicate a strong formal and informal interconnection between the firm and university. Indeed, it comes as something of a surprise to the reader when we find that Martin Wood is still employed at the university as late as 1969. More broadly, the firm grew within a labour pool generated by the university and its ancillary institutions. Scientists and engineers were recruited from AERE Harwell and later the Rutherford and Appleton Laboratory.

The firm existed within both a local and a global intellectual community. Throughout, the Woods were clearly keen to stay in touch with the broad development of new scientific discoveries, and as such form part of the global scientific community. (Commercial enterprise must, however, exist in competition with rival firms, trying to exploit the same scientific or technological advances that emerge. The firm's R & D strategy is crucial here—as will be discussed below.) But the firm also existed as an integral part of a more local community. Development contracts from Harwell, preferring the firm over larger rivals such a Arthur D. Little or Philips; the publicity generated for the firm through the publication of the results of this research; the close relationship with world-renowned Oxford scientists in NMR, such as Sir Rex Richards—all these indicate the way in which Oxford Instruments grew within a local network. The original market for scientific equipment that fostered the group, as noted in Chapter 14, continued to be important 'not only in terms of profits generated' but importantly as an 'incubator' for new technologies and products.

Indeed, the firm went on to generate its own centre of gravity, creating a cluster of firms in the Oxford region. That this could prove a burden as well as a benefit is illustrated by the problems that occur when rival firms threaten to set up in the area, or indeed when breakaway firms set up in competition, as happened in the case of Thor Cryogenics. This aside, regional clusters can clearly provide a net gain in terms of technology transfer, enhanced labour markets, and so on (Storper and Scott 1992). (They can, on the US model, also provide a focus for venture capital support in investment, though this has not noticeably been the case in Britain.) Yet, here again, we see Oxford Instruments working against the tide. While the success of Silicon Valley in California was centred on Stanford University, or the science park established around Cambridge enhanced enterprise in that region of the UK, Oxford could boast no counterpart. Here again we can see the Woods taking a proactive role through the establishment and efforts of The Oxford Trust, a limited though very important initiative aimed at enhancing the links between education and industry and helping small high-technology firms in the Oxford area. The Trust's focus was subsequently broadened to address the fundamental problems of science in education.

The successful link between the firm and the university from its earliest days is also reflected in the culture of research that permeates the history of Oxford Instruments. Many questions surround R & D development strategy—how much to do, when to concentrate efforts on new products, diverting resources away from production processes, and so on. As noted above, a focus on R & D can be offputting to investors without a long-term perspective. There are the burdens of freeriders to think of—examples of firms investing great efforts in research, only to see others capitalize on development. (One of the most notable modern examples of this was the graphic user interface developed by Xerox but subsequently marketed by Apple, then Microsoft). Markets may not appear in ways predicted, as in the case of the considerable efforts that the firm undertook in developing the Helios X-ray lithography synchrotron for IBM. Nevertheless, research-driven innovation continues to be the leitmotif of the company throughout. Sustaining such a culture can prove difficult in a growing enterprise—indeed there was a need to re-emphasize this in the early 1980s when Martin Wood announced a 'returning' point, persuading shareholders to approve a return to a less risk-averse attitude, even at the expense of short-term profits.

In sustaining an R & D culture, has Oxford Instruments gone against the trend of British industry? In terms of the national effort, there are innumerable accounts of the ways in which R & D has been neglected. These range from individual technologies pioneered in Britain but developed abroad—for example, aniline dyes, jet engines, monoclonal antibodies, liquid crystals (the computer and atomic energy are occasionally added to this list)—to a general failure in the percentage of GDP dedicated to R & D. Debates continue over the accuracy of historic estimates of R & D levels (Mowery and Rosenberg 1989; Edgerton and Horrocks 1994) and the ways in which governments may enhance R & D effort. Historians have also emphasized the ways in which the military prerogatives of the cold war distorted the national R & D effort, in particular the ways in which research such as at Harwell and the RRE was disconnected from industry, in contrast to the USA, where, owing to contracted-out R & D, military technologies may have had greater spin-off or synergy with the civil economy (Mowery and Rosenberg 1989; Coopey 1993; Bromberg 1999). Again the case of Oxford seems to confound this argument. Instead we have the picture of both Harwell and the RRE in a dynamic relationship with the private-sector Oxford

Instruments, helping to build an important advanced technology enterprise with a global presence.

In the 1960s the debate was complicated further by the assertions of Bruce Williams and others that R & D spend was not essential to national economic prosperity—that the Japanese, for example, did well on licensed and imported technology. Nevertheless the government attempted, through the newly created Ministry of Technology, to enhance and orchestrate the national R & D effort. This experiment was short lived and faded under the general political pressures that saw the Labour government defeated in 1970. British governments subsequently addressed the level and direction of R & D in Britain in only sporadic, piecemeal fashion (Gummett 1980; Vig 1986; Zuckerman 1988). (The Japanese economy has, of course, subsequently entered a mature phase and R & D levels have climbed to much higher levels, and R & D holds a central position in the strategy of Japanese corporations.) It is interesting to note that the role of government appears only fleetingly in the history of Oxford Instruments—for example, in the reference to the neglect of basic research during the Thatcher years. As with the provision of finance for SMEs, the role of government in fostering growth sectors has at best been tangential and uncoordinated.

The Oxford Model?

The impression left by the Oxford Instruments history is that the enterprise would thrive in spite of an unpromising environment. If we look for a keystone to the enterprise, it must lie with the entrepreneurial drive and enthusiasm of the founders. The combination of technical and managerial ability meant that the growing firm could take advantage of unfolding market opportunities, and beyond this could generate new markets where none previously existed. The history clearly demonstrates the need for a vibrant SME sector during times of industrial and technological change. As Audrey Wood notes, 'It was certainly a difficult industry for a novice company to be in, but a small group can learn fast and can usually be more flexible than a large hierarchical company.' But the Oxford group needed to grow beyond this. The history demonstrates the difficulties in sustaining the ethos and culture, and indeed the fleetfootedness of a 'novice' enterprise as it grows, but it clearly demonstrates the ways in which these difficulties can be overcome. The story outlined in this book remains a vivid illustration of some of the difficulties facing a growing high-technology company in post-war Britain, and supports some of the general arguments put forward to explain the relative decline of the British economy. But the story also stands in marked contrast to many of the issues usually raised by declinists, most notably the standard of management and entrepreneurship and the nature of the scientific–manufacturing nexus. The history is just that. The contexts that provided the enterprise with its opportunities were specific to a particular time. Beyond this, however, the history has many facets that should be instructive in terms of policy, but more importantly provides a model for others to emulate.

Glossary

atom the smallest part of an element that has an independent existence. It consists of a nucleus surrounded by much smaller orbiting electrons. The nucleus has a positive electric charge normally equal to the negative charge of the electrons. The nucleus consists of protons that carry the charge, and may also contain one or more neutrons that are equal in mass to the protons but have no charge

beam line an evacuated passage through which particles or radiation leaving an accelerator flow to the target; a beam line may contain focusing magnets, beam splitters, collimators, valves, and diagnostic devices

catalyst a substance that speeds up the rate of a chemical reaction but itself remains unchanged

centigrade C the temperature scale devised by Celsius (1701–44). (See box 2.1 and Fig. 2.2.)

chip a single crystal of a semiconductor, usually silicon, on which an integrated circuit is created

cold iron quadrupole a superconducting magnet using iron cores, which are cooled to the same low temperature as the superconductor

collimator a device to produce a parallel beam in a beam line

compensating coils extra coils fitted near the main coils of a magnet to compensate for some irregularity in the main field

computerized tomography CT a process by which images of slices through a patient can be obtained from X-ray exposures taken in a 180° arc round the body; a three-dimensional image is then constructed by computerized image analysis

cooling cycle the series of processes—compression, cooling, and expansion through which a gas is cycled to generate cooling at one point

core memory a type of computer memory that consists of an electrical circuit round a semiconductor or ferrite core

critical current this is the maximum current that a superconductor can carry without reverting to the normal resistive state of the material

cryocooler synonymous with mechanical refrigerator

cryostat a vessel that enables a sample to be maintained at a low temperature

crystallography the study of the geometrical form of crystals that assists in the determination of their molecular structure

current density the electric current divided by the cross-sectional area of the conductor through which it is flowing

die a block of hardened steel with a tapered hole through it, through which a wire can be pulled to reduce its diameter

diffusion pump a type of vacuum pump

digital electronics the field of electronics in which equipment handles electrical signals in discrete pulses

direct current DC an electric current in which the flow of electrons is only in one direction, as opposed to alternating current, AC, in which the current flows back and forth

electrocardiogram (ECG) a recording of the patterns resulting from the electrical activity associated with the contraction of the heart muscles

electrodes electrical terminals

electroencephalogram (EEG) a recording of the pattern of electrical signals resulting from activity in the brain

electrons constituents of all atoms; they are very light, negatively charged particles, which orbit round the nucleus

enzymes a large group of proteins, produced by living cells, that act as catalysts in many of the essential chemical reactions in living organisms

ferrite an oxide of iron and other metals that is strongly magnetic but an electrical insulator; used for light and efficient cores for transformers and electronic devices

field gradient the extent to which the field strength varies with distance in a certain direction

field modulation coils extra coils fitted near the main coils of a magnet to modulate the main field

frequency the number of cycles, oscillations, or vibrations that occur in a given period of time, normally one second

gallium arsenide a compound semiconductor

gyrotron a high power source of microwave radiation

helmholz coil a coil divided into two halves such that the gap between them is equal to the inner radius of the coil; this produces a homogeneous field at its centre

high-energy physics the branch of physics concerned with high-energy particles

high resolution the property of a measuring instrument that is able to detect very small variations in what it is measuring; the property of a processing system that yields information in great detail

homogeneity the degree of uniformity—e.g. magnetic field strength—throughout a certain volume

hormones chemicals produced by glands in higher animals that regulate many of the functions of the body

induced current the current that develops in a conductor or a circuit when the magnetic field within it changes

integrated circuit a miniature electronic circuit on a piece of semiconductor, usually silicon

ion an atom or group of atoms that has either lost one or more electrons, making it positively charged (cation), or has gained one or more electrons, making it negatively charged (anion)

isotopes variants of the same element, all having the same chemistry and number of protons but different numbers of neutrons and hence different atomic weights

Kelvin K the temperature scale named after Lord Kelvin (1824–1907). (See Box 2.1 and Fig. 2.2.)

lines of force imaginary lines in a magnetic field that enable the direction and strength of the field to be visualised

magnetic mirror machine a device used for containing a plasma (an ionized gas) in a nuclear fusion experiment

magnetic moment an intrinsic property that can be measured by the rotational force felt by a magnet or a magnetic molecule in a magnetic field of unit field strength

metabolites complex substances that take part in building up and breaking down chemicals and tissues within the body, with the absorption and release of energy; a metabolic pathway is a series of chemical reactions by which complex chemicals in the body are built up or broken down

microprocessors basic elements of electronic circuits and computers, which accept and deliver instructions according to how they are programmed

molecule a combination of two or more atoms chemically combined together; the smallest part of a compound that can exist and show the properties of that compound

multifilamentary conductor a conductor made up from a number of filaments of conductor as opposed to a single core conductor

nano-technology technology concerning devices having extremely small dimensions, measured in nanometres (a nanometre is a thousand millionth of a metre)

neutron a part of the nucleus of an atom with the same mass (in practical terms equivalent to weight) as a proton, but with no electric charge

niobium tin (NbSn) an inter-metallic compound of niobium and tin that is a very high performance superconductor, although brittle and difficult to fabricate. The scientific abbreviation is Nb_3Sn, showing the proportions of each element

niobium titanium (NbTi) an alloy of the two elements, niobium and titanium, that is the most commonly used superconductor

niobium zirconium (NbZr) an alloy of niobium and zirconium; it was the earliest superconductor to be formed into wire for winding practical magnets

nuclear magnetic resonance (NMR) see Box 7.1

optical lithography the normal essentially photographic process by which the pattern for an electrical circuit is transferred to the surface of a silicon chip

oscilloscope an electronic instrument that provides a visible image on a glass screen of rapidly changing electrical signals

permanent magnet a piece of magnetic material, usually an alloy of iron, in which the atoms, which are all individual atomic magnets, have been made to line up in the same direction and so collectively produce an external magnetic field

pH a measure of acidity

photo-multiplier a detector and amplifier for low levels of light

positron a particle similar to an electron but with a positive charge

proton a part of the nucleus of an atom that carries a positive electric charge

quadrupole magnet a magnet that focuses a beam of charged particles flowing through it

quantum theory the theory that grew up around Max Planck's concept of energy not being continuous but divided into discreet basic amounts or quanta

quench in this context a quench is the change that occurs when the wire in a magnet reverts from the superconducting to the normal resistive state

radio frequency (RF) frequency in the range used in radio transmission, which is 10kHz–100,000 MHz (10 thousand to 100,000 million cycles per second)

radio isotope an isotope of an element that is radioactive

rheostat a variable resistor that enables the flow of current to be varied

silicon wafer a thin slice of silicon cut from a cylindrical single crystal, wafers may be up to 25 centimetres in diameter and are used as the basic material for making semiconductor chips

solid state physics it involves studies of the nature and properties of solid materials, especially of semiconductors; recently the name 'Condensed Matter Physics' has been used, and includes the study of liquids

spectroscopy the study of matter and energy by analysing the spectrum of radiation that is characteristic of it

split coil a magnet coil that is split into two halves, to provide access to the centre in the equatorial plane

sputter deposition a process for depositing a thin uniform film of a metal onto a surface

superfluidity the property of liquid helium at very low temperatures that enables it to flow without friction

swage swaging is the process of forming a long bar of metal from a short billet by passing it through a series of mechanical hammers

synchrotron a particle accelerator that imparts energy to electrons or protons, either to carry out experiments in particle physics, or to use the beam of radiation, including high intensity X-rays, which can be extracted when the accelerating electron beam is forced to move in a curved path by a magnetic field

References

ACCOSP (1991), Advisory Council Central Oxfordshire Science Parks, *Oxford & Technology Transfer: The Role of Science Parks* (Oxford: The Oxford Trust and ACCOSP).

ARCHIE, C. (1993), 'Performance of the IBM Synchrotron X-ray Source for Lithography', *IBM Journal of Research and Development*, 37/3: 373–83.

BAGHAI, M., COLEY, S., and WHITE, D. (1999), *The Alchemy of Growth* (London: Orion Business Books).

BANNOCK, G. (1981), *The Economics of Small Firms* (Oxford: Blackwell).

BELL, E. (1992), 'Bringing together Regional Resources: The Oxford Trust and Technology Transfer in Oxfordshire', research paper (The Oxford Trust).

BELL, E., STOTT, M. A., and KINGHAM, D. R. (1994), *Oxfordshire Firms and Technology Transfer: Interaction between Small and Medium Sized Enterprises (SMEs) and Local Research Base* (Oxford: The Oxford Trust and Oxfordshire County Council).

BIRCH, D. (1979), *The Jobs Generation Process* (Cambridge, Mass.: MIT).

BITTERMAN, A. (1999) 'Magnetic Surgery', *Superconductor & Cryogenics*, Spring: 10–16.

BRADSTOCK, P. A. (1987), *A Discussion Paper Concerning the Science Park Issue in Oxfordshire* (Oxford: The Oxford Trust).

BRANKIN, P. (1993), 'Establishing a Sales Operation in Japan', *Engineering Management Journal*, 3/4: 185–91.

BROMBERG, J. L. (1999), *NASA and the Space Industry* (Baltimore: Johns Hopkins).

BULLOCK, M. (1983), *Academic Enterprise, Industrial Innovation, and the Development of High Technology Financing in the United States* (London: Brand Brothers & Co.).

BYGRAVE, W. D., and TIMMONS, J. A. (1992), *Venture Capital at the Crossroads* (Boston, Mass.: Harvard Business School Press).

CAIRNCROSS, A. (1992), *The British Economy since 1945* (Oxford: Blackwell).

CAPIE, F., and M. COLLINS, M. (1992), *Have the Banks Failed British Industry?* (London: IEA).

CATTERMOLE, M. J. G., and WOLFE, A. F. (1987), *Horace Darwin's Shop* (Bristol: Adam Hilger).

CLUGSTON, M. J. (1998) (ed.), *The New Penguin Dictionary of Science* (London: Penguin Books).

COGHLAN, A. (1994), 'City's Dash for Cash Crushes Research', *New Scientist*, 141/1913: 4.

Committee for High Field NMR (1998), 'A New Millennium Resource', mimeo (National Science Foundation).

Company Reporting (1993), *The 1993 UK R&D Scoreboard* (Edinburgh: Company Reporting Ltd.).

COOPEY, R. (1993), 'Industrial Policy in the White Heat of the Scientific Revolution', in R. Coopey, S. Fielding, and N. Tiratsoo, *The Wilson Governments, 1964–1970* (London: Pinter).

——and CLARKE, D. (1996), *3i: Fifty Years Investing in Industry* (Oxford: Oxford University Press).

DALTON, I. G. (1989), 'The Development of Technology Transfer Institutes at Heriot-Watt University Edinburgh, Scotland', unpublished paper (Edinburgh: Heriot-Watt University).

——(1995), 'Technology Transfer Institutes: A Novel and Successful Mechanism', unpublished paper (Edinburgh: Heriot-Watt University).

DE KLERK, D. (1965), *The Construction of High-Field Electromagnets* (Oxford: Newport Instruments Ltd.).

DEW-HUGHES, D. (1966), 'Hard Superconductors: Review Paper', *Materials Science and Engineering,* 1: 2–29.

EDGERTON, D. (1991), *England and the Aeroplane* (Manchester: Manchester University Press).

——and HORROCKS, S. (1994), 'British Industrial Research and Development before 1945', *Economic History Review*, 47/2: 213–38.

ELBAUM, B., and LAZONICK, W. (1986), *The Decline of the British Economy* (Oxford: Oxford University Press).

Electronic Business (1989), 'X-Ray Lithography: Wave of the Future?', 27 Nov., 26–35.

GARNSEY, E. (1997), 'Science-Based Enterprise: Threat or Opportunity?' *Physics World*, July: 15–17.

——and LAWTON SMITH, H. (1998), 'Proximity and Complexity in the Emergence of High Technology Industry: The Oxbridge Comparison', *Geoforum*, 29/4: 433–50.

GUMMETT, P. (1980), *Scientists in Whitehall* (Manchester: Manchester University Press).

HAKE, R. R., BERLINCOURT, T. G., and LESLIE, D. H. (1962), 'A 59-Kilogauss Niobium-Zirconium Superconducting Solenoid', in H. Kolm, B. Lax, F. Bitter, and R. Mills (eds.), *High Magnetic Fields: Proceedings of the International Conference on High Magnetic Fields* (New York: John Wiley & Sons), 341–3.

HAMEL, G., and PRAHALAD, C. K. (1994), *Competing for the Future* (Boston: Harvard Business School Press).

HANDY, C. (1994), *The Empty Raincoat* (London: Hutchinson).

HAWKSWORTH, D. G., McDOUGALL, I. L., BIRD, J. M., and BLACK, D. (1986), 'Considerations in the Design of MRI Magnets with Reduced Stray Fields', *MAG* 23/2: 1309.

HEINZ, P. (1991), 'X-Ray Lithography Strategies: Japan versus the US', *Microelectronics Manufacturing Technology*, Sept.: 6–7.

Her Majesty's Stationery Office (1966), *Interim Report of the Working Group on Manpower Parameters for Scientific Growth* (Swann Report) (London: HMSO).

——(1966), *Interim Report: Enquiry into the Flow of Candidates in Science and Technology into Higher Education* (Dainton Report) (London: HMSO).

——(1968), *The Flow into Employment of Scientists, Engineers and Technologists: Report of the Working Group on Manpower Parameters for Scientific Growth* (Swann Report) (London: HMSO).

HOBART ELLIS, R., Jr. (1968), 'Britons Seek Closer Relations between Industry and University', *Physics Today*, 21/1: 54–8.

HOPE J., and FRASER, R. (1999), *Beyond Budgeting: White Paper* (London: Beyond Budgeting Round Table/CAM-I).

HOWARD, W., and LOUIS, R. (1998), *The Oxford History of the Twentieth Century* (New York: Oxford University Press).

HULM, J. K., FRASER, M. J., RIEMERSMA, H., VENTURINO, A. J., and WIEM, R. E. (1962), 'A High-Field Niobium-Zirconium Superconducting Solenoid', in H. Kolm, B. Lax, F. Bitter, and R. Mills (eds.), *High Magnetic Fields: Proceedings of the International Conference on High Magnet Fields* (New York: John Wiley & Sons), 332–40.

HUTTON, W. (1995), *The State We're In* (London: Jonathan Cape).

INGHAM, G. (1984), *Capitalism Divided: The City and Industry in British Social Development* (London: Macmillan).

ISAACS, A., DAINTITH, J., and MARTIN, E. (1996) (eds.), *Concise Science Dictionary* (3rd edn., Oxford: Oxford University Press).

IVES, J. R., and WOODS, J. F. (1975), '4-Channel 24-Hour Cassette Recorder for Long-Term EEG Monitoring of Ambulatory Patients', *Electroencephalography and Clinical Neurophysiology*, 39: 88–92.

KENNEDY, W. P. (1987), *Industrial Structure, Capital Markets and the Origins of British Economic Decline* (Cambridge: Cambridge University Press).

KINDLEBERGER, C. P. (1996), *Manias, Panics and Crashes: A History of Financial Crises* (London: Macmillan).

KING, R. (1973), *Michael Faraday of the Royal Institution* (London: Royal Institution of Great Britain).

LAWTON SMITH, H. (1989), 'The Location and Development of Advanced Technology Industry in Oxfordshire in the Context of the Research Environment', D.Phil. thesis (Oxford).

——(1991), 'The Role of Incubators in the Local Industrial Development: The Cryogenics Industry in Oxfordshire', *Entrepreneurs & Regional Development*, 3/2: 175–94.

——and BRADSTOCK, P. (1986), '*Science Parks: A Review*', review paper (Oxford: The Oxford Trust).

LESOINE, L. G., and LEAVEY, J. A. (1998), 'IBM Advanced Lithography Facility: The First Five Years', *Solid State Technology*, 41/7: 101–12.

KAMMERLINGH ONNES, H. (1911), 'Further Experiments with Liquid Helium, D. On the Change of the Electrical Resistance of Pure Metals at Very Low Temperatures, etc. V. The Disappearance of the Resistance of Mercury', *Physical Laboratory Communications*,122b (Leyden University), 13–15.

MACARTHUR, B. (1993) (ed.), *The Penguin Book of Twentieth-Century Speeches* (London: Penguin Books).

MACKENZIE, D., and Spinardi, G. (1993), 'The Technological Impact of a Defence Research Establishment', in R. Coopey, M. Uttley, and G. Spinardi (eds.), *Defence Science and Technology: Adjusting to Change* (Reading: Harwood).

MCMASTERS, J. B. (1931–8), *History of the People of the United States* (8 vols.; New York: D. Appleton-Century).

MADDOCK, I. (1973), 'Can Science-Based Companies Survive?' *New Scientist*, 59/862: 565–70.

MALPAS, R. (1999), 'Business Needs Science', unpublished speech for 'Oxford University in Business', University of Oxford Science and Technology Day, 4 Nov.

MARSH, P. (1987), 'Little Businesses are Big Business', *New Scientist*, 113/1546: 46–9.

MAZARR, M. J. (1999), *Global Trends 2005* (Bloomsburg, Pa.: MacMillan Press).

MENDELSSOHN, K. (1963), 'Patterns of Superconductivity', *Cryogenics*, 3/3: 129–40.

MENDOZA, E. (1962), 'Superconductivity and Production of High Magnetic Fields', *Times Science Review*, Winter: 12–14.

MIHELL, D., KINGHAM, D., and STOTT, M. (1987), *The Development of the Biotechnology Sector in Oxfordshire: Implications for Public Policy* (Oxford: Oxford Innovation and Oxfordshire County Council).

MOWERY, D. C., and ROSENBERG, N. (1989), *Technology and the Pursuit of Economic Growth* (Cambridge: Cambridge University Press).

NEWTON, L. (1996), 'Regional Bank–Industry Relations during the Mid-19th Century', *Business History*, 38: 64–83.

Nippon Steel (1993), *Nippon: The Land and its People* (Tokyo: Gakuseisha Publishing).

OGLE, P. L. (1985), 'New Era of Limits Leaves Mark on Program at Annual RSNA Meeting', *Diagnostic Imaging*, 51/4: 87.

The Oxford Trust (1990a), *CuriOXity: Through the Technological Barriers* (Oxford: The Oxford Trust).

—— (1990b). *Scintilla*, 3 (Oxford: The Oxford Trust).

—— (1991a), *CuriOXity: A Review* (Oxford: The Oxford Trust).

—— (1991b), *Oxford & Technology Transfer: The Role of Science Parks* (Oxford: The Oxford Trust).

—— (1991c), *Schools/Industry Grant: A Review* (Oxford: The Oxford Trust).

PATERSON, R. A. H. (1975), 'Seasonal Reduction of Slow-Wave Sleep at an Antarctic Coastal Station', *Lancet*, 1/7904: 468–9.

PETER, L. J., and HULL, R. (1970), *The Peter Principle* (New York: Bantam Books).

PETERS, T. J., and WATERMAN, R. H. (1982), *In Search of Excellence* (New York: Harper & Row).

PRESSNELL, L. (1956), *Country Banking in the Industrial Revolution* (Oxford: Oxford University Press).

REYNOLDS, K., QUICK, S., and DAY, L. (1998), 'Oxford Magnet Technology: The 483 Order Fulfillment Project', private report submitted for the Siemens Engineering Excellence Award (1998 winner).

ROLLIN, B. V. (1946), 'Nuclear Magnetic Resonance and Spin Lattice Equilibrium', *Nature*, 158: 669–70.

RUBINSTEIN, W. D. (1994), *Capitalism, Culture and Decline in Britain, 1750–1990* (London: Routledge).

SALTER, L. C., AUTLER, S. H., KOLM, H. H., ROSE, D. J., and GOOEN, K. (1962), 'A Niobium-Tin Superconducting Magnet', in H. Kolm, B. Lax, F. Bitter, and R. Mills (eds.), *High Magnetic Fields: Proceedings of the International Conference on High Magnetic Fields* (New York: John Wiley & Sons), 341–3.

Segal Quince & Partners (1985), *The Cambridge Phenomenon* (Cambridge: Segal Quince & Partners).

SMITH, D., and SMITH, R. (1985), *The Economics of Militarism* (London: Pluto).

Snow, C. P. (1959), *The Two Cultures and the Scientific Revolution* (Oxford: Oxford University Press).

STANWORTH, J., and GRAY, C. (1991), *Bolton 20 Years On: The Small Firm in the 1990s* (London: Paul Chapman).

STORPER, M., and SCOTT, A. (1992), *Pathways to Industrialization and Regional Development* (London: Routledge).

STOTT, F. D. (1977), 'Ambulatory Monitoring', *British Journal of Clinical Equipment*, 2/2: 61–8.

TIBBETTS, R. (1992), 'The Role of Small Firms in Developing and Commercializing Novel Advances in Scientific Instrumentation and Equipment: Lessons from the U.S. and NSF Small Business Innovation Research (SBIR) Program', invited paper for the international workshop Equipping Science for the 21st Century, Amsterdam.

University of Cambridge (1969), 'Report on Relations between the University and Industry (Mott Report)', *Cambridge University Reporter*, 22 Oct.

University of Oxford (1996), 'Intellectual Property Policy Statement of Enactment of Changes to be Made to the Legislation Governing Intellectual Property Generated with the University', *Oxford University Gazette*, 31 Oct.

VIG, N. J. (1986), *Science and Technology in British Politics* (London: Pergamon).

WATSON, T. J., Jr., and PETRE, P. (1990), *Father, Son, & Co.* (New York: Bantam Press).

Wellcome Trust (1998), *Wellcome Witnesses to Twentieth Century Medicine*, ii (London: Wellcome Trust).

WIENER, M. J. (1985), *English Culture and the Decline of the Industrial Spirit 1850–1980* (Harmondsworth: Penguin Books; first published, 1981).

WILLIAMS, P. (1991), 'Time and the City: Short Termism in the UK', Stockton Lecture (London Business School).

WILLIS, T., KINGHAM, D., and STAFFORD, J. (1996), 'Oxfordshire's Motor Sport Industry: Building on Local Strengths', report commissioned by Heart of England Training and Enterprise Council and Oxford Innovation.

WILSON, A. D. (1993), 'X-ray Lithography in IBM, 1980–1992, the Development Years', *IBM Journal of Research and Development*, 37/3: 299–318.

WILSON, M. N. (1983), *Superconducting Magnets* (New York: Oxford University Press).

WOOD, M. (1962), 'Some Aspects of Design of Superconducting Solenoids', *Cryogenics*, 2/5: 297–300.

—— (1999), 'Phoenixes to the Rescue', unpublished speech for 'Challenges of Growth', the 2nd Cambridge Enterprise Conference.

Working Group on Innovation (1992), *Interim Report* (London: Working Group on Innovation).

ZUCKERMAN, S. (1988), *From Apes to Warlords* (London: Hamilton).

Index